INVENTORY CONTROL

Second Edition

Recent titles in the **INTERNATIONAL SERIES IN OPERATIONS RESEARCH & MANAGEMENT SCIENCE**
Frederick S. Hillier, Series Editor, *Stanford University*

Maros/ *COMPUTATIONAL TECHNIQUES OF THE SIMPLEX METHOD*
Harrison, Lee & Neale/ *THE PRACTICE OF SUPPLY CHAIN MANAGEMENT: Where Theory and Application Converge*
Shanthikumar, Yao & Zijm/ *STOCHASTIC MODELING AND OPTIMIZATION OF MANUFACTURING SYSTEMS AND SUPPLY CHAINS*
Nabrzyski, Schopf & Węglarz/ *GRID RESOURCE MANAGEMENT: State of the Art and Future Trends*
Thissen & Herder/ *CRITICAL INFRASTRUCTURES: State of the Art in Research and Application*
Carlsson, Fedrizzi, & Fullér/ *FUZZY LOGIC IN MANAGEMENT*
Soyer, Mazzuchi & Singpurwalla/ *MATHEMATICAL RELIABILITY: An Expository Perspective*
Chakravarty & Eliashberg/ *MANAGING BUSINESS INTERFACES: Marketing, Engineering, and Manufacturing Perspectives*
Talluri & van Ryzin/ *THE THEORY AND PRACTICE OF REVENUE MANAGEMENT*
Kavadias & Loch/*PROJECT SELECTION UNDER UNCERTAINTY: Dynamically Allocating Resources to Maximize Value*
Brandeau, Sainfort & Pierskalla/ *OPERATIONS RESEARCH AND HEALTH CARE: A Handbook of Methods and Applications*
Cooper, Seiford & Zhu/ *HANDBOOK OF DATA ENVELOPMENT ANALYSIS: Models and Methods*
Luenberger/ *LINEAR AND NONLINEAR PROGRAMMING, 2nd Ed.*
Sherbrooke/ *OPTIMAL INVENTORY MODELING OF SYSTEMS: Multi-Echelon Techniques, Second Edition*
Chu, Leung, Hui & Cheung/ *4th PARTY CYBER LOGISTICS FOR AIR CARGO*
Simchi-Levi, Wu & Shen/ *HANDBOOK OF QUANTITATIVE SUPPLY CHAIN ANALYSIS: Modeling in the E-Business Era*
Gass & Assad/ *AN ANNOTATED TIMELINE OF OPERATIONS RESEARCH: An Informal History*
Greenberg/ *TUTORIALS ON EMERGING METHODOLOGIES AND APPLICATIONS IN OPERATIONS RESEARCH*
Weber/ *UNCERTAINTY IN THE ELECTRIC POWER INDUSTRY: Methods and Models for Decision Support*
Figueira, Greco & Ehrgott/ *MULTIPLE CRITERIA DECISION ANALYSIS: State of the Art Surveys*
Reveliotis/ *REAL-TIME MANAGEMENT OF RESOURCE ALLOCATIONS SYSTEMS: A Discrete Event Systems Approach*
Kall & Mayer/ *STOCHASTIC LINEAR PROGRAMMING: Models, Theory, and Computation*
Sethi, Yan & Zhang/ *INVENTORY AND SUPPLY CHAIN MANAGEMENT WITH FORECAST UPDATES*
Cox/ *QUANTITATIVE HEALTH RISK ANALYSIS METHODS: Modeling the Human Health Impacts of Antibiotics Used in Food Animals*
Ching & Ng/ *MARKOV CHAINS: Models, Algorithms and Applications*
Li & Sun/ *NONLINEAR INTEGER PROGRAMMING*
Kaliszewski/ *SOFT COMPUTING FOR COMPLEX MULTIPLE CRITERIA DECISION MAKING*
Bouyssou et al/ *EVALUATION AND DECISION MODELS WITH MULTIPLE CRITERIA: Stepping stones for the analyst*
Blecker & Friedrich/ *MASS CUSTOMIZATION: Challenges and Solutions*
Appa, Pitsoulis & Williams/ *HANDBOOK ON MODELLING FOR DISCRETE OPTIMIZATION*
Herrmann/ *HANDBOOK OF PRODUCTION SCHEDULING*

* *A list of the early publications in the series is at the end of the book* *

INVENTORY CONTROL

Second Edition

Sven Axsäter

Springer

Sven Axsäter
Lund University
Lund, Sweden

Library of Congress Control Number: 2006922871

ISBN-10: 0-387-33250-2 (HB)   ISBN-10: 0-387-33331-2 (e-book)
ISBN-13: 978-0387-33250-5 (HB)   ISBN-13: 978-0387-33331-1 (e-book)

Printed on acid-free paper.

© 2006 by Springer Science+Business Media, LLC
All rights reserved. This work may not be translated or copied in whole or in part without the written permission of the publisher (Springer Science + Business Media, LLC, 233 Spring Street, New York, NY 10013, USA), except for brief excerpts in connection with reviews or scholarly analysis. Use in connection with any form of information storage and retrieval, electronic adaptation, computer software, or by similar or dissimilar methodology now know or hereafter developed is forbidden.

The use in this publication of trade names, trademarks, service marks and similar terms, even if the are not identified as such, is not to be taken as an expression of opinion as to whether or not they are subject to proprietary rights.

Printed in the United States of America.

9 8 7 6 5 4 3 2 1

springer.com

CONTENTS

Preface xv

1 INTRODUCTION 1
   1.1 Importance and objectives of inventory control 1
   1.2 Overview and purpose of the book 2
   1.3 Framework 5
   References 5

2 FORECASTING 7
   2.1 Objectives and approaches 7
   2.2 Demand models 8
      2.2.1 Constant model 9
      2.2.2 Trend model 9
      2.2.3 Trend-seasonal model 10
      2.2.4 Choosing demand model 10
   2.3 Moving average 11
   2.4 Exponential smoothing 12
      2.4.1 Updating procedure 12
      2.4.2 Comparing exponential smoothing to a moving average 13
      2.4.3 Practical considerations and an example 14
   2.5 Exponential smoothing with trend 16
      2.5.1 Updating procedure 16
      2.5.2 Practical considerations and an example 17
   2.6 Winters' trend-seasonal method 18
      2.6.1 Updating procedure 18
      2.6.2 Practical considerations and an example 20
   2.7 Using regression analysis 21
      2.7.1 Forecasting demand for a trend model 21
      2.7.2 Practical considerations and an example 23
      2.7.3 Forecasts based on other factors 24
      2.7.4 More general regression models 25
   2.8 Sporadic demand 26

2.9 Box-Jenkins techniques 27
2.10 Forecast errors 29
    2.10.1 Common error measures 29
    2.10.2 Updating MAD or σ^2 30
    2.10.3 Determining the standard deviation as a function of demand 32
    2.10.4 Forecast errors for other time periods 32
    2.10.5 Sales data instead of demand data 34
2.11 Monitoring forecasts 34
    2.11.1 Checking demand 35
    2.11.2 Checking that the forecast represents the mean 35
2.12 Manual forecasts 36
References 37
Problems 38

3 COSTS AND CONCEPTS 43
3.1 Considered costs and other assumptions 44
    3.1.1 Holding costs 44
    3.1.2 Ordering or setup costs 44
    3.1.3 Shortage costs or service constraints 45
    3.1.4 Other costs and assumptions 45
3.2 Different ordering systems 46
    3.2.1 Inventory position 46
    3.2.2 Continuous or periodic review 47
    3.2.3 Different ordering policies 48
        3.2.3.1 (R, Q) policy 48
        3.2.3.2 (s, S) policy 49
References 50

4 SINGLE-ECHELON SYSTEMS: DETERMINISTIC LOT SIZING 51
4.1 The classical economic order quantity model 52
    4.1.1 Optimal order quantity 52
    4.1.2 Sensitivity analysis 54
    4.1.3 Reorder point 54
4.2 Finite production rate 55
4.3 Quantity discounts 56
4.4 Backorders allowed 59
4.5 Time-varying demand 61
4.6 The Wagner-Whitin algorithm 63

4.7 The Silver-Meal heuristic 66
4.8 A heuristic that balances holding and ordering costs 68
4.9 Exact or approximate solution 70
References 70
Problems 72

5 SINGLE-ECHELON SYSTEMS: REORDER POINTS 77

5.1 Discrete stochastic demand 77
   5.1.1 Compound Poisson demand 77
   5.1.2 Logarithmic compounding distribution 80
   5.1.3 Geometric compounding distribution 82
   5.1.4 Smooth demand 83
   5.1.5 Fitting discrete demand distributions in practice 85
5.2 Continuous stochastic demand 85
   5.2.1 Normally distributed demand 85
   5.2.2 Gamma distributed demand 86
5.3 Continuous review (R, Q) policy - inventory level distribution 88
   5.3.1 Distribution of the inventory position 88
   5.3.2 An important relationship 90
   5.3.3 Compound Poisson demand 90
   5.3.4 Normally distributed demand 91
5.4 Service levels 94
5.5 Shortage costs 95
5.6 Determining the safety stock for given S_1 96
5.7 Fill rate and ready rate constraints 97
   5.7.1 Compound Poisson demand 97
   5.7.2 Normally distributed demand 98
5.8 Fill rate - a different approach 99
5.9 Shortage cost per unit and time unit 101
   5.9.1 Compound Poisson demand 101
   5.9.2 Normally distributed demand 103
5.10 Shortage cost per unit 106
5.11 Continuous review (s, S) policy 107
5.12 Periodic review - fill rate 109
   5.12.1 Basic assumptions 110
   5.12.2 Compound Poisson demand - (R, Q) policy 111
   5.12.3 Compound Poisson demand - (s, S) policy 112
   5.12.4 Normally distributed demand - (R, Q) policy 113
5.13 The newsboy model 114
5.14 A model with lost sales 117

5.15 Stochastic lead-times 119
  5.15.1 Two types of stochastic lead-times 119
  5.15.2 Handling sequential deliveries independent of the lead-time demand 120
  5.15.3 Handling independent lead-times 122
  5.15.4 Comparison of the two types of stochastic lead-times 123
  References 124
  Problems 126

6 SINGLE-ECHELON SYSTEMS: INTEGRATION - OPTIMALITY 129
6.1 Joint optimization of order quantity and reorder point 129
  6.1.1 Discrete demand 129
    6.1.1.1 (R, Q) policy 130
    6.1.1.2 (s, S) policy 132
  6.1.2 An iterative technique 133
  6.1.3 Fill rate constraint - a simple approach 135
6.2 Optimality of ordering policies 137
  6.2.1. Optimality of (R, Q) policies when ordering in batches 138
  6.2.2 Optimality of (s, S) policies 140
6.3 Updating order quantities and reorder points in practice 140
  References 145
  Problems 146

7 COORDINATED ORDERING 149
7.1 Powers-of-two policies 150
7.2 Production smoothing 154
  7.2.1 The Economic Lot Scheduling Problem (ELSP) 155
    7.2.1.1 Problem formulation 155
    7.2.1.2 The independent solution 156
    7.2.1.3 Common cycle time 157
    7.2.1.4 Bomberger's approach 159
    7.2.1.5 A simple heuristic 160
    7.2.1.6 Other problem formulations 163
  7.2.2 Time-varying demand 163
    7.2.2.1 A generalization of the classical dynamic lot size problem 163
    7.2.2.2 Application of mathematical programming approaches 170
  7.2.3 Production smoothing and batch quantities 170

7.3 Joint replenishments 172
  7.3.1 A deterministic model 173
    7.3.1.1 Approach 1. An iterative technique 174
    7.3.1.2 Approach 2. Roundy's 98 percent approximation 176
  7.3.2 A stochastic model 180
References 181
Problems 183

8 MULTI-ECHELON SYSTEMS: STRUCTURES AND ORDERING POLICIES 187
8.1 Inventory systems in distribution and production 188
  8.1.1 Distribution inventory systems 188
  8.1.2 Production inventory systems 189
  8.1.3 Repairable items 192
  8.1.4 Lateral transshipments in inventory systems 192
  8.1.5 Inventory models with remanufacturing 194
8.2 Different ordering systems 195
  8.2.1 Installation stock reorder point policies and KANBAN policies 196
  8.2.2 Echelon stock reorder point policies 197
  8.2.3 Comparison of installation stock and echelon stock policies 198
  8.2.4 Material Requirements Planning 204
  8.2.5 Ordering system dynamics 213
References 215
Problems 217

9 MULTI-ECHELON SYSTEMS: LOT SIZING 221
9.1 Identical order quantities 222
  9.1.1 Infinite production rates 222
  9.1.2 Finite production rates 223
9.2 Constant demand 225
  9.2.1 A simple serial system with constant demand 225
  9.2.2 Roundy's 98 percent approximation 230

9.3 Time-varying demand 236
- 9.3.1 Sequential lot sizing 236
- 9.3.2 Sequential lot sizing with modified parameters 238
- 9.3.3 Other approaches 240
- 9.3.4 Concluding remarks 241

References 242
Problems 243

10 MULTI-ECHELON SYSTEMS: REORDER POINTS 247

10.1 The Clark-Scarf model 248
- 10.1.1 Serial system 249
- 10.1.2 The Clark-Scarf approach for a distribution system 256

10.2 The METRIC approach for distribution systems 261

10.3 Two exact techniques 266
- 10.3.1 Disaggregation of warehouse backorders 266
- 10.3.2 A recursive procedure 267

10.4 Optimization of ordering policies 271

10.5 Batch-ordering policies 273
- 10.5.1 Serial system 273
- 10.5.2 Distribution system 276
  - 10.5.2.1 Some basic results 277
  - 10.5.2.2 METRIC type approximations 278
  - 10.5.2.3 Disaggregation of warehouse backorders 279
  - 10.5.2.4 Following supply units through the system 280
  - 10.5.2.5 Practical considerations 280

10.6 Other assumptions 281
- 10.6.1 Guaranteed service model approach 281
- 10.6.2 Coordination and contracts 283
  - 10.6.2.1 The newsboy problem with two firms 284
  - 10.6.2.2 Wholesale-price contract 285
  - 10.6.2.3 Buyback contract 286

References 287
Problems 291

11 IMPLEMENTATION 295

11.1 Preconditions for inventory control 295
- 11.1.1 Inventory records 296
  - 11.1.1.1 Updating inventory records 296
  - 11.1.1.2 Auditing and correcting inventory records 297
- 11.1.2 Performance evaluation 298

11.2 Development and adjustments 299
    11.2.1 Determine the needs 300
    11.2.2 Selective inventory control 301
    11.2.3 Model and reality 302
    11.2.4 Step-by-step implementation 303
    11.2.5 Simulation 304
    11.2.6 Short-run consequences of adjustments 305
    11.2.7 Education 306
References 307

APPENDIX 1
ANSWERS AND HINTS TO PROBLEMS 309

APPENDIX 2
NORMAL DISTRIBUTION TABLES 321

INDEX 325

Preface

Modern information technology has created new possibilities for more sophisticated and efficient control of supply chains. Most organizations can reduce their costs associated with the flow of materials substantially. Inventory control techniques are very important components in this development process. A thorough understanding of relevant inventory models is a prerequisite for successful implementation. I hope that this book will be a useful tool in acquiring such an understanding.

The book is primarily intended as a course textbook. It assumes that the reader has a good basic knowledge of mathematics and probability theory, and is therefore most suitable for industrial engineering and management science/operations research students. The book can be used both in undergraduate and more advanced graduate courses.

About fifteen years ago I wrote a Swedish book on inventory control. This book is still used in courses in production and inventory control at several Swedish engineering schools and has also been appreciated by many practitioners in the field. Positive reactions from many readers made me contemplate writing a new book in English on the same subject. Encouraging support of this idea from the Springer Editors Fred Hillier and Gary Folven finally convinced me to go ahead with that project six years ago.

The resulting first edition of this book was published in 2000 and contained quite a lot of new material that was not included in its Swedish predecessor. It has since then been used in quite a few university courses in different parts of the world, and I have received many positive reactions. Still some readers have felt that the book was too compact and some have asked for additional topics. Some of those who have used the book as a textbook have also requested more problems to be solved by the students.

The Springer Editors Fred Hillier and Gary Folven finally convinced me to publish a Second Edition of my book. This new edition is quite different from the previous one. The text has been expanded by more than 50 percent. My main goal has been to make the new book more suitable as a textbook. There are eleven chapters compared to six in the previous version. The explanations of different results are more detailed, and a considerable number of exercises have been added. I have also included several new topics. The additions include: alternative forecasting techniques, more material on different stochastic demand processes and how they can be fitted to empirical data, generalized treatment of single-echelon periodic review systems, capacity constrained lot sizing, short sections on lateral transshipments and on remanufacturing, coordination and contracts.

When working with the book I have been much influenced by other textbooks and various scientific articles. I would like to thank the authors of these books and papers for indirectly contributing to my book.

There are also a number of individuals that I would like to thank. Before I started to work on the revision, Springer helped me to arrange a review process, where a number of international scholars were asked to suggest suitable changes in the book. These scholars were: Shoshana Anily, Tel Aviv University, Saif Benjaafar, University of Minnesota, Eric Johnson, Dartmouth College, George Liberopoulos, University of Thessaly, Suresh Sethi, University of Texas-Dallas, Jay Swaminathan, University of North Carolina, Ruud Teunter, Lancaster University, Geert-Jan Van Houtum, Eindhoven University of Technology, Luk Van Wassenhove, INSEAD, and Yunzeng Wang, Case Western Reserve University. Some of them had used the first edition of the book in their classes. The review process resulted in most valuable suggestions for improvements, and I want to thank all of you very much.

Several colleagues of mine at Lund University have helped me a lot. I would especially like to mention Johan Marklund for much and extremely valuable help with both editions, and Kaj Rosling (now at Växjö University) for his important suggestions concerning the first edition. Furthermore, Jonas Andersson, Peter Berling, Fredrik Olsson, Patrik Tydesjö, and Stefan Vidgren have reviewed the manuscript at different stages and offered valuable suggestions which have improved this book considerably. Thank you so much.

Finally, I would also like to thank the Springer people: Fred Hillier, Gary Folven, and Carolyn Ford for their support, and Sharon Bowker for polishing my English.

Sven Axsäter

1 INTRODUCTION

1.1 Importance and objectives of inventory control

For more or less all organizations in any sector of the economy, *Supply Chain Management*, i.e., the control of the material flow from suppliers of raw material to final customers, is a crucial problem. The strategic importance of this area is today fully recognized by top management. The total investment in inventories is enormous, and the control of capital tied up in raw material, work-in-progress, and finished goods offers a very important potential for improvement. Scientific methods for inventory control can give a significant competitive advantage. This book deals with a wide range of different inventory models that can be used when developing inventory control systems.

Advances in information technology have drastically changed the possibilities to apply efficient inventory control techniques. Furthermore, the recent progress in research has resulted in new and more general methods that can reduce the supply chain costs substantially. The field of inventory control has indeed changed during the last decades. It used to mean application of simple decision rules, which essentially could be carried out manually. Modern inventory control is based on quite advanced and complex decision models, which may require considerable computational efforts.

Inventories cannot be decoupled from other functions, for example purchasing, production, and marketing. As a matter of fact, the objective of inventory control is often to balance conflicting goals. One goal is, of course, to keep stock levels down to make cash available for other purposes. The purchasing manager may wish to order large batches to get volume discounts. The production manager similarly wants long production runs to avoid time-consuming setups. He also prefers to have a large raw material inventory to avoid stops in production due to missing materials. The marketing manager would like to have a high stock of finished goods to be able to provide customers a high service level.

It is seldom trivial to find the best balance between such goals, and that is why we need inventory models. In most situations some stocks are required. The two main reasons are *economies of scale* and *uncertainties*. Economies of scale mean that we need to order in *batches*. Uncertainties in supply and demand together with lead-times in production and transportation inevitably create a need for *safety stocks*. Still, most organizations can reduce their inventories without increasing other costs by using more efficient inventory control tools.

There are important inventory control problems in all supply chains. For those who are working with logistics and supply chains, it is difficult to think of any qualification that is more essential than a thorough understanding of basic inventory models.

1.2 Overview and purpose of the book

The main purpose of this book is that it should be useful as a course textbook. The structure of the book is illustrated in Figure 1.1.

After this introduction we consider different *forecasting techniques* in Chapter 2. We focus on methods like exponential smoothing and moving average procedures for estimating the future demand from historical demand data. We also provide techniques for evaluating the size of forecast errors.

Chapters 3 - 6 deal with basic inventory problems for a *single installation* and *items that can be handled independently*. More precisely, Chapter 3 presents various basic concepts. Chapter 4 deals with *deterministic lot sizing* and Chapter 5 with *safety stocks* and *reorder points*. In Chapter 6 we discuss *integration* and *optimality*.

The contents in Chapters 2 - 6 provide the foundation for an efficient standard inventory control system, which can include:

- A forecasting module, which periodically updates demand forecasts and evaluates forecast errors.

- A module for determination of reorder points and order quantities.

- Continuous or periodic monitoring of inventory levels and outstanding orders. Triggering of suggested orders when reaching the reorder points.

INTRODUCTION 3

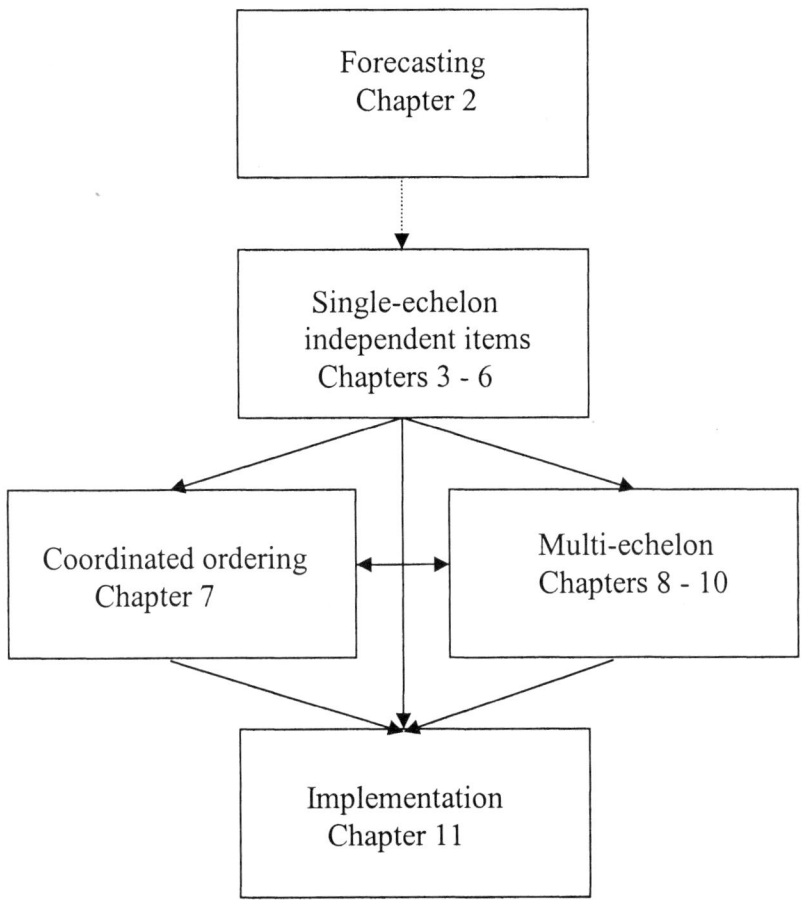

Figure 1.1 Structure of the book.

In Chapter 7 we leave the assumption of independent items and consider <u>coordinated replenishments</u>. Both production smoothing models and so-called joint replenishment problems are analyzed.

Chapters 8 - 10 focus on multi-echelon inventory systems, i.e., on several installations which are coupled to each other. The installations can represent, for example, stocks of raw materials, components, work-in-process, and final products in a production system, or a central warehouse and a number of retailers in a distribution system. In Chapter 8 we consider *structures* and *ordering policies*. Chapter 9 deals with *lot sizing* and Chapter 10 with *safety stocks* and *reorder points*.

Finally, in Chapter 11 we discuss various practical problems in connection with *implementation* of inventory control systems.

Over the years a substantial number of excellent books and overview papers dealing with various inventory control topics have been published. A selection of these publications is listed at the end of this chapter. A natural question then is why this book is needed. To explain this, note first that this book is different from most other books because it also covers very recent advances in inventory theory, for example new techniques for multi-echelon inventory systems and Roundy's 98 percent approximation. Furthermore, this book is also different from most other books because it assumes a reader with a good basic knowledge of mathematics and probability theory. This makes it possible to present different inventory models in a compact and hopefully more efficient way. The book attempts to explain fundamental ideas in inventory modeling in a simple but still rigorous way. However, to simplify, several models are less general than they could have been.

Because the book assumes a good basic knowledge of mathematics and probability theory, it is most suitable for industrial engineering and management science/operations research students. It can be used in a basic undergraduate course, and/or in a more advanced graduate course.

Chapter 2 may be omitted in a course which is strictly focused on inventory control. If it is included, it should probably be the first part of the course. Chapters 3 - 6 should precede Chapters 7 - 10. Chapter 7 can either precede or succeed Chapters 8 - 10. Chapter 11 should come at the end of the course.

An *undergraduate course* can, for example, be based on the following parts of the book: Sections 2.1 - 2.6, Sections 2.10 - 2.12, Chapters 3 - 4, Section 5.1.1, Section 5.2.1, Sections 5.3 - 5.8, Section 5.13, Section 6.3, Section 7.2.1, Section 8.1, Sections 8.2.1 - 8.2.2, Sections 8.2.4 - 8.2.5, Section 9.1, Section 9.2.1, Chapter 11.

For students that have taken the suggested undergraduate course, or a corresponding course, a *graduate course* can build on a selection of the remaining parts of the book, e.g., Sections 5.1.2 - 5.1.5, Section 5.2.2. Sections 5.9 - 5.12, Sections 5.14 - 5.15, Sections 6.1 - 6.2, Section 7.1, Section 7.3, Section 8.2.3, Sections 9.2 - 9.3, Chapter 10.

A graduate course for students that have no prior knowledge of inventory control but a good mathematical background should include most of the material suggested for the undergraduate course, but can exclude some of the sections suggested for the graduate course.

Another purpose of this book is to describe and explain efficient inventory control techniques for practitioners, and in that way simplify and promote implementation in practice. The book can, e.g., be used as a *handbook* when implementing and adjusting inventory control systems.

1.3 Framework

Models and methods in this book are based on the cost structure that is most common in industrial applications. We consider holding costs including opportunity costs of alternative investments, ordering or setup costs, and shortage costs or service level constraints. We will not deal with, for example, inventory problems related to financial speculation, i.e., when the value of an item can be expected to increase, or with aggregate planning models for smoothing production in case of seasonal demand variations. The interaction with production is recognized through setup costs but also in some models by explicit capacity constraints. The book does not cover production planning settings that are not directly related to inventory control.

The models considered in the book assume that the basic conditions for inventory control are given, for example in the form of demand distributions, lead-times, service requirements, and holding and ordering costs. In practice, most of these conditions can be changed at least in the long run. There are, consequently, many important questions concerning inventories that are related to the structure and organization of the inventory control system. Such questions may concern evaluation of investments to reduce setup costs, or whether the customers should be served through a single-stage or a multi-stage inventory system. Although we do not treat such questions directly, it is important to note that a correct evaluation must always be based on inventory models of the type considered in this book. The question is always whether the savings in inventory-related costs are larger than the costs for changing the structure of the system.

References

Brown, R. G. 1967. *Decision Rules for Inventory Management*, Holt, Rinehart and Winston, New York.

Chikán, A. Ed. 1990. *Inventory Models*, Kluwer Academic Publishers, Boston.

De Kok, A. G., and S. C. Graves. Eds. 2003. *Supply Chain Management: Design, Coordination and Operation*, Handbooks in OR & MS, Vol.11, North Holland, Amsterdam.

Graves, S. C., A. Rinnooy Kan, and P. H. Zipkin. Eds. 1993. *Logistics of Production and Inventory*, Handbooks in OR & MS, Vol.4, North Holland, Amsterdam.

Hadley, G., and T. M. Whitin. 1963. *Analysis of Inventory Systems*, Prentice-Hall, Englewood Cliffs, NJ.

Hax, A., and D. Candea. 1984. *Production and Inventory Management*, Prentice-Hall, Englewood Cliffs, NJ.

Johnson, L. A., and D. C. Montgomery. 1974. *Operations Research in Production Planning, Scheduling, and Inventory Control*, Wiley, New York.

Love, S. F. 1979. *Inventory Control*, McGraw-Hill, New York.

McClain, J. O., and L. J. Thomas. 1980. *Operations Management: Production of Goods and Services*, Prentice-Hall, Englewood Cliffs, NJ.

Muller, M. 2003. *Essentials of Inventory Management*, AMACOM, New York.

Naddor, E. 1966. *Inventory Systems*, Wiley, New York.

Nahmias, S. 1997. *Production and Operations Analysis*, 3rd edition, Irwin, Boston.

Orlicky, J. 1975. *Material Requirements Planning*, McGraw-Hill, New York.

Plossl, G. W., and O. W. Wight. 1985. *Production and Inventory Control*, Prentice-Hall, Englewood Cliffs, NJ.

Porteus, E. L. 2002. *Foundations of Stochastic Inventory Theory*, Stanford University Press.

Sherbrooke, C. C. 2004. *Optimal Inventory Modeling of Systems*, 2nd edition, Kluwer Academic Publishers, Boston.

Silver, E. A., D. F. Pyke, and R. Peterson. 1998. *Inventory Management and Production Planning and Scheduling*, 3rd edition, Wiley, New York.

Tersine, R. J. 1988. *Principles of Inventory and Materials Management*, 3rd edition, North-Holland, New York.

Veinott, A. 1966. The Status of Mathematical Inventory Theory, *Management Science*, 12, 745-777.

Vollman, T. E., W. L. Berry, and D. C. Whybark. 1997. *Manufacturing Planning and Control Systems*, 4th edition, Irwin, Boston.

Wagner, H. M. 1962. *Statistical Management of Inventory Systems*, Wiley, New York.

Zipkin, P. H. 2000. *Foundations of Inventory Management*, McGraw-Hill, Singapore.

2 FORECASTING

There are two main reasons why an inventory control system needs to order items some time before customers demand them. First, there is nearly always a *lead-time* between the ordering time and the delivery time. Second, due to certain ordering costs, it is often necessary to order in *batches* instead of unit for unit. This means that we need to look ahead and forecast the future demand. A demand forecast is an estimated average of the demand size over some future period. But it is not enough to estimate the average demand. We also need to determine how uncertain the forecast is. If the forecast is more uncertain, a larger safety stock is required. Consequently, it is also necessary to estimate the forecast error, which may be represented by the standard deviation or the Mean Absolute Deviation (*MAD*).

2.1 Objectives and approaches

In this chapter we shall consider forecasting methods that are suitable in connection with inventory control. Typical for such forecasts is that they concern a relatively short time horizon. Very seldom is it necessary to look more than one year ahead. In general, there are then two types of approaches that may be of interest:

- Extrapolation of historical data

When extrapolating historical data, the forecast is based on previous demand data. The available techniques are grounded in statistical methods for analysis of time series. Such techniques are easy to apply and use in computerized inventory control systems. It is no problem to regularly update forecasts for thousands of items, which is a common requirement in connection with practical inventory control. Extrapolation of historical data is the most common and important approach to obtain forecasts over a short horizon, and we shall devote the main part of this chapter to such techniques.

- Forecasts based on other factors

It is very common that the demand for an item depends on the demand for some other items. Consider, for example, an item that is used exclusively as a component when assembling some final products. It is then often natural to first forecast the demand for these final products, for example by extrapolation of historical data. Next we determine a production plan for the products. The demand for the considered component is then obtained directly from the production plan. This technique to "forecast" demand for dependent items is used in *Material Requirements Planning* (*MRP*) that is dealt with in Section 8.2.4.

But there are also other factors that might be reasonable to consider when forecasting demand. Assume, for example, that a sales campaign is just about to start or that a competing product is introduced on the market. Clearly this can mean that historical data are no longer representative when looking ahead. It is normally difficult to take such factors into account in the forecasting module of a computerized inventory control system. It is therefore usually most practical to adjust the forecast manually in case of such special events.

It is also possible, at least in principle, to use other types of dependencies. A forecast for the demand of ice cream can be based on the weather forecast. Consider, as another example, forecasting of the demand for a spare part that is used as a component in certain machines. The demand for the spare part can be expected to increase when the machines containing the part as a component are getting old. It is therefore reasonable to look for dependencies between the demand for the spare part and previous sales of the machines. As another example we can assume that the demand during a certain month will increase with the advertising expenditure the previous month. Such dependencies could be determined from historical data by regression analysis. (See Section 2.7.) Applications of such techniques are, however, very limited.

2.2 Demand models

Extrapolation of historical data is, as mentioned, the most common approach when forecasting demand in connection with inventory control. To deter-

FORECASTING

mine a suitable technique, we need to have some idea of how to model the stochastic demand. In principle, we should try to determine the model from analysis of historical data. In practice this is very seldom done. With many thousands of items, this initial work does not seem to be worth the effort in many situations. In other situations there are not enough historical data. A model for the demand structure is instead determined intuitively. In general, the assumptions are very simple.

2.2.1 Constant model

The simplest possible model means that the demands in different periods are represented by independent random deviations from an average that is assumed to be relatively stable over time compared to the random deviations. Let us introduce the notation:

x_t = demand in period t,
a = average demand per period (assumed to vary slowly),
ε_t = independent random deviation with mean zero.

A constant model means that we assume that the demand in period t can be represented as

$$x_t = a + \varepsilon_t. \tag{2.1}$$

Many products can be represented well by a constant model, especially products that are in a mature stage of a product life cycle and are used regularly. Examples are consumer products like toothpaste, many standard tools, and various spare parts. In fact, if we do not expect a trend or a seasonal pattern, it is in most cases reasonable to assume a constant model.

2.2.2 Trend model

If the demand can be assumed to increase or decrease systematically, it is possible to extend the model by also considering a linear trend. Let

a = average demand in period 0,
b = trend, that is the systematic increase or decrease per period (assumed to vary slowly).

A trend model means that the demand is modeled as:

$$x_t = a + bt + \varepsilon_t . \tag{2.2}$$

During a product life cycle there is an initial growth stage and a phase-out stage at the end of the cycle. During these stages it is natural to assume that the demand follows a trend model with a positive trend in the growth stage and a negative trend in the phase-out stage.

2.2.3 Trend-seasonal model

Let

F_t = seasonal index in period t (assumed to vary slowly).

If, for example, $F_t = 1.2$, this means that the demand in period t is expected to be 20 percent higher due to seasonal variations. If there are T periods in one year, we must require that for any T consecutive periods $\sum_{k=1}^{T} F_{t+k} = T$. When using a multiplicative trend-seasonal demand model it is assumed that the demand can be expressed as

$$x_t = (a + bt)F_t + \varepsilon_t . \tag{2.3}$$

By setting $b = 0$ in (2.3) we obtain a constant-seasonal model.

In (2.3) it is assumed that the seasonal variations increase and decrease proportionally with increases and decreases in the level of the demand series. In most cases this is a reasonable assumption. An alternative assumption could be that the seasonal variations are additive.

Many products have seasonal demand variations. For example the demand for ice cream is much larger during the summer than in the winter. Some products, like various Christmas decorations, are only sold during a very short period of the year. Still, the number of items with seasonal demand variations is usually very small compared to the total number of items. A seasonal model is only meaningful if the demand follows essentially the same pattern year after year.

2.2.4 Choosing demand model

When looking at the three demand models considered, it is obvious that (2.2) is more general than (2.1), and that (2.3) is more general than (2.2). It may then appear that it should be most advantageous to use the most general model (2.3). This is, however, not true. A more general demand model cov-

FORECASTING   11

ers a wider class of demands, but on the other hand, we need to estimate more parameters. Especially if the independent deviations are large, it may be very difficult to determine accurate estimates of the parameters, and it can therefore be much more efficient to use a simple demand model with few parameters. A more general model should be avoided unless there is some evidence that the generality will give certain advantages.

It is important to understand that the independent deviations ε_t cannot be forecasted, or in other words, the best forecast for ε_t is always zero. Consequently, if the independent deviations are large there is no possibility to avoid large forecast errors. Consider the constant model (2.1). It is obvious that the best forecast is simply our best estimate of a. In (2.2) the best forecast for the demand in period t is similarly our best estimate of $a + bt$, and in (2.3) our best forecast is the estimate of $(a + bt)F_t$.

In some situations it may be interesting to use more general demand models than (2.1) - (2.3). (See Section 2.9.) This would, however, require a detailed statistical analysis of the demand structure. In practice this is rarely done in connection with inventory control.

One practical problem is that it is quite often difficult to measure demand, since only sales are recorded. If historical sales, instead of historical demands, are used for forecasting demand, considerable errors may occur in situations where a relatively large portion of the total demand is lost due to shortages. (See Section 2.10.5.)

2.3 Moving average

Assume that the underlying demand structure is described by the constant model (2.1). Since the independent deviations ε_t cannot be predicted, we simply want to estimate the constant a. If a were completely constant the best estimate would be to take the average of all observations of x_t. But a can be expected to vary slowly. This means that we need to focus on the most recent values of x_t. The idea of the moving average technique is to take the average over the N most recent values. Let

$\hat{a}_t$ = estimate of a after observing the demand in period t,
$\hat{x}_{t,\tau}$ = forecast for period $\tau > t$ after observing the demand in period t.

We obtain:

$$\hat{x}_{t,\tau} = \hat{a}_t = (x_t + x_{t-1} + x_{t-2} + ... + x_{t-N+1})/N. \qquad (2.4)$$

Note that the forecast demand is the same for any value of $\tau > t$. This is, of course, because we are assuming a constant demand model.

The value of N should depend on how slowly we think that a is varying, and on the size of the stochastic deviations ε_t. If a is varying more slowly and the stochastic deviations are larger, we should use a larger value of N. This will limit the influence of the stochastic deviations. On the other hand, if a is varying more rapidly and the stochastic variations are small, we should prefer a small value of N, which will allow us to follow the variations in a in a better way.

If we use one month as our period length and set $N = 12$, the forecast is the average over the preceding year. This may be an advantage if we want to prevent seasonal variations from affecting the forecast.

2.4 Exponential smoothing

2.4.1 Updating procedure

When using exponential smoothing instead of a moving average, the forecast is updated differently. The result is in many ways similar, though. We are again assuming a constant demand model, and we wish to estimate the parameter a. To update the forecast in period t, we use a linear combination of the previous forecast and the most recent demand x_t,

$$\hat{x}_{t,\tau} = \hat{a}_t = (1-\alpha)\hat{a}_{t-1} + \alpha x_t, \tag{2.5}$$

where $\tau > t$ and

$\alpha =$ smoothing constant ($0 < \alpha < 1$).

Due to the constant demand model the forecast is again the same for any future period.

Note that the updating procedure can also be expressed as

$$\hat{x}_{t,\tau} = \hat{a}_t = \hat{a}_{t-1} + \alpha(x_t - \hat{a}_{t-1}). \tag{2.6}$$

We have assumed that $0 < \alpha < 1$ although it is also possible to use $\alpha = 0$ and $\alpha = 1$. The value $\alpha = 0$ means simply that we do not update the forecast, while $\alpha = 1$ means that we choose the most recent demand as our forecast.

2.4.2 Comparing exponential smoothing to a moving average

To be able to compare exponential smoothing to a moving average, we can express the forecast in the following way:

$$\hat{a}_t = (1-\alpha)\hat{a}_{t-1} + \alpha x_t = (1-\alpha)((1-\alpha)\hat{a}_{t-2} + \alpha x_{t-1}) + \alpha x_t$$

$$= \alpha x_t + \alpha(1-\alpha)x_{t-1} + (1-\alpha)^2 \hat{a}_{t-2} = ... = \alpha x_t + \alpha(1-\alpha)x_{t-1}$$

$$+ \alpha(1-\alpha)^2 x_{t-2} + ... + \alpha(1-\alpha)^n x_{t-n} + (1-\alpha)^{n+1} \hat{a}_{t-n-1}. \quad (2.7)$$

Let us now compare (2.7) to (2.4). In (2.4) the N last period demands all have the weight $1/N$. In (2.7) we have, in principle, positive weights for all previous demands, but the weights are decreasing exponentially as we go backwards in time. This is the reason for the name exponential smoothing. The sum of the weights is still unity.[1] When using a moving average, a larger value of N means that we put relatively more emphasis on old values of demand. When applying exponential smoothing, a small value of α will give essentially the same effect.

When using a moving average according to (2.4) the forecast is based on the demands in periods t, t - 1, ..., t - N + 1. The ages of these data are respectively 0, 1, ..., and N - 1 periods. The weights are all equal to $1/N$. The average age is therefore $(N - 1)/2$ periods. To be able to compare the parameter N to the smoothing constant α, we shall also determine the average age of the data when using exponential smoothing according to (2.5), or equivalently (2.7). We obtain:[2]

$$\alpha 0 + \alpha(1-\alpha)1 + \alpha(1-\alpha)^2 2 + ... = \alpha(1-\alpha)S'(1-\alpha) = (1-\alpha)/\alpha, \quad (2.8)$$

---

[1] Let $0 \leq x < 1$ and consider the infinite geometric sum $S(x) = 1 + x + x^2 + x^3...$ Note that $S(x) = 1 + x \cdot S(x)$. This implies that $S(x) = 1/(1-x)$. The sum of the weights in (2.7) is $\alpha \cdot S(1-\alpha) = 1$.

[2] Let $0 \leq x < 1$ and consider the infinite geometric sum $S'(x) = 1 + 2x + 3x^2 ..$ $= 1 + x + x^2 + ... + x(1 + 2x + 3x^2 ...) = S(x) + x \cdot S'(x)$. This implies that $S'(x) = S(x)/(1-x) = 1/(1-x)^2$.

and we can conclude that the forecasts are based on data of the "same average age" if

$$(1-\alpha)/\alpha = (N-1)/2, \qquad (2.9)$$

or equivalently when

$$\alpha = 2/(N+1). \qquad (2.10)$$

Consider, for example, a moving average that is updated monthly with $N = 12$. This means that each month in the preceding year has weight 1/12. Consider then an exponential smoothing forecast that is also updated monthly. A value of α "corresponding" to $N = 12$ is according to (2.10) obtained as $\alpha = 2/(12 + 1) = 2/13 \approx 0.15$.

2.4.3 Practical considerations and an example

If the period length is one month, it is common in practice to use a smoothing constant α between 0.1 and 0.3. Table 2.1 shows the weights for different previous demands for $\alpha = 0.1$ and $\alpha = 0.3$.

Table 2.1 Weights for demand data in exponential smoothing

Period	Weight	$\alpha = 0.1$	$\alpha = 0.3$
t	α	0.100	0.300
$t-1$	$\alpha(1-\alpha)$	0.090	0.210
$t-2$	$\alpha(1-\alpha)^2$	0.081	0.147
$t-3$	$\alpha(1-\alpha)^3$	0.073	0.103
$t-4$	$\alpha(1-\alpha)^4$	0.066	0.072

We can see from Table 2.1 that the forecasting system will react much faster if we use $\alpha = 0.3$. On the other hand, the stochastic deviations will affect the forecast more compared to when $\alpha = 0.1$. When choosing α we always have to compromise.

If the forecast is updated more often, for example each week, a smaller α should be used. To see how much smaller we can apply (2.10). Assume that we start with a monthly update and that we use the value of α "corresponding" to a moving average with $N = 12$, i.e., $\alpha \approx 0.15$. When changing to weekly forecasts it is natural to change N to 52. The "corresponding" value of α is obtained from (2.10) as $\alpha = 2/(52 + 1) \approx 0.04$.

When starting to forecast according to (2.5) in some period t, an initial forecast to be used as $\hat{a}_{t-1}$ is needed. We can use some simple estimate of the average period demand. If no such estimate is available, it is possible to start with $\hat{a}_{t-1} = 0$, since $\hat{a}_{t-1}$ will not affect the forecast in the long run, see (2.7). However, especially for small values of α, it can take a long time until the forecasts are reliable. If it is necessary to start with a very uncertain initial forecast, it may be a good idea to use a rather large value of α to begin with, since this will reduce the influence of the initial forecast.

Example 2.1 The demand for an item usually fluctuates considerably. A moving average or a forecast obtained by exponential smoothing gives essentially an average of more recent demands. The forecast cannot, as we have emphasized before, predict the independent stochastic deviations. Table 2.2 shows some typical demand data and the corresponding exponential smoothing forecasts with $\alpha = 0.2$. It is assumed that the forecast after period 2 is $\hat{a}_2 = 100$.

Table 2.2 Forecasts obtained by exponential smoothing with $\alpha = 0.2$. Initial forecast $\hat{a}_2 = 100$.

Period	Demand in period t, x_t	Forecast at the end of period t, $\hat{a}_t$
3	72	94
4	170	110
5	67	101
6	95	100
7	130	106

In Table 2.2 the forecast immediately after period 3 is obtained by applying (2.5).

$$\hat{a}_3 = 0.8 \cdot 100 + 0.2 \cdot 72 = 94.4,$$

which is rounded to 94 in Table 2.2. Note that when determining $\hat{a}_3$ the demands in future periods are not known. Therefore at this stage, $\hat{a}_3$ serves as our forecast for any future period. After period 4 the forecast is again updated

$$\hat{a}_4 = 0.8 \cdot 94.4 + 0.2 \cdot 170 = 109.52.$$

If we compare exponential smoothing to a moving average there are some obvious but minor advantages with exponential smoothing. Because the average a, which we wish to estimate is assumed to vary slowly, it is reasonable to use larger weights for the most recent demands as is done in exponential smoothing. As we have discussed before, however, a moving average over a full year may be advantageous if we want to eliminate the influence of seasonal variations on the forecast. It is also interesting to note that with exponential smoothing we only need to keep track of the previous forecast and the most recent demand.

In practice, exponential smoothing (or possibly a moving average) is, in general, a suitable technique for most items. But there is also usually a need for other methods for relatively small groups of items for which it is feasible and interesting to follow up trends and/or seasonal variations.

2.5 Exponential smoothing with trend

2.5.1 Updating procedure

Let us now instead assume that the demand follows a trend model according to (2.2). To forecast demand we need to estimate the two parameters a and b, compared to only a in case of a constant model. As before, we cannot predict the independent deviations ε_t. There are different techniques for estimating a and b. We shall here consider a method that was first suggested by Holt (1957). (Another technique based on linear regression is described in Section 2.7.) Estimates of a and b are successively updated according to (2.11) and (2.12).

$$\hat{a}_t = (1-\alpha)(\hat{a}_{t-1} + \hat{b}_{t-1}) + \alpha x_t, \qquad (2.11)$$

$$\hat{b}_t = (1-\beta)\hat{b}_{t-1} + \beta(\hat{a}_t - \hat{a}_{t-1}), \qquad (2.12)$$

where α and β are smoothing constants between 0 and 1.

The "average" $\hat{a}_t$ corresponds to period t, i.e., the period for which we have just observed the demand. The forecast for a future period, $t + k$ is obtained as

$$\hat{x}_{t,t+k} = \hat{a}_t + k \cdot \hat{b}_t. \qquad (2.13)$$

The most important difference compared to simple exponential smoothing according to (2.5) is that the forecasts for future periods are no longer the same. Note that the trend, i.e., the change per period, can just as well be negative.

The method means that $\hat{a}_t$ is always adjusted to fit the present period. With this in mind, (2.11) is essentially equivalent to (2.5). As long as x_t is unknown, $\hat{a}_{t-1} + \hat{b}_{t-1}$ is our best estimate for the mean demand in period t. We determine $\hat{a}_t$ as a linear combination of this estimate and the new demand x_t. The average difference between two consecutive values of $\hat{a}_t$ should in the long run be equal to the trend. Therefore, we use these differences in (2.12) to update the trend by exponential smoothing.

It can be shown that if the demand is a linear function without stochastic variations, the forecast will, in the long run independent of the initial values, estimate the future demand exactly. When using simple exponential smoothing this is not the case. See Problem 2.4.

2.5.2 Practical considerations and an example

The idea behind exponential smoothing with trend is to be able to follow systematic linear changes in demand better. As with exponential smoothing, larger values of the smoothing constants α and β will mean that the forecasting system reacts faster to changes but will also make the forecasts more sensitive to stochastic deviations. When choosing values in practice it can be recommended to have a relatively low value of β since errors in the trend can give serious forecast errors for relatively long forecast horizons. Note that the trend is multiplied by k in (2.13). It is therefore very unfortunate if pure stochastic variations are interpreted as a trend. When updating the forecast monthly, typical values of the smoothing constants may be $\alpha = 0.2$ and $\beta = 0.05$. When the forecasting system is initiated it is usually reasonable to set the trend to 0 and, as in exponential smoothing, let the initial $\hat{a}$ be equal to some estimate of the average period demand. If the initial values are very uncertain it can be reasonable, also for exponential smoothing with trend, to use extra large smoothing constants in an initial phase.

Example 2.2 We consider the same demand data as in Example 2.1. Table 2.3 illustrates the forecasts when applying exponential smoothing with trend and looking one and five periods ahead, respectively. The smoothing constants are $\alpha = 0.2$ and $\beta = 0.1$. At the end of period 2, $\hat{a}_2 = 100$ and $\hat{b}_2 = 0$.

Table 2.3 Forecasts obtained by exponential smoothing with trend. The smoothing constants are $\alpha = 0.2$ and $\beta = 0.1$, and the initial forecast $\hat{a}_2 = 100$, $\hat{b}_2 = 0$.

Period t	Demand in period t, x_t	Forecast for period $t+1$ at the end of period t, $\hat{a}_t + \hat{b}_t$	Forecast for period $t+5$ at the end of period t, $\hat{a}_t + 5\hat{b}_t$
3	72	94	92
4	170	110	114
5	67	102	102
6	95	100	100
7	130	107	109

In period 3 we obtain from (2.11) and (2.12)

$$\hat{a}_3 = 0.8 \cdot (100 + 0) + 0.2 \cdot 72 = 94.4,$$

$$\hat{b}_3 = 0.9 \cdot 0 + 0.1 \cdot (94.4 - 100) = -0.56.$$

Our forecast for period 4 is then 94.4 - 0.56 = 93.84 ≈ 94. At the end of period 4 we obtain the real demand 170. Applying (2.11) and (2.12) again we get

$$\hat{a}_4 = 0.8 \cdot (94.4 - 0.56) + 0.2 \cdot 170 = 109.072,$$

$$\hat{b}_4 = 0.9 \cdot (-0.56) + 0.1 \cdot (109.072 - 94.4) = 0.9632.$$

The forecasts in Table 2.3 are rounded to integers.

2.6 Winters' trend-seasonal method

Let us now assume that the demand is modeled by a multiplicative trend-seasonal model, see (2.3). Such a model is usually only used for a few items with very clear seasonal variations like Christmas decorations.

2.6.1 Updating procedure

Winters (1960) suggested the following technique that can be seen as a generalization of exponential smoothing with trend. We shall describe how the parameters a, b, and F_t are updated. Note first that in (2.3), $a + bt$ represents the development of demand if we disregard the seasonal variations. When we record the demand x_t in period t, we can similarly interpret $x_t / \hat{F}_t$ as the demand without seasonal variations. At this stage we have not updated $\hat{F}_t$ with respect to the new observation x_t. In complete analogy with (2.11) and (2.12) we now have

$$\hat{a}_t = (1-\alpha)(\hat{a}_{t-1} + \hat{b}_{t-1}) + \alpha(x_t / \hat{F}_t), \qquad (2.14)$$

$$\hat{b}_t = (1-\beta)\hat{b}_{t-1} + \beta(\hat{a}_t - \hat{a}_{t-1}), \qquad (2.15)$$

It remains to update the seasonal indices. We first determine

$$\hat{F}'_t = (1-\gamma)\hat{F}_t + \gamma(x_t / \hat{a}_t), \qquad (2.16)$$

and

$$\hat{F}'_{t-i} = \hat{F}_{t-i} \text{ for } i = 1, 2,..., T\text{-}1, \qquad (2.17)$$

where $0 < \gamma < 1$ is another smoothing constant. Recall that T is the number of periods per year. Note that (2.17) means that the other indices are left unchanged in this initial step. We must also require, however, that the sum of T consecutive seasonal indices is equal to T. Therefore, we need to normalize all indices

$$\hat{F}_{t-i} = \hat{F}'_{t-i} (T / \sum_{k=0}^{T-1} \hat{F}'_{t-k}) \text{ for } i = 0, 1, \ldots, T\text{-}1. \qquad (2.18)$$

These indices are also applied to future periods until the indices are updated the next time. For example,

$$\hat{F}_{t-i+kT} = \hat{F}_{t-i} \text{ for } i = 0, 1, \ldots, T\text{-}1, \text{ and } k = 1, 2, \ldots. \qquad (2.19)$$

The forecast for period $t + k$ is obtained as

$$\hat{x}_{t,t+k} = (\hat{a}_t + k \cdot \hat{b}_t)\hat{F}_{t+k}. \qquad (2.20)$$

An alternative is to use manually set seasonal indices. In this case we do not need to go through (2.16) - (2.19).

The corresponding constant seasonal model is obtained if (2.14) and (2.15) are replaced by

$$\hat{a}_t = (1-\alpha)\hat{a}_{t-1} + \alpha(x_t/\hat{F}_t), \qquad (2.21)$$

and (2.20) by

$$\hat{x}_{t,t+k} = \hat{a}_t \hat{F}_{t+k}. \qquad (2.22)$$

2.6.2 Practical considerations and an example

Example 2.3 To illustrate the computations we shall go through a complete updating of all parameters. Assume that we are dealing with monthly updates, i.e., that $T = 12$. The smoothing constants are $\alpha = 0.2$, $\beta = 0.05$, and $\gamma = 0.2$. Assume that the last update took place in period 23 and that this update resulted in the following parameters: $\hat{a}_{23} = 9$, $\hat{b}_{23} = 1$, $\hat{F}_{12} = \hat{F}_{13} = \hat{F}_{14} = 1.2$, $\hat{F}_{15} = \hat{F}_{16} = \hat{F}_{17} = \hat{F}_{18} = 1$, $\hat{F}_{19} = 0.4$ and, $\hat{F}_{20} = \hat{F}_{21} = \hat{F}_{22} = \hat{F}_{23} = 1$. Note that the sum of the seasonal indices equals 12. At this stage $\hat{F}_{24} = \hat{F}_{12} = 1.2$ according to (2.19).

In period 24 we record the demand $x_{24} = 7$. Applying (2.14) - (2.16) we get

$$\hat{a}_{24} = 0.8(9+1) + 0.2(7/1.2) = 9.167$$

$$\hat{b}_{24} = 0.95 \cdot 1 + 0.05(9.167 - 9) = 0.958$$

$$\hat{F}'_{24} = 0.8 \cdot 1.2 + 0.2 \cdot (7/9.167) = 1.113$$

We obtain $\sum_{i=13}^{24} \hat{F}'_i = 11.913$. By applying (2.18) we get the updated normalized indices for periods 13-24 as $\hat{F}_{13} = \hat{F}_{14} = 1.209$, $\hat{F}_{15} = \hat{F}_{16} = \hat{F}_{17} = \hat{F}_{18} = 1.007$, $\hat{F}_{19} = 0.403$, $\hat{F}_{20} = \hat{F}_{21} = \hat{F}_{22} = \hat{F}_{23} = 1.007$, and $\hat{F}_{24} = 1.121$.

The forecast for period 26 is obtained from (2.20) as

FORECASTING

$$\hat{x}_{24,26} = (9.167 + 2 \cdot 0.958) \cdot 1.209 = 13.40,$$

where we apply $\hat{F}_{26} = \hat{F}_{14}$ according to (2.19).

When using a trend-seasonal method, which also updates the seasonal indices, it is quite often difficult to distinguish systematic seasonal variations from independent stochastic deviations. The problem is that we have so many parameters to estimate. The indices may then become very uncertain. Sometimes it can therefore be more efficient to estimate the indices in other ways. For example, if a group of items can be expected to have very similar seasonal variations, it may be advantageous to estimate the indices from the total demand for the whole group of items. By doing so we can limit the influence of the purely stochastic deviations. Assume, for example, that we have a number of items that can all be classified as Christmas decorations. By aggregating historical demand data for all items in this group, it may be possible to estimate seasonal indices that are quite accurate for all items in the group together. We can, for example, use Winter's procedure as described above for determining the indices for the aggregate data. Furthermore, we can usually assume that the obtained indices are also reasonably accurate for the individual items in the group. So when determining forecasts for individual items we regard these aggregate seasonal indices as given.

In general, it can also be recommended that only items with very obvious seasonal variations be accepted as seasonal items.

2.7 Using regression analysis

2.7.1 Forecasting demand for a trend model

Assume once again a demand model with trend according to (2.2). One technique to forecast the future demand has been described in Section 2.5. Let us consider an alternative technique. Assume that we have just obtained the demand in period t and that we wish to base the forecast on the N most recent observations $x_t, x_{t-1}, \ldots, x_{t-N+1}$. One possibility is to use simple linear regression, which means that we fit a line $y_{t+k} = \hat{a}_t + \hat{b}_t k$ to the observations so that the sum of the squared errors is minimized for $k = -N+1, -N+2, \ldots, 0$. We can then use y_{t+k} for $k > 0$ as our forecast for period $t + k$, i.e., we set $\hat{x}_{t,t+k} = y_{t+k}$.

Let us now determine $\hat{a}_t$ and $\hat{b}_t$. We wish to minimize

$$\sum_{k=-N+1}^{0}(x_{t+k}-\hat{a}_t-\hat{b}_t k)^2. \tag{2.23}$$

Setting the derivatives with respect to $\hat{a}_t$ and $\hat{b}_t$ equal to zero we obtain

$$-2\sum_{k=-N+1}^{0}(x_{t+k}-\hat{a}_t-\hat{b}_t k)=0, \tag{2.24}$$

$$-2\sum_{k=-N+1}^{0}k(x_{t+k}-\hat{a}_t-\hat{b}_t k)=0. \tag{2.25}$$

Using the notation

$$\bar{k}=\frac{1}{N}\sum_{k=-N+1}^{0}k=-(N-1)/2, \tag{2.26}$$

$$\bar{x}=\frac{1}{N}\sum_{k=-N+1}^{0}x_{t+k}, \tag{2.27}$$

we get from (2.24)

$$\hat{a}_t=\bar{x}-\hat{b}_t\bar{k}, \tag{2.28}$$

and by inserting in (2.25)

$$\hat{b}_t=\frac{\sum_{k=-N+1}^{0}k(x_{t+k}-\bar{x})}{\sum_{k=-N+1}^{0}(k^2-k\bar{k})}=\frac{\sum_{k=-N+1}^{0}(k-\bar{k})(x_{t+k}-\bar{x})}{\sum_{k=-N+1}^{0}(k-\bar{k})^2}. \tag{2.29}$$

Finally we get $\hat{a}_t$ from (2.28).

FORECASTING 23

2.7.2 Practical considerations and an example

This technique based on linear regression can be seen as a generalization of a moving average (Section 2.3) to a demand model with trend, in the same way as we can see exponential smoothing with trend (Section 2.5) as a generalization of exponential smoothing (Section 2.4). When determining $\hat{a}_t$ and $\hat{b}_t$ by linear regression, we give the same weight to the N most recent known demands as we do when determining a moving average. Furthermore, if instead of fitting a line to the observations, we fit a constant $\hat{a}_t$ so that the sum of the squared errors is minimized for $k = -N+1, -N+2, ..., 0$, we will obtain a moving average. See Problem 2.5. Normally it is reasonable to give more weight to more recent observations as is done when applying exponential smoothing with trend, which has also become a more common forecasting technique.

Example 2.4 Let us go back to the data in Example 2.2 where we applied exponential smoothing with trend. Assume that the demands in periods 3-7 are given. The corresponding values of k are $k = -4, -3, ..., 0$. See Figure 2.1.

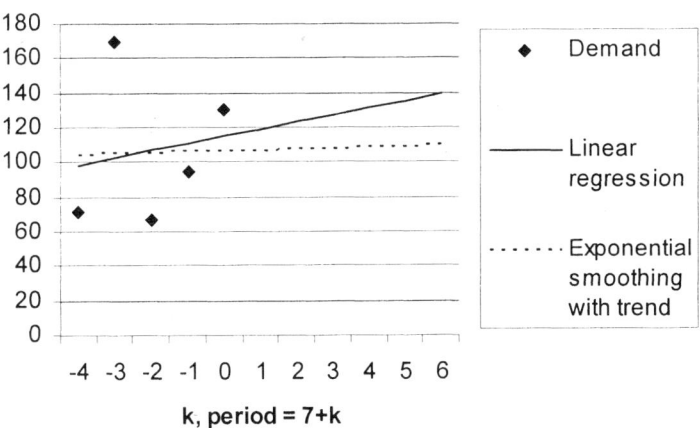

Figure 2.1 Forecasts (for $k > 0$) with linear regression and exponential smoothing with trend.

In Example 2.2 we updated an initial forecast $\hat{a}_2 = 100$ and $\hat{b}_2 = 0$ successively. In period 7 we got $\hat{a}_7 = 106.1571$ and $\hat{b}_7 = 0.567955$. See Figure 2.1, where the corresponding forecasts $\hat{x}_{7,7+k} = \hat{a}_7 + \hat{b}_7 k$ are illustrated. In the figure we also show the line for $k \leq 0$, although these values are not used because we already know the demands. We have also used linear regression (2.26) - (2.29) based on the demands in periods 3, 4, ..., 7 to determine $\hat{a}_7 = 115$ and $\hat{b}_7 = 4.1$. The corresponding forecast is also illustrated in Figure 2.1. Again we also show the line for $k \leq 0$, although these values are not used as forecasts.

In this example we have only used five demands, $N = 5$, when determining the linear regression forecast. For the considered demand variations, which are rather typical, it is normally better to use a considerably larger N ($N \geq 10$) to get a more stable forecast. It is easy to see from Figure 2.1 that an occasional very high or very low demand can affect the forecast considerably.

2.7.3 Forecasts based on other factors

The application of linear regression for forecasting demand in case of a trend model is fairly limited. Usually exponential smoothing with trend is preferred. Of more practical interest is to use linear regression when the demand depends on one or more other factors that are known when making the forecast.

Consider as an example a spare part that is exclusively used as a replacement when maintaining a certain machine. The maintenance activities are carried out by the company selling the spare part, and the customers have to order the maintenance two months in advance. In such a case it is natural to assume that the demand for the spare part in the next month is essentially proportional to the known number of machines undergoing maintenance in that month. Let

x_t = demand in month t,
z_t = number of machines undergoing maintenance in month t.

It is then natural to assume the following demand model

$$x_t = a + bz_t + \varepsilon_t, \qquad (2.30)$$

where a and b are constants and ε_t are independent random variations with mean zero. Note now that we can fit a line $y_{t+k} = \hat{a}_t + \hat{b}_t z_{t+k}$ to the N most recent observations so that the sum of the squared errors is minimized exactly as we did in Section 2.7.1, where we dealt with the special case $z_{t+k} = $

k. We can then use y_{t+k} for $k > 0$ as our forecast for period $t + k$, i.e., we set $\hat{x}_{t,t+k} = y_{t+k}$. In complete analogy with Section 2.7.1 we get

$$\hat{b}_t = \frac{\sum_{k=-N+1}^{0}(z_{t+k} - \bar{z})(x_{t+k} - \bar{x})}{\sum_{k=-N+1}^{0}(z_{t+k} - \bar{z})^2}, \qquad (2.31)$$

and

$$\hat{a}_t = \bar{x} - \hat{b}_t \bar{z}, \qquad (2.32)$$

where $\bar{x}$ and $\bar{z}$ are obtained as

$$\bar{x} = \frac{1}{N}\sum_{k=-N+1}^{0} x_{t+k}, \qquad (2.33)$$

$$\bar{z} = \frac{1}{N}\sum_{k=-N+1}^{0} z_{t+k}. \qquad (2.34)$$

Note that we must require that not all z_{t+k} ($k = -N + 1, -N + 2,\ldots,0$) are identical. In that case $\hat{b}_t z_{t+k}$ is constant and there is no unique solution for $\hat{a}_t$ and $\hat{b}_t$. One optimal solution is obtained by setting $\hat{b}_t = 0$. From (2.32) we then get $\hat{a}_t = \bar{x}$, i.e., simply a moving average.

2.7.4 More general regression models

Several generalizations of the models considered in Sections 2.7.1-2.7.3 are possible. We can include more than one factor by using multiple regression, i.e., the demand model may be expressed as

$$x_t = a + \sum_{j=1}^{J} b_j z_{j,t} + \varepsilon_t. \qquad (2.35)$$

Such models are, however, not common when forecasting demand in connection with inventory control.

Furthermore, some nonlinear relationships can easily be transformed into linear models. Assume, for example, that we wish to choose positive parameters $\hat{a}_t$ and $\hat{b}_t$ so that the function

$$y_{t+k} = \hat{a}_t \hat{b}_t^{z_{t+k}} \tag{2.36}$$

closely follows some previous demand data. If we take logarithms of both sides we get

$$\log y_{t+k} = \log \hat{a}_t + (\log \hat{b}_t) z_{t+k}, \tag{2.37}$$

and we can then use simple linear regression to determine $\log \hat{a}_t$ and $\log \hat{b}_t$.

2.8 Sporadic demand

Sometimes an item is demanded very seldom, while the quantity that is demanded by a customer may be relatively large. It may, for example, turn out that customers order only about twice per year. If we use exponential smoothing the resulting forecast will decrease in periods without demand and go up again when a customer arrives. If we use a small smoothing constant the forecast will still be relatively stable, but on the other hand, it will react very slowly to demand changes. Croston (1972) has suggested a simple technique to handle such a situation. The forecast is only updated in periods with positive demand. In case of a positive demand, two averages are updated by exponential smoothing: the size of the positive demand, and the time between two periods with positive demand. This gives a more stable forecast and also a better feeling for the structure of the demand. Let as before x_t be the demand in period t. Define also

- k_t = the stochastic number of periods since the preceding positive demand,
- $\hat{k}_t$ = estimated average of the number of periods between two positive demands at the end of period t,
- $\hat{d}_t$ = estimated average of the size of a positive demand at the end of period t,
- $\hat{a}_t$ = estimated average demand per period at the end of period t.

We update $\hat{k}_t$ and $\hat{d}_t$ in the following way:

(i) $x_t = 0$
$$\hat{k}_t = \hat{k}_{t-1},$$
$$\hat{d}_t = \hat{d}_{t-1}, \quad (2.38)$$

(ii) $x_t > 0$
$$\hat{k}_t = (1-\alpha)\hat{k}_{t-1} + \alpha k_t,$$
$$\hat{d}_t = (1-\beta)\hat{d}_{t-1} + \beta x_t, \quad (2.39)$$

where $0 < \alpha, \beta < 1$ are smoothing constants.

We get the forecast for the demand per period as

$$\hat{a}_t = \hat{d}_t / \hat{k}_t. \quad (2.40)$$

2.9 Box-Jenkins techniques

The stochastic variations in the demand models that we have considered, (2.1) - (2.3), are assumed to be independent. This is a major simplification. It is, however, easy to realize that situations exist when this is not true. If there are only a few large customers we can sometimes expect demands in consecutive periods to be negatively correlated. A high demand in one period can indicate that several of the customers have replenished, and it is reasonable to expect that the demand in the next period will be a little lower. However, there are also situations with positively correlated demand. A high demand in one period may mean that the product is exposed to more potential customers, and a high demand can also be expected in the next period.

Forecasting techniques that can handle correlated stochastic demand variations and other more general demand processes have been developed by Box and Jenkins (1970). See also Box et al. (1994).

A general non-seasonal demand model is known as an AutoRegressive Integrated Moving Average (ARIMA) model. There is a large variety of such models. It is common to use the notation ARIMA(p, d, q) where

AR: p = order of autoregressive part,
I: d = degree of first differencing involved,
MA: q = order of the moving average part.

Consider as an example an ARIMA(p, 0, q) model.

$$x_t = a + b_1 x_{t-1} + b_2 x_{t-2} + \ldots + b_p x_{t-p} + \varepsilon_t + c_1 \varepsilon_{t-1} + c_2 \varepsilon_{t-2} + \ldots + c_q \varepsilon_{t-q}. \tag{2.41}$$

In (2.41) ε_i are passed errors that, combined with previous demands, are used as explanatory variables, while b_i and c_i are constants. The errors are assumed to be independent and have zero mean. The forecasting technique can be divided into two steps. In each period all constants are first estimated from historical data, and next the model is used to determine forecasts for future periods. Assume that we want to forecast the demand in period t, and that we know the outcome in all previous periods. The error in period t, ε_t, is then not known and is replaced by zero since it has zero mean. Otherwise we can use observed values of previous demands and errors. If we also want to forecast the demand in period $t+1$ we increase the subscripts in (2.41) by one throughout. To get the forecast for x_{t+1}, we also use our forecast for x_t and we set the unknown errors ε_{t+1} and ε_t equal to zero.

More extensive computations are needed to use such a technique. Furthermore, a relatively large record of historical data must be kept available. This, in connection with inventory control, can usually only be motivated for a few very important products. Various computer programs exist for fitting ARIMA models.

The demand model in (2.41) has $d = 0$. Because of that it is also equivalently denoted ARMA(p, q). Similarly, if $p = 0$ we can equivalently say IMA(d, q). If $p = d = 0$ we get a MA(q) model.

Let us now consider an ARIMA(p, 1, q) model.

$$x'_t = a + b_1 x'_{t-1} + b_2 x'_{t-2} + \ldots + b_p x'_{t-p} + \varepsilon_t + c_1 \varepsilon_{t-1} + c_2 \varepsilon_{t-2} + \ldots + c_q \varepsilon_{t-q}. \tag{2.42}$$

The only difference compared to (2.41) is that x_i is now replaced by $x'_i = x_i - x_{i-1}$. If $d = 2$, x'_i is replaced by $x''_i = x'_i - x'_{i-1}$.

To illustrate the richness of the class of ARIMA models let us consider a simple example, ARIMA(0, 1, 1) with $a = 0$ and $c_1 = -(1 - \alpha)$,

$$x_{t+1} = x_t + \varepsilon_{t+1} - (1 - \alpha) \varepsilon_t, \tag{2.43}$$

i.e., after observing the demand in period t our forecast for period $t + 1$ is

$$\hat{a}_t = x_t - (1 - \alpha) \varepsilon_t. \tag{2.44}$$

FORECASTING 29

Because $x_t = \hat{a}_{t-1} + \varepsilon_t$, we can rewrite (2.44) as

$$\hat{a}_t = \hat{a}_{t-1} + \alpha\varepsilon_t = \hat{a}_{t-1} + \alpha(x_t - \hat{a}_{t-1}) = (1-\alpha)\hat{a}_{t-1} + \alpha x_t. \qquad (2.45)$$

So we can conclude that exponential smoothing is just a special case of an ARIMA model.

In practice it is usually sufficient to consider values of p, d, q in the range 0, 1, 2. This will simplify the model identification. Still it is possible to cover a very large set of practical forecasting situations.

It is also possible to add seasonality to ARIMA models. See e.g., Makridakis et al. (1998).

2.10 Forecast errors

2.10.1 Common error measures

So far we have only discussed how we can estimate the mean of the future demand. But it is not sufficient only to know the mean. To be able to determine a suitable safety stock we also need to know how uncertain the forecast is, i.e., how large the forecast errors tend to be.

The most common way to describe variations around the mean is through the *standard deviation*. Let X be a stochastic variable with mean $m = E(X)$. The standard deviation σ is defined as

$$\sigma = \sqrt{E(X-m)^2}, \qquad (2.46)$$

and σ^2 is denoted the *variance*. In connection with forecast errors it is, by old tradition, not common to estimate σ or σ^2 directly. Instead the *Mean Absolute Deviation (MAD)* is estimated. *MAD* is the expected value of the absolute deviation from the mean

$$MAD = E|X - m|. \qquad (2.47)$$

The original reason that *MAD* is estimated instead of σ or σ^2 was that this simplified the computations. Today it is no problem to estimate σ or σ^2 directly, but still most forecasting systems first evaluate *MAD*. It is obvious that *MAD* and σ in most cases give a very similar picture of the variations around the mean. It is also possible to relate them to each other. A common assumption is that the forecast errors are *normally* distributed. In that case it is easy to show that

$$\sigma = \sqrt{\pi/2} \, MAD \approx 1.25 \, MAD. \tag{2.48}$$

This relationship is very often used in connection with forecasting, also in situations when it is less natural to assume that the forecast errors are normally distributed. The approximation $\sqrt{\pi/2} \approx 1.25$ is in no way needed but is still quite often used in practice.

2.10.2 Updating MAD or σ^2

We shall now describe how to successively update *MAD* or σ^2. Let MAD_t be the estimate of *MAD* after period t. At the end of period $t-1$ we obtained from the forecasting system a forecast for period t, $\hat{x}_{t-1,t}$. At this stage we could regard this as a "mean" for the stochastic demand in period t, x_t. (Of course, this is not always true. Very often there are substantial systematic errors in the forecast.) After period t we know x_t and the corresponding absolute deviation from the "mean", $|x_t - \hat{x}_{t-1,t}|$. It is, in general, assumed that these absolute variations can be seen as independent random deviations from a mean which varies relatively slowly, i.e., that they follow a constant model according to (2.1). It is then natural to update the average absolute error, MAD_t, by exponential smoothing, see (2.5). The forecast for the absolute error at the end of period t is consequently determined as

$$MAD_t = (1-\alpha)MAD_{t-1} + \alpha |x_t - \hat{x}_{t-1,t}|, \tag{2.49}$$

where $0 < \alpha < 1$ is a smoothing constant (not necessarily the same as in (2.5)).

Since the observed absolute deviations usually vary quite a lot, it is common to use a relatively small smoothing constant like $\alpha = 0.1$ in the case of monthly updates.

An alternative to (2.49) would be to update MAD_t as a moving average.

When we have determined MAD_t we use (2.48) to get the corresponding standard deviation.

$$\sigma_t = \sqrt{\pi/2} \, MAD_t \approx 1.25 \, MAD_t. \tag{2.50}$$

FORECASTING

Example 2.5 We consider the same demand data and the same forecasting technique (exponential smoothing with trend) as in Example 2.2. Table 2.4 shows the updated values of MAD_t when using $\alpha = 0.1$ and the initial value $MAD_2 = 20$.

Table 2.4 Updated values of MAD_t with $\alpha = 0.1$ and initial value $MAD_2 = 20$. The forecasts are obtained by exponential smoothing with trend. The smoothing constants are $\alpha = 0.2$ and $\beta = 0.1$, and the initial forecast $\hat{a}_2 = 100$, $\hat{b}_2 = 0$, see Example 2.2.

Period	Demand in period t, x_t	Forecast for period $t+1$ at the end of period t, $\hat{a}_t + \hat{b}_t$	Mean Absolute Deviation at the end of period t, MAD_t
3	72	94	20.8
4	170	110	26.3
5	67	102	28.0
6	95	100	25.9
7	130	107	26.3

Note first that the forecast for period 3 in period 2 was $\hat{a}_2 + \hat{b}_2 = 100$. In period 3 we then obtain from (2.49)

$$MAD_3 = 0.9 \cdot 20 + 0.1 \cdot |72 - 100| = 20.8,$$

and similarly in period 4

$$MAD_4 - 0.9 \cdot 20.8 + 0.1 \cdot |170 - 93.84| = 26.336.$$

Note that when updating MAD we use forecasts that are not rounded, e.g., 94.4 - 0.56 = 93.84 instead of 94, see Example 2.2.

A less common technique is to update the variance in the same way

$$\sigma_t^2 = (1-\alpha)\sigma_{t-1}^2 + \alpha(x_t - \hat{x}_{t-1,t})^2. \quad (2.51)$$

When starting to update MAD or σ^2 we need an initial value. This value is easy to obtain if historical data of forecast errors are available. If such data are not available it can be reasonable to determine initial values as functions of the estimated demand. See Section 2.10.3. As we have discussed before concerning forecasts, it may also be a good idea to use a relatively large smoothing constant during a short initial period if the initial *MAD* is very uncertain.

The determination of forecast errors in connection with inventory control is normally carried out as described in this section. However, other possibilities exist, for example, in connection with linear regression. See also Section 2.10.3.

2.10.3 Determining the standard deviation as a function of demand

Sometimes when a forecasting system is initiated we do not have a sufficient number of historical data per item to evaluate initial values of *MAD* or σ. Still, we may have some idea of the size of the average demand $\hat{a}$. In such a situation it is common to determine an initial value of σ as a function of the average demand. We can get a corresponding *MAD* from (2.50). Such functions are sometimes also used as an alternative to the updating procedures (2.49) or (2.51). This can be reasonable in case of very low demand, when it may be difficult to get good estimates by using (2.49) or (2.51). Examples of such items can be certain types of spare parts.

The forecast errors are nearly always increasing with the average demand, but the relative errors are decreasing. When determining σ as a function of the average demand $\hat{a}$, a reasonable and common approach is to use the form

$$\sigma = k_1 \hat{a}^{k_2}, \tag{2.52}$$

or equivalently

$$\log \sigma = \log k_1 + k_2 \log \hat{a}, \tag{2.53}$$

so we can evaluate $\log k_1$ and k_2 by linear regression. The same function is used for many items, so we do not need to go back in time much to get a relatively good estimate. As before, we can get a corresponding *MAD* from (2.50).

2.10.4 Forecast errors for other time periods

Our determination of MAD_t and σ_t in Section 2.10.2 concerns the forecast error when we look one period ahead. The error when forecasting the demand

FORECASTING   33

in a more distant period is usually larger. In principle we could update the error for such a forecast similarly. This is not often done in practice, though. It is most common to use MAD_t as the expected absolute error not only for the demand in period $t + 1$, but also for the demand in period $t + k$ for $k > 1$. This means that we usually are underestimating the errors for such forecasts.

The estimates MAD_t and σ_t represent the mean absolute deviation and the standard deviation of the demand during one period. If σ_t, for example, is updated monthly, the time period implicitly considered is one month. In connection with inventory control it is very common, though, that we need to consider a shorter or a longer period, e.g., a certain lead-time. Assume that after updating σ_t in period t, we need to determine the standard deviation of the forecast error over L time periods, $\sigma(L)$, where L can be any positive number, i.e., not necessarily integral. If, as discussed above, we disregard that the errors usually increase when forecasting the demand for a more distant period, and also assume that forecast errors are independent over time, we obtain:

$$\sigma(L) = \sigma_t \sqrt{L} . \qquad (2.54)$$

Example 2.6 Assume that MAD_t is updated each month and that the most recent value is $MAD_t = 40$. From (2.50) we obtain $\sigma_t \approx 1.25 \cdot 40 = 50$. In case of independence over time, the standard deviation over two months is obtained as $\sigma(2) = 50 \cdot 2^{1/2} \approx 71$, and over 0.5 month as $\sigma(0.5) = 50 \cdot 0.5^{1/2} \approx 35$.

The determination of $\sigma(L)$ in (2.54) is very common. It is important, though, to recall that it is based on the assumptions discussed above. For a long lead-time the standard deviation is therefore often underestimated. One way to reduce the error is to use a forecast period that is of about the same length as the majority of the lead-times. This will limit the possible error when applying (2.54). If the lead times vary a lot it may also be reasonable to update the forecast and the forecast error for different periods.

Another possibility is to use the following slightly more general relationship instead of (2.54)

$$\sigma(L) = \sigma_t L^c , \qquad (2.55)$$

where $0.5 \leq c \leq 1$. Consider first the assumption that the forecast errors are independent. The value $c = 0.5$ corresponds to independence over time and gives (2.54), while the other extreme $c = 1$ corresponds to forecast errors that are completely correlated over time. However, we can also use $c > 0.5$ to compensate for disregarding that the forecast error for a more distant period tends to be larger.

2.10.5 Sales data instead of demand data

In practice it is quite often difficult to measure the real demand, because if a demand cannot be met this results in a lost sale, which is not accounted for. If it is sales instead of demand that is observed, the forecast is of course also for the sales. This means a systematic error that is not considered by the techniques discussed in Sections 2.10.1-2.10.4. If we keep a high service level there are very few lost sales and the error can more or less be disregarded. But situations also exist when this error may be quite large. An obvious possibility to reduce the error is to try to account for demands that can not be met. When a customer asks for some item that is not in stock, this should be accounted for as a demand. Another possibility is to adjust the forecasting technique. Consider a certain period and assume that there are 15 units in stock in the beginning of the period, and that there are no incoming deliveries during the period. Assume also that demands that cannot be met directly are lost. If the stock on hand at the end of the period is 5, we know that the period demand was 10. However, if the stock on hand at the end of the period is 0, we only know that the demand was at least 15. There are forecasting techniques that take this into account and can provide more accurate forecasts. See Nahmias (1994).

2.11 Monitoring forecasts

The forecasting methods that we have described will, in general, work excellently for most items. However, a forecasting system will normally comprise several thousand items. Various problems will occur for a few of these items in each period. These problems may depend on errors in the input to the system or on a less appropriate forecasting technique. This means that there is a need to follow up and check the forecasts. It would be extremely time consuming to do this manually for all items. It is therefore usually suitable to let the forecasting system itself perform certain automatic tests to check whether an item should go through a detailed manual examination. These tests are similar to techniques used in connection with statistical quality control. The tests are performed each time the forecasts are updated and result in a list of items that should be checked in more detail. If necessary the forecasting procedure can be reinitiated for some of these items with new initial values and possibly new parameters, e.g., smoothing constants.

FORECASTING

Two types of tests that are very often done automatically are to check whether the input, i.e., the new demand, is reasonable, and to evaluate whether the forecast represents the mean adequately.

2.11.1 Checking demand

If the forecasting system works well, *MAD* and the corresponding standard deviation σ illustrate how much the demand can be expected to deviate from the forecast. Assume that that in period *t*-1 we obtained a certain forecast for period t, $\hat{x}_{t-1,t}$ and a certain MAD_{t-1}. If $\hat{x}_{t-1,t}$ can be regarded as the mean and the deviations of the demand from the forecast are normally distributed, we can determine the probability that the forecast error in period t is within *k* standard deviations as,

$$P(|x_t - \hat{x}_{t-1,t}| < k\sigma_{t-1}) = \Phi(k) - \Phi(-k), \qquad (2.56)$$

where $\Phi(\cdot)$ is the cumulative normal distribution function. See Appendix 2. It is common to use the test

$$|x_t - \hat{x}_{t-1,t}| < k_1 MAD_{t-1} = k_1 \sigma_{t-1} / \sqrt{\pi/2}, \qquad (2.57)$$

with $k_1 = 4$ to check whether x_t is "reasonable". (When checking x_t it is appropriate to use MAD_{t-1} instead of MAD_t, which has been affected by the demand that we are checking.) By applying (2.56) with $k = 4/\sqrt{\pi/2} \approx 3.19$, we can see that the probability that the test should be satisfied is approximately 99.8 percent under normal conditions. Consequently, if (2.57) is not satisfied there is probably some error in the new demand or in the forecast, alternatively, an event with a very low probability has occurred. Any way it is reasonable to check the item manually. If it turns out that there is an error, it should be corrected. After that the forecasting system for the item is restarted with new initial values.

2.11.2 Checking that the forecast represents the mean

The determination of MAD_t in (2.49) means that we evaluate the average absolute error. It is also common to update the average error in a similar way. Let

z_t = estimate of the average error in period *t*.

In complete analogy with (2.49) we have

$$z_t = (1-\alpha)z_{t-1} + \alpha(x_t - \hat{x}_{t-1,t}), \tag{2.58}$$

where $0 < \alpha < 1$ is a smoothing constant. If the forecast works as a correct mean, positive and negative forecast errors should be of about the same size in the long run, and z_t can be expected to be relatively close to zero. On the other hand, if z_t differs significantly from zero we have reason to believe that the forecast is biased. When updating z_t it is natural to use zero as the initial value and to have a relatively small smoothing constant like $\alpha = 0.1$ in case of monthly updates. A common test is

$$|z_t| < k_2 MAD_t. \tag{2.59}$$

What is a suitable value of the parameter k_2? Comparing (2.58) to (2.49), it is evident that we should expect $|z_t| < MAD_t$. A reasonable initial value of k_2 may be 0.5. If this leads to too many error signals it can be reasonable to increase k_2. Similarly, if there are very few error signals we can tighten the control by using a smaller k_2.

Systematic errors that are detected by (2.59) can have different explanations. One possibility is that the forecasting method is inadequate. For example, if demand has a trend and we are using simple exponential smoothing, there will always be a systematic error. Another common reason is that a large change in the average demand has occurred. It will then take a long time for the forecast to approach the new demand level. If such a situation is detected by (2.59), we can improve the forecasts by restarting the system with a new, more accurate initial forecast.

The tests (2.57) and (2.59) should, in general, be thought of as a basis for manual adjustments of the forecasting system. It is also possible to design an *adaptive* forecasting system that makes various adjustments automatically. The system can, for example, change forecasting method and smoothing constants after performing various tests. See e. g., Gardner and Dannenbring (1980). Such systems are very seldom used in practice, though.

2.12 Manual forecasts

Most of the forecasting methods we have discussed in this chapter have the purpose of extrapolating historical demands as well as possible. There are situations though, when such forecasting techniques are less suitable. There can be known factors that will affect the future demand, but which have not affected the previous demand. In some cases it is possible to model the influence of such factors as described in Section 2.7.3. However, in general, it is both difficult and impractical to take such factors into account within a computerized forecasting system. Usually it is much more practical to let manual forecasts replace automatic forecasts. A forecasting system should therefore be designed so that such replacements are easy.

Normally, manual forecasts are only needed for a limited number of items over relatively short times. It is never a good idea to use a manual forecast if the forecast simply means an extrapolation of historical data. Such forecasts are obtained much more easily and with fewer errors by using the various methods that have been described earlier in this chapter.

Examples of situations when manual forecasts could be considered are:

- price changes
- sales campaigns
- conflicts that affect demand
- new products without historical data
- new competitive products on the market
- new regulations

A special problem with manual forecasts is that they sometimes have systematic errors because of optimistic or pessimistic attitudes by the forecasters.

References

Box, G. E. P., and G. M. Jenkins. 1970. *Time Series Analysis, Forecasting and Control*, Holden-Day, San Francisco.

Box, G. E. P., G. M. Jenkins, and G. C. Reinsell. 1994. *Time Series Analysis, Forecasting and Control*, 3rd edition, Prentice-Hall, Englewood Cliffs, N.J.

Croston, J. D. 1972. Forecasting and Stock Control for Intermittent Demands, *Operational Research Quarterly*, 23, 3, 289-303.

Gardner, E. S. 1985. Exponential Smoothing: the State of the Art, *Journal of Forecasting*, 4, 1-28.

Gardner, E. S., and D. G. Dannenbring. 1980. Forecasting with Exponential Smoothing: Some Guidelines for Model Selection, *Decision Sciences*, 11, 370-383.

Holt, C. C. 1957. Forecasting Seasonals and Trends by Exponentially Weighted Moving Averages, ONR Memorandum No. 52, Carnegie Institute of Technology, Pittsburgh, Pennsylvania.

Makridakis, S., S. C. Wheelwright, and R. J. Hyndman. 1998. *Forecasting: Methods and Applications*, 3rd edition, Wiley, New York.

Montgomery, D. C., L. A. Johnson, and J. S. Gardiner. 1990. *Forecasting and Time Series Analysis*, 2nd edition, McGraw-Hill, New York.

Nahmias, S. 1994. Demand Estimation in Lost Sales Inventory Systems, *Naval Research Logistics*, 41, 739-757.

Nahmias, S. 1997. *Production and Operations Analysis*, 3rd edition, Irwin, Boston.

Silver, E. A., D. F. Pyke, and R. Peterson. 1998. *Inventory Management and Production Planning and Scheduling*, 3rd edition, Wiley, New York.

Winters, P. R. 1960. Forecasting Sales by Exponentially Weighted Moving Averages, *Management Science*, 6, 324-342.

Problems

2.1*   Assume the following demands:

Period	1	2	3	4	5	6	7	8	9	10
Demand	460	452	458	470	478	480	498	500	490	488

Determine the moving average with $N = 3$ after observing the demands in periods 3-10.

2.2   Determine $\hat{a}_5$, $\hat{a}_6$, and $\hat{a}_7$ in Example 2.1.

2.3   Determine $\hat{a}_i$ and $\hat{b}_i$ for $i = 5, 6, 7$ in Example 2.2.

2.4*   Assume that the demand in period t is $c + d \cdot t$, where c and d are any constants.

a) Apply exponential smoothing with trend starting at $t = 1$ with $\hat{a}_0 = 0$ and $\hat{b}_0 = 0$. Using

$$\begin{pmatrix} 1-\alpha & 1-\alpha \\ -\alpha\beta & 1-\alpha\beta \end{pmatrix}^n \to 0, \text{ as } n \to \infty$$

prove that the forecast error will approach 0 in the long run.

b) Apply exponential smoothing starting at $t = 1$ with $\hat{a}_0 = 0$. Prove that the forecast error will approach d/α in the long run.

---

* Answer and/or hint in Appendix 1.

FORECASTING

2.5 a) Assume that we determine a constant forecast $\hat{a}_t$ at time t by minimizing the sum of the quadratic deviations from the demands $x_t, x_{t-1}, ..., x_{t-N+1}$. Show that the resulting forecast is a moving average.

b) Assume that we determine a constant forecast $\hat{a}_t$ at time t by minimizing the sum of the weighted quadratic deviations for all previous demands $x_t, x_{t-1}, ...$. The weight for $(x_{t-k} - \hat{a}_t)^2$ is w^k where $0 < w < 1$. Show that this is equivalent to exponential smoothing.

2.6 Assume that the demand follows a constant model according to (2.1), where a is completely constant and the standard deviation of ε_t is σ.

a) What is the standard deviation of the forecast error in the long run for a moving average?

b) What is the standard deviation of the forecast error in the long run for an exponential smoothing forecast?

c) Show that if we set the standard deviations in a) and b) equal we get the relationship (2.10), i.e., $\alpha = 2/(N+1)$.

2.7* A company uses Winters' trend-seasonal method for an item. The forecast is updated each quarter. Each quarter has a seasonal index. All smoothing constants are equal to 0.2. After the update at the end of 2004 we have:

$\hat{a}_{04.4} = 1000.00$

$\hat{b}_{04.4} = 10.00$

$\hat{x}_{04.4,05.1} = 808.00$

$\hat{x}_{04.4,05.2} = 1020.00$

$\hat{x}_{04.4,05.3} = 1648.00$

$\hat{x}_{04.4,05.4} = 624.00$

The demand in the first quarter 2005 was 795, and in the second quarter 1023. Update the forecast twice. What is then the forecasts for the coming four quarters?

2.8 Assume that the demand is updated each period by Winters' trend-seasonal method. The smoothing constants are $\alpha = 0.2$, $\beta = 0.1$, $\gamma = 0.2$. There are $T = 7$ equally long periods during a year. Assume further that the last update took place at the end of period 13 and resulted as follows

$\hat{a}_{13} = 5, \hat{b}_{13} = 1$

$\hat{F}_7 = \hat{F}_8 = 1.4, \hat{F}_9 = \hat{F}_{10} = \hat{F}_{11} = 1, \hat{F}_{12} = \hat{F}_{13} = 0.6$

The observed demand in period 14 is 3. Carry out the update and determine the forecast for period 17.

2.9 Consider the following demand data.

Period	1	2	3	4	5	6	7	8	9	10
Demand	430	452	428	470	478	480	498	500	610	488

Use the trend method in Section 2.7.1 to obtain forecasts for periods 11, 12, and 13. Use all data, i.e., the forecast is determined after observing the demand in period 10 and $N = 10$.

2.10 Consider the example described in Section 2.7.3 dealing with a spare part that is exclusively used as a replacement when maintaining a certain machine. The maintenance activities are carried out by the company selling the spare part, and the customers have to order the maintenance two months in advance. Consider the following data showing demand and the known number of machines undergoing maintenance.

Month	1	2	3	4	5	6	7	8
Demand	66	45	77	78	64	79		
No of machines	110	97	150	143	125	160	151	172

Use linear regression to forecast demand in months 7 and 8.

2.11 Consider the following demand data.

Period	1	2	3	4	5	6	7	8	9	10
Demand	0	0	0	0	22	0	0	0	27	0

Use Croston's method to update a demand forecast as described in Section 2.8. At the end of period 0 after a positive demand in this period we have $\hat{k}_0 = 6$ and $\hat{d}_0 = 31$. Use $\alpha = \beta = 0.2$.

2.12* Consider the ARIMA(0, 2, 2) model

$$x_t = 2x_{t-1} - x_{t-2} + \varepsilon_t - (2-\alpha-\alpha\beta)\varepsilon_{t-1} + (1-\alpha)\varepsilon_{t-2}.$$

Show that this model is equivalent to exponential smoothing with trend.

2.13* Consider the following monthly demands:

Month	1	2	3	4	5	6
Demand	718	745	767	728	788	793

FORECASTING   41

  a) Determine forecasts by exponential smoothing. Update also *MAD*. At the end of month 0 the forecast was 730 and *MAD* was 20. Use $\alpha = 0.2$ for both updates.
  b) Determine instead forecasts by exponential smoothing with trend. The initial forecast is $\hat{a}_0 = 730$ and $\hat{b}_0 = 0$. Use $\alpha = \beta = 0.2$. What is the forecast for month 5 at the end of month 2?

2.14* A company is using exponential smoothing with trend to forecast the demand for a certain product. The smoothing constants are $\alpha = 0.2$ and $\beta = 0.1$. *MAD* is updated with exponential smoothing with $\alpha = 0.1$. At the end of month 4 the mean and trend were 220 and 10, respectively, and *MAD* was 35. The demand in month 5 is 250. Update forecast and *MAD* and estimate mean and standard deviation for the total demand during the three months 6-8. The forecast errors are independent over time.

2.15 The demand for a product has a clear trend. The company producing the product is therefore determining its forecasts by applying exponential smoothing with trend. At the end of March the updated mean was 50 and the trend was 13. *MAD* was 2. The smoothing constants for exponential smoothing with trend are $\alpha = 0.2$ and $\beta = 0.05$. *MAD* is updated by exponential smoothing with $\alpha = 0.1$. The demands during April, May, and June are 57, 64, and 73 respectively. Update forecast and *MAD* during these months. Determine also a forecast for the total demand during July and August and the standard deviation for this period. (Forecast errors during different periods are assumed to be independent.)

2.16 The forecast for an item is determined by exponential smoothing with $\alpha = 0.2$. When updating *MAD*, $\alpha = 0.1$. After the update at the end of period 6 the forecast is 320 and *MAD* = 82. We have just observed the demand in period 7, which is 290.

  a) Update forecast and *MAD*.
  b) Consider the first half of period 10. After the update in a), what is the forecast and the standard deviation for the demand during this half period. Forecast errors during different periods are assumed to be independent.
  c) Assume that you wish to change forecasting method to a moving average, so that the forecast is still based on data with the same average age. How would you choose N?

2.17 When estimating the forecast error it is common to use that *MAD* is coupled to the standard deviation as

$$\sqrt{\pi/2} \cdot MAD = \sigma$$

under the assumption of Normal distribution. Prove this.

2.18 Both the forecast and *MAD* are updated monthly by exponential smoothing. Both smoothing constants are equal to 0.1. After the update that took place at the end of March the forecast was 132 and *MAD* was 42.

a) In April the demand is 92. Update forecast and *MAD*.
b) In this forecast, what is the weight for the demand in January?

2.19 Consider the following monthly demands:

Month	1	2	3	4	5	6
Demand	668	745	778	728	634	789

a) Determine forecasts by exponential smoothing. At the end of month 0 the forecast was 730. Use $\alpha = 0.2$.
b) Update also *MAD*. At the end of month 0 *MAD* was 30. Use again $\alpha = 0.2$. What is the estimate for σ at the end of period 6?
c) For comparison, use also the updating procedure (2.51). Set the initial $\sigma = \sqrt{\pi/2} \cdot MAD = \sqrt{\pi/2} \cdot 30$. Use again $\alpha = 0.2$. What is now the estimate for σ at the end of period 6?

2.20 Consider (2.49) and (2.58). Assume that we use the same α and that $|z_0| \leq MAD_0$. Prove that this implies $|z_t| \leq MAD_t$ for all $t > 0$.

3 COSTS AND CONCEPTS

In Chapters 3-6 we consider a large and very important class of inventory problems, for which, in general, we can offer satisfactory solutions that can be used in practice relatively easily. This class of systems is characterized by two qualities:

- Different items can be controlled independently.

- The items are stocked at a single location, i.e., not in a multi-stage inventory system.

The first condition is very often satisfied, but there are also many important counter examples. When orders are issued to a factory for example, we usually wish to obtain a smooth production load. This means that we need to coordinate orders for different items and the condition is evidently not satisfied. Another example is when we get a discount that is based on the total order sum for several items. In that case we wish to order these items at the same time. We shall deal with such situations in Chapter 7.

There are also many practical settings where the second condition is not satisfied. For example, when distributing an item over a large geographical region, it is common to use an inventory system with a central warehouse together with local stocks close to the final customers. Such situations are considered in Chapters 8-10.

Typical settings where the two conditions are satisfied can be found at many wholesalers. In production there is very often a certain coupling between different stocks. Sometimes, though, this coupling is so weak that it can be disregarded.

3.1 Considered costs and other assumptions

The models that we deal with are based on certain standard assumptions. In general, we consider the following costs or constraints. (See also Berling, 2005, Silver et al., 1998, and Zipkin, 2000.)

3.1.1 Holding costs

By holding stock we have an opportunity cost for capital tied up in inventory. This part of the holding cost should, in principle, be closely related to the return on an alternative investment. It is not necessarily equal to the expected return though, because we also need to take into account financial risks associated with alternative investments. The capital cost is usually regarded to be the dominating part of the holding cost. Other parts can be material handling, storage, damage and obsolescence, insurance, and taxes. All costs that are variable with the inventory level should be included. Consider, for example, the costs for storage space. In some situations with owned warehouses, these costs may be essentially fixed. This means that they should not be included in the holding cost. In other situations, however, it may be possible to increase or decrease rented warehouse space in the short run, and the costs should consequently be included in the holding cost.

The holding cost per unit and time unit is often determined as a percentage of the unit value. Note that this percentage can usually not be the same for all types of items. It should, for example, be high for computers due to obsolescence. In general, the percentage should be considerably higher than the interest rate charged by the bank.

An alternative to including the financial costs in the holding cost is to consider the net present value, i.e., the discounted value, of all expenditures and revenues. See, for example, Grubbström (1980). In this book we will only consider standard holding costs.

3.1.2 Ordering or setup costs

There are usually fixed costs associated with a replenishment (independent of the batch size). In production, common reasons are setup and learning costs. But there are also administrative costs associated with the handling of orders. Other fixed costs can occur in connection with transportation and material handling. The setup cost may be especially high if an expensive capacity constrained machine has to stop during the setup. In some situations, though, most of the setup can take place outside the machine, which reduces the costs.

COSTS AND CONCEPTS   45

When ordering from an outside supplier there are also various fixed costs for an order, like costs for order forms, authorization, receiving, inspection, and handling of invoices from the supplier.

3.1.3 Shortage costs or service constraints

If an item is demanded and cannot be delivered due to a shortage, various costs can occur. There are situations when a customer agrees to wait while his order is backlogged, but also situations when the customer chooses some other supplier. If the customer order is backlogged, there are often extra costs for administration, price discounts for late deliveries, material handling, and transportation. If the sale is lost, the contribution of the sale is also lost. In any case, it usually means a loss of good will that may affect the sales in the long run. Most of these costs are difficult to estimate.

Shortage costs in production are, in general, even more difficult to estimate. If a component is missing for example, this can cause a chain of negative consequences like delays, rescheduling, etc.

There are also situations when shortage costs are easy to evaluate. Assume, for example, that a missing component can be bought at a higher price in a store next door. We can then use the additional cost as our shortage cost.

Because shortage costs are so difficult to estimate, it is very common to replace them by a suitable service constraint. Of course it is also difficult to determine an adequate service level, but yet this is regarded to be somewhat simpler in many practical situations. Still, the motivation for a service constraint is nearly always some underlying shortage cost.

3.1.4 Other costs and assumptions

The above costs are those that are most common in inventory models. Although we shall in most cases limit our attention to these costs, it is evident that it can also be relevant to consider other costs.

As discussed in Section 3.1.1 the holding cost is usually determined as a percentage of the unit cost, which is normally obtained as the cost figure given by accounting. Otherwise it is in most cases assumed that the unit cost is not affected by inventory control decisions concerning e.g., order quantity and reorder point. There are some obvious exceptions, though. One example is when a large order quantity can affect the unit cost through a discount. See Section 4.3.

Other costs that should be considered are the costs of operating the inventory control system. This includes costs for data acquisition and computation, as well as costs for training the involved personnel.

In this book, we shall generally, assume that demand is backordered and not lost in case of a shortage. Models with backordering are usually much simpler, and unless shortages are very common, can also be used as an approximation in case of lost sales. A model with lost sales is considered in Section 5.14.

3.2 Different ordering systems

3.2.1 Inventory position

The purpose of an inventory control system is to determine *when* and *how much* to order. This decision should be based on the stock situation, the anticipated demand, and different cost factors.

When talking about the stock situation, it is natural to think of the physical stock *on hand*. But an ordering decision can not be based only on the stock on hand. We must also include the *outstanding orders* that have not yet arrived, and *backorders*, i.e., units that have been demanded but not yet delivered. In inventory control the stock situation is therefore characterized by the *inventory position*:

inventory position = stock on hand + outstanding orders - backorders.

If the customers can *reserve* units for later delivery, the reserved units are usually also subtracted from the inventory position. This means that the inventory control handles the reserved units in the same way as units that have been ordered for immediate delivery. This may be reasonable unless the delivery time is too distant, which will result in unnecessary holding costs. If the delivery time is far away, a reasonable policy is to subtract the reserved units from the inventory position first when the remaining time until delivery is less than a certain time limit. The customers will then, with a high probability, still get their orders in time. We shall from now on disregard reservations.

Although the ordering decisions are based on the inventory position, holding and shortage costs will depend on the *inventory level*:

inventory level = stock on hand - backorders.

In some situations the holding costs should also include holding costs for outstanding orders. The resulting additional cost term is easy to evaluate because the average outstanding orders is easily obtained as the average lead-time demand.

3.2.2 Continuous or periodic review

An inventory control system can be designed so that the inventory position is monitored continuously. As soon as the inventory position is sufficiently low, an order is triggered. We denote this *continuous review*. The triggered order will be delivered after a certain lead-time. Let

L = lead-time.

The lead-time is the time from the ordering decision until the ordered amount is available on shelf. It is not only the transit time from an external supplier or the production time in case of an internal order. It also includes, for example, order preparation time, transit time for the order, administrative time at the supplier, and time for inspection after receiving the order.

An alternative to continuous review is to consider the inventory position only at certain given points in time. In general, the intervals between these reviews are constant and we talk about *periodic review*. Let

T = review period, i. e., the time interval between reviews.

Both alternatives have advantages as well as disadvantages. Continuous review will reduce the needed safety stock. When using a continuous review system the inventory position when ordering should guard against demand variations during the lead-time L. With periodic review the uncertain time is instead $T + L$. Consider a review when no order is triggered. The next possibility to order is T time units ahead, and consequently, the time to the next possible delivery is $T + L$. Therefore, the inventory position must guard against demand variations during this time.

Periodic review has advantages, especially when we want to coordinate orders for different items (see Chapter 7). Although modern information technology has reduced the costs for inspections considerably, periodic review will also reduce the costs for the inventory control system. This is especially true for items with high demand. For items with low demand it does not cost much more to use continuous review. The advantages of continuous review are also usually larger for such items. In practice it is therefore common to use continuous review for items with low demand, and periodic review for items with higher demand.

Periodic review with a short review period T is, of course, very similar to continuous review.

3.2.3 Different ordering policies

The two most common ordering policies in connection with inventory control are often denoted (R, Q) policy and (s, S) policy.

3.2.3.1 (R, Q) policy

When the inventory position declines to or below the reorder point R, a batch quantity of size Q is ordered. (If the inventory position is sufficiently low it may be necessary to order more than one batch to get above R. The considered policy is therefore sometimes also denoted (R, nQ) policy.) If demand is continuous, or one unit at a time, we will always hit the reorder point exactly in case of continuous review.

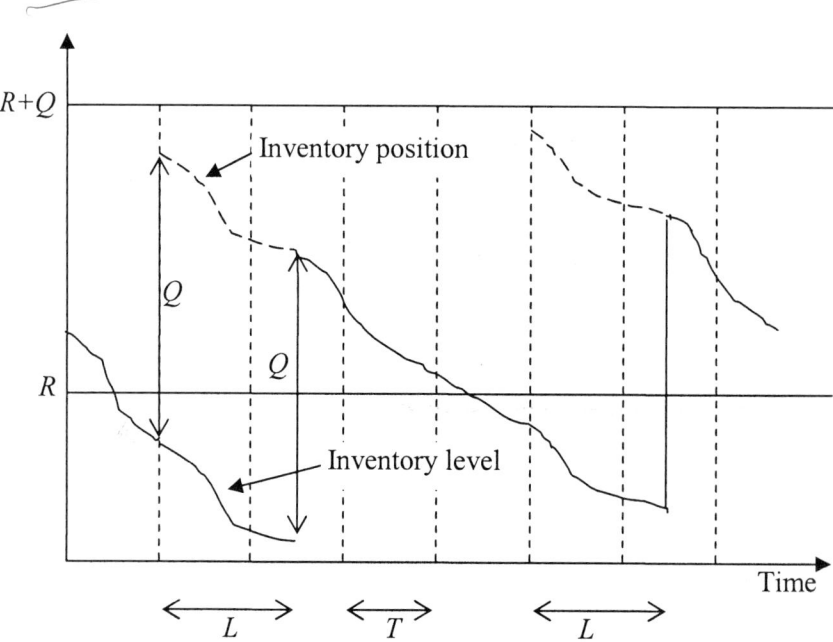

Figure 3.1 (R, Q) policy with periodic review. Continuous demand.

In case of periodic review, or if the triggering demand is for more than one unit, it will often occur that the inventory position is below R when ordering. In this case we will not reach the position $R + Q$ after ordering. Figure 3.1 illustrates this for a periodic review (R, Q) policy. Note that the development of the inventory position is completely independent of the lead-time L.

COSTS AND CONCEPTS

An ordering policy that is very similar to an (R, Q) policy is a so-called KANBAN policy. With a KANBAN policy there are N containers, each containing Q units of the item and with a card on the bottom. When a container becomes empty, the card (a KANBAN) is used as an order for Q units. $N-1$ of the KANBAN cards are always associated with full containers, which are either in stock or ordered but not yet delivered. So the inventory position is $(N-1)Q$ plus the number of units in the container that at present is satisfying the demand. When this container is empty an order is triggered. A KANBAN policy is consequently very similar to an (R, Q) policy with $R = (N-1)Q$. But if there are already N outstanding orders, i.e., no stock on hand, no more orders can be triggered since no KANBANs are available. We can therefore interpret a KANBAN policy as an (R, Q) policy where backorders are not subtracted from the inventory position.

3.2.3.2 (s, S) policy

This policy is also similar to an (R, Q) policy. The reorder point is denoted by s. When the inventory position declines to or below s, we order up to the maximum level S. The difference compared to an (R, Q) policy is that we no longer order multiples of a given batch quantity. If we always hit the reorder point exactly (continuous review and continuous demand), the two policies are equivalent provided $s = R$ and $S = R + Q$. But if we do not always hit the reorder point exactly, the equivalence does not hold. Figure 3.2 illustrates such a situation with periodic review. Since we order when the inventory position is below the reorder point, we would not have reached the inventory position $R + Q$ after ordering in case of an (R, Q) system. See Figure 3.1.

In case of periodic review, it may happen with either policy that we do not reach the reorder point at a certain inspection. In that case no order is triggered. A variation of the (s, S) policy denoted S policy, or order-up-to-S policy, or *base stock policy*, will always order unless the period demand is zero. An S policy simply means that we order up to S independent of the inventory position. An S policy is equivalent to an (s, S) policy with $s = S - 1$ and to an (R, Q) policy with $R = S - 1$ and $Q = 1$. In case of continuous review, a single demand may similarly not trigger an order when using an (R, Q) policy or an (s, S) policy. A policy which always immediately orders the demanded number of units is also an S policy, but is in connection with continuous review often denoted $(S - 1, S)$ policy.

For single-echelon systems it is possible to show that the optimal policy, under very general assumptions, is of the (s, S) type. See Section 6.2. From a theoretical point of view it should therefore be advantageous to use (s, S) policies. However, the cost differences are, in general, very small, and in practice it is often easier to use an (R, Q) policy with a fixed batch quantity.

Independent of which policy we choose, it remains to determine the control parameters R and Q, or s and S. We shall deal with this problem in Chapters 4 – 6.

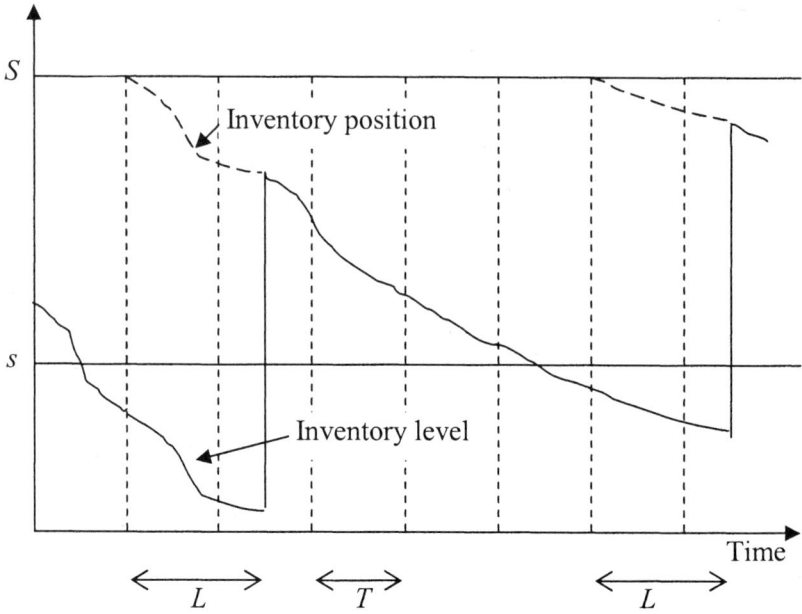

Figure 3.2 (s, S) policy with periodic review. Continuous demand.

References

Berling, P. 2005. On Determination of Inventory Cost Parameters, Ph. D. Thesis, Lund University.

Grubbström, R. W. 1980. A Principle for Determining the Correct Capital Costs of Work-in-Progress and Inventory, *International Journal of Production Research*, 18, 259-271.

Silver, E. A., D. F. Pyke, and R. Peterson. 1998. *Inventory Management and Production Planning and Scheduling*, 3rd edition, Wiley, New York.

Zipkin, P. H. 2000. *Foundations of Inventory Management*, McGraw-Hill, Singapore.

4 SINGLE-ECHELON SYSTEMS: DETERMINISTIC LOT SIZING

When using an (R, Q) policy we need to determine the two parameters R and Q. We shall first consider the determination of the batch quantity Q. (When using an (s, S) policy, Q essentially corresponds to $S - s$.) We shall assume that the future demand is deterministic and given. If the lead-time is constant, it does then not affect the problem and we can therefore just as well assume that the lead-time $L = 0$. The only difference in case of a positive lead-time is that we need to order L time units earlier. See also the discussion in Section 4.1.3. (If the demand is stochastic, we cannot disregard the lead-time.)

Deterministic demand may seem to be a very unrealistic assumption. In most situations there are stochastic variations in demand. It turns out, though, that the assumption of deterministic demand is in general very reasonable. First of all there exist situations when a company is really facing deterministic demand. An example is a company delivering according to a long-range contract. More important is that it is also often feasible to use deterministic lot sizing in case of stochastic demand. The determination of Q should, even in a stochastic case, essentially mean that we balance ordering and holding costs. Although a larger Q will, in fact, reduce the need for safety stock, this is of minor importance. A standard procedure in practice is therefore to first replace the stochastic demand by its mean and use a deterministic model to determine Q. Given Q, a stochastic model is then used in a second step to determine the reorder point R. Under relatively general assumptions it is possible to show that this approximate procedure will give a cost increase compared to optimum that is always lower than 11.8 percent

(See Axsäter, 1996 and Zheng, 1992). In general, the cost increase is much lower than that.

4.1 The classical economic order quantity model

4.1.1 Optimal order quantity

The most well-known result in the whole inventory control area may be the classical economic order quantity formula. This simple result has had and still has an enormous number of practical applications. It was first derived by Harris (1913), but Wilson (1934) is also recognized in connection with this model. The model is based on the following assumptions:

- Demand is constant and continuous.
- Ordering and holding costs are constant over time.
- The batch quantity does not need to be an integer.
- The whole batch quantity is delivered at the same time.
- No shortages are allowed.

We shall use the following notation:

h = holding cost per unit and time unit,
A = ordering or setup cost,
d = demand per time unit,
Q = batch quantity,
C = costs per time unit.

Since no safety stock is needed and no shortages are allowed the inventory level will vary over time as in Figure 4.1, i.e., a batch is delivered exactly when the preceding batch is finished.

The relevant costs are the costs that vary with the batch quantity Q, i.e., the holding costs and the ordering costs. We obtain

$$C = \frac{Q}{2}h + \frac{d}{Q}A. \qquad (4.1)$$

The first term in (4.1) represents the holding costs, which we obtain as the average stock, $Q/2$, multiplied by the holding cost h. The average number of orders per time unit is d/Q, and multiplying by the ordering cost A, we obtain the ordering costs per time unit in the second term.

SINGLE-ECHELON – LOT SIZING

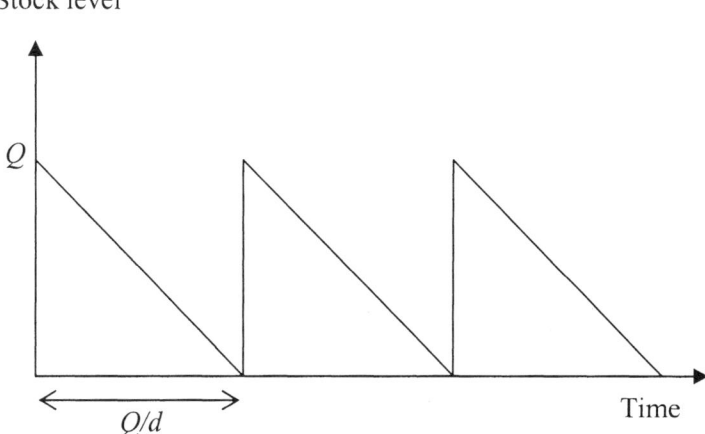

Figure 4.1 Development of inventory level over time.

The cost function C is obviously convex in Q, and we can therefore obtain the optimal Q from the first order condition

$$\frac{dC}{dQ} = \frac{h}{2} - \frac{d}{Q^2} A = 0. \tag{4.2}$$

Solving for Q we obtain the economic order quantity:

$$Q^* = \sqrt{\frac{2Ad}{h}}. \tag{4.3}$$

If we insert (4.3) in (4.1) we get

$$C^* = \sqrt{\frac{Adh}{2}} + \sqrt{\frac{Adh}{2}} = \sqrt{2Adh}, \tag{4.4}$$

i.e., for the optimal order quantity the holding costs happen to be exactly equal to the ordering costs.

In (4.1) we have expressed the costs as a function of the order quantity Q. An alternative is to express the costs in terms of the cycle time $T = Q/d$. Obviously, $T^* = Q^*/d$.

4.1.2 Sensitivity analysis

How important is it to use the optimal order quantity? From (4.1), (4.3), and (4.4) we obtain

$$\frac{C}{C^*} = \frac{Q}{2}\sqrt{\frac{h}{2Ad}} + \frac{1}{2Q}\sqrt{\frac{2Ad}{h}} = \frac{1}{2}\left(\frac{Q}{Q^*} + \frac{Q^*}{Q}\right). \qquad (4.5)$$

From (4.5) it is evident that the relative cost increase C/C^* when using the batch quantity Q instead of the optimal Q^* is a simple function of Q/Q^*. It turns out that even quite large deviations from the optimal order quantity will give very limited cost increases. For example, if $Q/Q^* = 3/2$ (or 2/3) we get $C/C^* = 1.08$ from (4.5), i.e., the cost increase is only 8 percent. It is also interesting to note that the costs are even less sensitive to errors in the cost parameters. For example, if we use a value of the ordering cost A that is 50 percent higher than the correct ordering cost, we can see from (4.3) that the resulting relative error in the batch quantity is $Q/Q^* = (3/2)^{1/2} = 1.225$ and the relative cost increase is only about 2 percent. Consequently, we can conclude that the choice of cost parameters when using the classical economic order quantity is not that critical. It should be noted, however, that other more complex situations exist when even small errors in the batch quantities can affect the costs dramatically. We shall discuss such situations in Chapter 7.

Example 4.1 Consider an item with unit cost $100, and holding cost rate per year corresponding to 20 percent of the value. The constant demand is 300 units per year. The ordering cost is $200. We get $h = 0.20 \cdot 100 = \$20$ per year. $A = \$200$ and $d = 300$ per year. Applying (4.3) we obtain the optimal batch quantity $Q^* = (2Ad/h)^{1/2} = (2 \cdot 200 \cdot 300/20)^{1/2} = 77.5$ units.

In practice Q often has to be an integer. Assume that $n < Q^* < n + 1$. Since the cost function (4.1) is convex, it is obvious that the best integral Q is either n or $n + 1$. We can see from (4.5) that we should choose the lower value $Q = n$ if $Q^*/n \leq (n + 1)/Q^*$.

4.1.3 Reorder point

Assume that we want to use a continuous review (R, Q) policy under the considered simple assumptions. How should we choose the reorder point R? If the lead-time $L = 0$ we simply choose $R = 0$. This means that we order when the inventory position is equal to zero and we get the delivery immedi-

SINGLE-ECHELON – LOT SIZING

ately. Note that when $L = 0$ the inventory position is always equal to the inventory level. Assume then a lead-time $L > 0$. As we have pointed out above, we only need to order the batch L time units earlier. This means that we need to change the reorder point to $R = Ld$, i.e., we add the deterministic lead-time demand. Assume for example, that this results in $R = 2.5Q$. An order is triggered when the inventory position is equal to R, while the inventory level will never exceed Q. When an order is triggered, there are two outstanding orders and the inventory level is equal to $Q/2$, i.e., the inventory position is no longer equal to the inventory level.

4.2 Finite production rate

If there is a finite production rate the whole batch is not delivered at the same time. The batch is instead delivered continuously with a certain production rate. It is quite easy to modify the basic model to encompass such successive deliveries. The inventory will now vary as in Figure 4.2.
Let

p = production rate $(p > d)$.

In each cycle we produce during the time Q/p. Consider the first cycle in Figure 4.2. During the time Q/p the stock is increasing at the rate $p - d$. The maximum level is $(Q/p)(p-d) = Q(1 - d/p)$. After the time Q/p the stock is decreasing with rate d.

The only difference compared to the classical model (4.1) is that the average inventory level is now $Q(1 - d/p)/2$ instead of $Q/2$ and we obtain:

$$C = \frac{Q(1-d/p)}{2}h + \frac{d}{Q}A. \qquad (4.6)$$

The optimal solution is then

$$Q^* = \sqrt{\frac{2Ad}{h(1-d/p)}}. \qquad (4.7)$$

It is easy to show that (4.5) is also valid in this case. Note furthermore that if $p \to \infty$, (4.7) degenerates to the classical economic order quantity formula (4.3).

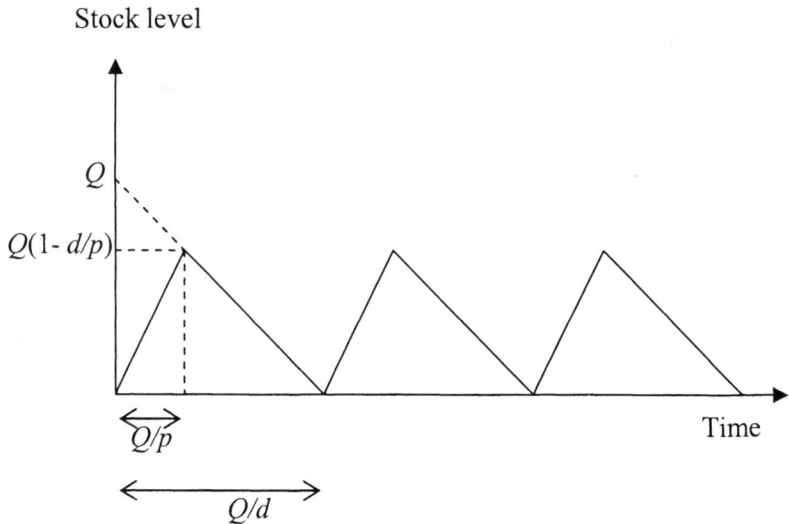

Figure 4.2 Development of the inventory level over time with finite production rate.

4.3 Quantity discounts

In general, it is relatively easy to include other costs that vary with the order quantity in lot sizing models. We shall illustrate this by considering the simple model in Section 4.1.1 with an additional assumption. If the order quantity is sufficiently large we get an <u>all-units discount</u> on the purchase price. Let

v = price per unit for $Q < Q_0$, i.e., the normal price,

v' = price per unit for $Q \geq Q_0$, where $v' < v$.

Note that we assume that we get the whole order at the lower price if $Q \geq Q_0$. It is also possible to handle the case when there is a discount only for units above the breakpoint Q_0. See Benton and Park (1996).

The holding cost will depend on the unit cost and we assume that it has the following structure:

$h = h_0 + rv$ for $Q < Q_0$,

$h' = h_0 + rv'$ for $Q \geq Q_0$,

SINGLE-ECHELON – LOT SIZING

i.e., we assume that the holding costs consist of two parts, a capital cost obtained by applying the interest rate r on the unit cost, and other out-of-pocket holding costs h_0 independent of the price per unit. As before, A is the ordering cost and d the constant demand per time unit.

Since the purchasing costs now depend on the order quantity they must be included in the objective function used to determine Q. We obtain

$$C = dv + \frac{Q}{2}(h_0 + rv) + \frac{d}{Q}A \quad \text{for } Q < Q_0, \tag{4.8}$$

$$C = dv' + \frac{Q}{2}(h_0 + rv') + \frac{d}{Q}A \quad \text{for } Q \geq Q_0. \tag{4.9}$$

Figure 4.3 illustrates the resulting cost function.

We shall use two steps to get the optimal solution:

1. First we consider (4.9) without the constraint $Q \geq Q_0$. We obtain

$$Q' = \sqrt{\frac{2Ad}{h_0 + rv'}} \tag{4.10}$$

and

$$C' = \sqrt{2Ad(h_0 + rv')} + dv'. \tag{4.11}$$

Note now that since $v' < v$, (4.9) provides a lower bound for (4.8) with $Q < Q_0$. Therefore if $Q' \geq Q_0$, (4.10) and (4.11) give the optimal solution, i.e., $Q^* = Q'$, and $C^* = C'$.

2. If $Q' < Q_0$ as in Figure 4.3 we need to optimize the cost function (4.8)

$$Q'' = \sqrt{\frac{2Ad}{h_0 + rv}}. \tag{4.12}$$

and

$$C'' = \sqrt{2Ad(h_0 + rv)} + dv. \tag{4.13}$$

Since $v > v'$ we know that Q'' is smaller than Q' so we have $Q'' < Q' < Q_0$. Clearly, (4.12) and (4.13) provide the lowest possible cost without a discount. Because of the convexity of (4.9) and that $Q' < Q_0$, we know that the lowest cost with a discount is

$$C(Q_0) = dv' + \frac{Q_0}{2}(h_0 + rv') + \frac{d}{Q_0} A . \qquad (4.14)$$

The optimal solution in this case is consequently obtained as the minimum of (4.13) and (4.14).

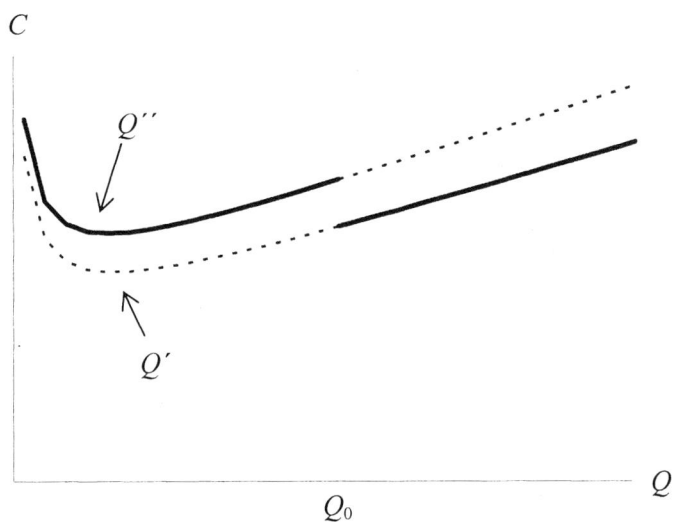

Figure 4.3 Costs for different values of Q.

Example 4.2 Assume that $v = \$100$ and $v' = \$95$ for $Q \geq Q_0 = 100$. Furthermore $h_0 = \$5$ per unit and year, and $r = 0.2$, i.e., $h = \$25$ and $h' = \$24$ per unit and year. The constant demand is $d = 300$ per year, and the setup cost $A = \$200$.

In the first step we get $Q' = 70.71$ and $C' = 30197$ from (4.10) and (4.11). Since $Q' < Q_0$ we need to go to the second step. From (4.12) and (4.13) we get $Q'' = 69.28$ and $C'' = 31732$. Applying (4.14) we obtain $C(100) = 30300$. Consequently $Q^* = Q_0 = 100$ is the optimal order quantity.

SINGLE-ECHELON – LOT SIZING

4.4 Backorders allowed

So far we have assumed that backorders are not allowed. Backorders can make sense in a deterministic model, though. Therefore we shall now allow backorders and introduce a certain penalty cost proportional to the waiting time for the customer.

b_1 = shortage penalty cost per unit and time unit.

Otherwise the assumptions are exactly the same as when we derived the classical economic order quantity in Section 4.1.1.

The backorder penalty costs per time unit are b_1 times the average number of backorders, since each unit backordered during one time unit means that some customer is waiting during that same time. Recall that the inventory level is the stock on hand minus backorders. The backorder cost considered has a structure that is very similar to the holding cost. The only difference is that the backorder cost is charged when the inventory level is negative and the holding cost when it is positive. Note that we can similarly interpret the holding cost h as a penalty cost per unit and time unit for units waiting to be demanded. Each unit on hand is waiting for its demand to occur. So the total of such penalty costs (holding costs) per unit of time are obtained as h times the average inventory on hand. (There are other types of shortage costs, as will be discussed in Section 5.5.)

Note also that because the average number of backorders equals the average customer waiting time per unit of time, the average number of backorders divided by the demand d is the average waiting time per unit demanded. This relationship is valid in much more general stochastic settings and is then denoted Little's formula. See also Section 10.2. Similarly, the average inventory on hand divided by d is the average waiting time for a unit to be demanded.

Allowing backorders means that some units are delivered in stock after they have been demanded. Let us define

x = fraction of demand that is backordered.

This means that when a batch Q is delivered the fraction x of the batch is used to cover backorders that have occurred before the batch was delivered. We wish to optimize both x and Q. The development of the inventory level is very similar to that in Figure 4.1. See Figure 4.4.

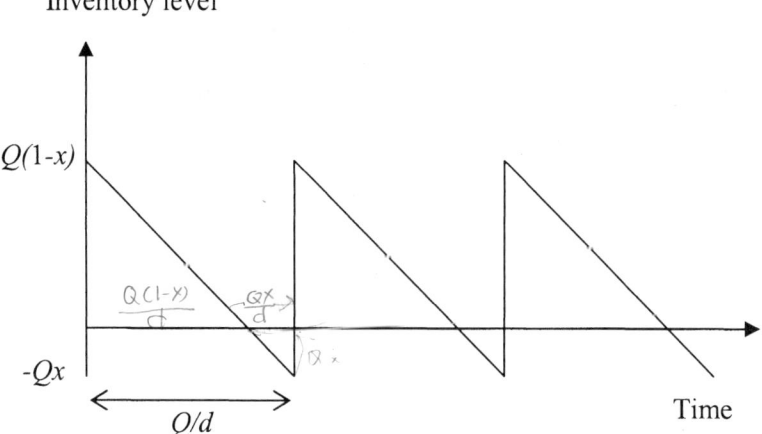

Figure 4.4 Development of inventory level over time with backorders.

The costs are now obtained as

$$C = \frac{Q(1-x)^2}{2}h + \frac{Qx^2}{2}b_1 + \frac{d}{Q}A. \quad (4.15)$$

In (4.15) we use that the ratio of time with positive stock is (1-x) and with backorders x. When the inventory level is positive, the average stock on hand is $Q(1-x)/2$, but when the inventory level is negative, the stock on hand is zero. So we obtain the average stock on hand as $(1-x)Q(1-x)/2 + x \cdot 0 = Q(1-x)^2/2$. In complete analogy the average number of backorders is $Qx^2/2$.

The cost is convex in x, and it is easy to see that the optimal x is independent of Q. By setting the derivative equal to zero we obtain

$$x^* = \frac{h}{h+b_1}, \quad (4.16)$$

and inserting in (4.15) we get

$$C(x^*) = \frac{Q}{2}\frac{hb_1}{h+b_1} + \frac{d}{Q}A. \quad (4.17)$$

In complete analogy with (4.3) and (4.4) we then obtain the optimal batch quantity and costs as

$$Q^* = \sqrt{\frac{2Ad(h+b_1)}{hb_1}}, \tag{4.18}$$

$$C^* = \sqrt{\frac{2Ad\,hb_1}{(h+b_1)}}. \tag{4.19}$$

Note that the batch quantity (4.18) is larger and the optimal costs (4.19) are lower than the corresponding batch quantity and costs for the classical economic order quantity model (4.3) and (4.4). The costs are lower because we have more flexibility when allowing backorders. We get the cost function (4.1) if we force x to be zero.

It is easy to check that (4.5) is still valid. Note also that the solution will approach that of the classical economic lot size problem as $b_1 \to \infty$.

Example 4.3 Consider the same data as in Example 4.1 with the additional assumption that backorders are allowed with a backorder cost $b_1 = \$100$ per unit and year. The constant demand is $d = 300$ units per year. The ordering cost is $A = \$200$ and the holding cost $h = \$20$ per unit and year. Applying (4.16) we get $x^* = 20/120 = 1/6$ and from (4.18) we get the optimal batch quantity $Q^* = (2 \cdot 200 \cdot 300 \cdot 120/2000)^{1/2} = 84.85$ units.

4.5 Time-varying demand

So far we have only considered constant demand. Time-varying demand is quite common, though. The variations can depend on different reasons. For example, an item may be used as a component when producing some product that is only produced on certain known occasions. Another possibility is production on a contract, which requires that certain quantities are delivered on specified dates. There can also be known seasonal variations in demand. Note that we are still considering deterministic demand, i.e., all variations are known in advance. We can therefore still disregard the lead-time.

When dealing with lot sizing for time-varying demand, it is generally assumed that there are a finite number of discrete time steps, or periods. A period may be, for example, one day or a week. We know the demand in each period, and for simplicity, it is assumed that the period demand takes place at the beginning of the period. There is no initial stock. When delivering a batch, the whole batch is delivered at the same time. The holding cost and the ordering cost are constant over time. No backorders are allowed. We shall use the following notation:

T = number of periods,
d_i = demand in period i, $i = 1, 2, ..., T$, (it is assumed that $d_1 > 0$, since otherwise we can just disregard period 1),
A = ordering cost,
h = holding cost per unit and time unit.

We wish to choose batch quantities so that the sum of the ordering and holding costs is minimized. The problem considered is usually denoted the *classical dynamic lot size problem*. Since the demand is varying over time, we can no longer expect the optimal lot sizes to be constant.

Example 4.4 Let $T = 4$, $d_1 = 10$, $d_2 = 20$, $d_3 = 10$, $d_4 = 30$, $h = 2$, and $A = 150$. Consider the batch that should be delivered at the beginning of period 1. (Note that we always need a delivery in period 1.) It is important to realize that only a few values of the batch size are possible in an optimal solution. For example, a first batch size of 35 units can be disregarded. This batch size would cover the first two periods and half of the demand in period 3. But since a new batch is needed anyway in period 3, it would thus be advantageous to add 5 units to the second batch and reduce the first batch to 30. By doing so we can reduce the stock in periods 1 and 2 by 5 units, and this means, of course, lower holding costs. We can conclude that a batch should always cover the demand in an integer number of periods. Consequently, the first batch can only be 10, 30, 40, or 70. But in this example we can also rule out batch size 70, since this batch size implies that the demand in period 4 is stocked during periods 1 - 3, incurring the holding costs $3 \cdot 2 \cdot 30 = 180$. This means that it is cheaper to cover the demand in period 4 by a separate delivery that will incur the ordering cost 150 but eliminate the considered holding costs.

Let us summarize the conclusions from Example 4.4 as two properties of the optimal solution:

Property 1 *A replenishment must always cover the demand in an integer number of consecutive periods.*

Property 2 *The holding costs for a period demand should never exceed the ordering cost.*

SINGLE-ECHELON – LOT SIZING 63

Property 1 is often denoted the *zero-inventory property*, meaning that a delivery should only take place when the inventory is zero. It can be shown that Property 1 is valid in a more general setting, see e.g., Problem 4.15. A consequence of Property 1 is that the optimal solution has a relatively simple structure.

We will only deal with the case when backorders are not allowed. It is also common to consider a more general version of the problem with backorders.

4.6 The Wagner-Whitin algorithm

The dynamic lot size problem described in Section 4.5 is relatively easy to solve exactly. The most common approach is to use *Dynamic Programming*. This was first suggested by Wagner and Whitin (1958) and their solution is usually denoted the Wagner-Whitin algorithm.

Let us introduce the following notation:

f_k = minimum costs over periods 1, 2, ..., k, i.e., when we disregard periods $k + 1$, $k + 2$, ..., T,

$f_{k,t}$ = minimum costs over periods 1, 2, ..., k, given that the last delivery is in period t ($1 \leq t \leq k$).

Note first that

$$f_k = \min_{1 \leq t \leq k} f_{k,t}, \qquad (4.20)$$

since the last delivery must occur in some period in the optimal solution. It is also obvious that $f_0 = 0$ and that $f_1 = f_{1,1} = A$. This is so because with just one period we get a setup cost but no holding costs. Recall that the period demand is assumed to occur at the beginning of a period.

Assume now that we know f_{t-1} for some $t > 0$. It is then easy to obtain $f_{k,t}$ for $k \geq t$ as

$$f_{k,t} = f_{t-1} + A + h(d_{t+1} + 2d_{t+2} + ... + (k-t)d_k). \qquad (4.21)$$

Because we have a delivery in period t, the minimum costs for periods 1, 2, ..., t - 1 must be f_{t-1}. The cost in period t is the setup cost A. Recall again that the demands are assumed to take place at the beginning of the periods. This means that d_t will not cause any holding costs. The demand in period $t + 1$

incurs the holding costs hd_{t+1} since the quantity d_{t+1} is kept in stock during period t. The demand in period $t + 2$ is similarly kept in stock during two periods, t and $t + 1$, and incurs the holding costs $2hd_{t+2}$, etc.

Assume now that we know $f_1, f_2, ..., f_{k-1}$, i.e., that we have solved the problem for 1, 2, ..., $k - 1$ periods. We can then determine $f_{k,t}$ for $1 \leq t \leq k$ from (4.21). Next we apply (4.20) to get f_k and the problem is also solved for k periods. After that we are ready to use the same procedure for $k + 1$ periods, etc. From Property 2 in Section 4.5 we know that if $h(j - t)d_j > A$ we do not need to determine $f_{k,t}$ for $k \geq j$ when applying (4.21). Furthermore, it is obvious that we only need to consider $k \leq T$.

After going through all T periods we obtain the optimal solution as follows. First we consider (4.20) for $k = T$. The minimum value is the optimal cost, and the minimizing t is the period with the last delivery. Denote the minimizing t by t'. Since the last delivery is in period t', the solution for periods 1, 2, ..., $t'- 1$ must also be optimal. So we can again consider (4.20) for $k = t' - 1$. The optimal last delivery for this problem, say t'', is the second last delivery for the total problem. Next we consider (4.20) for $k = t'' - 1$, and we continue in this way until the minimizing t equals one, since we have then obtained all periods with deliveries.

We shall illustrate the computations with an example.

Example 4.5 The demands in $T = 10$ periods are given in the first two rows of Table 4.1. The ordering cost is $A = \$300$ and the holding cost $h = \$1$ per unit and period.

In Table 4.1 we start by setting $f_{1,1} = 300$. Next we use (4.21) to determine $f_{2,1} = f_{1,1} + 1 \cdot 60 = 360$, $f_{3,1} = f_{2,1} + 2 \cdot 90 = 540$, etc. When we come to $f_{6,1}$ we note that the holding cost associated with the demand in period 6 is $500 > A = 300$ and the column is therefore complete due to Property 2. It is clear that $f_1 = f_{1,1} = 300$. Applying (4.21) we obtain $f_{2,2}, f_{3,2}$, etc. After completing this column we apply (4.20) to get $f_2 = \min \{f_{2,1}, f_{2,2}\} = \min \{360, 600\} = 360$. Note that these two numbers are in the underlined diagonal in Table 4.1. After completing the next column we have to optimize over the next diagonal to the right to get $f_3 = \min\{540, 690, 660\} = 540$. We continue in this way until the table is complete. We get the minimum costs $f_{10} = \min\{1550, 1630, 1570, 1550, 1770\} = 1550$ by applying (4.20). It is obvious that the minimum costs are obtained both for $t' = 9$ and $t' = 6$. Let us first consider the solution with $t' = 9$. Since the last delivery is in period 9, the problem over 8 periods must also be solved optimally. By looking in the corresponding diagonal we get $t'' = 6$. Next we consider the five period problem and obtain $t''' = 3$. Finally we get $t'''' = 1$ and the solution is complete. It is easy to verify that the other optimal solution with $t' = 6$ simply means that the last two batches are combined into a single batch. The two optimal solutions are given in Table 4.2.

Table 4.1 Solution, $f_{k,t}$, of Example 4.5.

Period t	1	2	3	4	5	6	7	8	9	10
d_t	50	60	90	70	30	100	60	40	80	20
$k=t$	300	600	660	840	1030	1090	1370	1450	1530	1770
$k=t+1$	360	690	730	870	1130	1150	1410	1530	1550	
$k=t+2$	540	830	790	1070	1250	1230	1570	1570		
$k=t+3$	750	920	1090	1250	1370	1470	1630			
$k=t+4$	870		1330	1410		1550				
$k=t+5$			1530							

Table 4.2 Optimal batch sizes in Example 4.5.

Period t	1	2	3	4	5	6	7	8	9	10
Solution 1	110		190			200			100	
Solution 2	110		190			300				

The Wagner-Whitin algorithm was generalized to the backlogging case by Zangwill (1966).

Consider Table 4.1. There are T columns. Furthermore, there may also be as much as T cells in a column. This means that the computational complexity is $O(T^2)$, i.e., the computation time may be proportional to T^2. Recent more efficient methods for solving the dynamic lot size problem are given in Federgruen and Tzur (1991, 1994, 1995) and Wagelmans et al. (1992). The computational complexity of these techniques is $O(T\text{Log}(T))$.

In practice it is very common to consider dynamic lot size problems in a *rolling horizon* environment. In each period we want to determine the lot sizes in the first period. The real horizon is infinite but we consider a finite horizon as an approximation. The question then arises whether it is possible to replace an infinite horizon by a sufficiently long finite horizon, such that we still get the optimal solution in the first period. Several such planning horizon results have been derived. See e.g., Wagner and Whitin (1958), Kunreuther and Morton (1973), Lundin and Morton (1975), and Federgruen and Tzur (1994, 1995). Such planning horizons are generally quite long, though. It may even happen that no planning horizon exists. The following example from Bhaskaran and Sethi (1988) illustrates this. See also Section 4.9.

Example 4.6 The demands over an infinite horizon are given as $d_1 = 5$, $d_2 = 4$, $d_3 = 5$, $d_4 = 4$, and $d_i = 5$, for $i \geq 5$. Furthermore, $A = 1$ and $h = 0.05$. Let T be a considered finite horizon. It can be shown that the optimal initial batch varies as follows: for $T = 3, 6, 9, 12, \ldots$ it is 14, for $T = 4, 7, 10, 13, \ldots$ it is 18, and for $T = 5, 8, 11, 14, \ldots$ it is 9. So if we increase or decrease the horizon T by one time unit the initial batch is changing.

4.7 The Silver-Meal heuristic

Although the Wagner-Whitin algorithm and other exact methods are in most circumstances computationally feasible, in practice it is much more common to use simple heuristics to obtain an approximate solution. One of the best and most well-known methods is the Silver-Meal heuristic (Silver and Meal, 1973).

The Silver-Meal heuristic is, like most other lot sizing heuristics, a sequential method. When determining the delivery in period 1 we consider successively the demands in periods 2, 3, ... (Recall that we know that the delivery batch in period 1 must at least cover the demand in period 1). When considering period 2 we use a simple test to decide whether this period demand should be added to the delivery batch in the first period. If this is the case we continue with period 3, etc. Assume that this procedure shows that the next delivery will come in period k. The procedure is then repeated with period k as the "first period".

The idea of the Silver-Meal heuristic is to choose to have a new delivery when the average per period costs increase for the first time. This means that the first delivery shall cover n periods and the next delivery take place in period $n + 1$ if

$$\frac{A + h\sum_{j=2}^{k}(j-1)d_j}{k} \leq \frac{A + h\sum_{j=2}^{k-1}(j-1)d_j}{k-1}, \quad 2 \leq k \leq n, \quad (4.22)$$

and

$$\frac{A + h\sum_{j=2}^{n+1}(j-1)d_j}{n+1} > \frac{A + h\sum_{j=2}^{n}(j-1)d_j}{n}. \quad (4.23)$$

SINGLE-ECHELON – LOT SIZING

Example 4.7 We consider the same problem as in Example 4.5, i.e., $A = \$300$ and $h = \$1$ per unit and period. The demands are given in Table 4.3.

Table 4.3 Demands in Examples 4.7 and 4.8.

Period t	1	2	3	4	5	6	7	8	9	10
d_t	50	60	90	70	30	100	60	40	80	20

If the delivery in period 1 covers only the demand in period 1 the cost for this period is the ordering cost $A = 300$. We now successively consider the per period costs for deliveries that cover 2, 3, ... periods.

2 periods $(300 + 60)/2 = 180 < 300$,

3 periods $(300 + 60 + 2 \cdot 90)/3 = 180 \leq 180$,

4 periods $(300 + 60 + 2 \cdot 90 + 3 \cdot 70)/4 = 187.5 > 180$, which means a new delivery in period 4.

The same procedure is now applied with period 4 as the first period.

2 periods $(300 + 30)/2 = 165 < 300$,

3 periods $(300 + 30 + 2 \cdot 100)/3 = 176.67 > 165$, which means a new delivery in period 6.

Starting with period 6 as the first period we obtain

2 periods $(300 + 60)/2 = 180 < 300$,

3 periods $(300 + 60 + 2 \cdot 40)/3 = 146.67 \leq 180$,

4 periods $(300 + 60 + 2 \cdot 40 + 3 \cdot 80)/4 = 170 > 146.67$, which means a new delivery in period 9.

Starting with period 9 as the first period we obtain

2 periods $(300 + 20)/2 = 160 < 300$.

The corresponding solution is given in Table 4.4.

Table 4.4 Solution with the Silver-Meal heuristic.

Period t	1	2	3	4	5	6	7	8	9	10
Quantity	200			100		200			100	

The total costs are obtained as $C = 3 \cdot 180 + 2 \cdot 165 + 3 \cdot 146.67 + 2 \cdot 160 = 1630$ (3 periods with average period cost 180, 2 periods with average period cost 165, etc.). This is about 5% higher than the optimal costs obtained by the Wagner-Whitin algorithm. Note that in this case we are very close to obtaining the optimal solution. If, for example, the demand in period 3 had been 91 instead of 90, the heuristic would have suggested a delivery in period 3 and we would have obtained the optimal solution.

In most situations the cost increase is only about 1 - 2%, see e.g., Baker (1989). Still it has been demonstrated that the relative error can be arbitrarily large, see Axsäter (1982, 1985). It is straightforward to extend the logic of the procedure to more general problems, e.g., to include discounts.

When using the Silver-Meal heuristic we consider the costs per period. An alternative is to consider the costs per unit. This means that we are dividing by the total number of units instead of the number of periods in our comparisons. This alternative procedure is denoted the Least Unit Cost heuristic. It has been shown that the Silver-Meal heuristic, in general, outperforms the Least Unit Cost heuristic (Baker, 1989).

4.8 A heuristic that balances holding and ordering costs

When we derived the classical economic order quantity formula in Section 4.1.1 it turned out that in the optimal solution, holding costs and ordering costs are exactly equal. This is no longer true for the dynamic lot size problem, but it is reasonable to expect that holding and ordering costs are of about the same size. This is the basis for the following algorithm, which is a simplified version of a technique suggested by De Matteis and Mendoza (1968).

Let the first delivery quantity cover n periods, where n is determined by the condition

$$h\sum_{j=2}^{n}(j-1)d_j \leq A < h\sum_{j=2}^{n+1}(j-1)d_j , \qquad (4.24)$$

SINGLE-ECHELON – LOT SIZING

i.e., we choose to have a new delivery as soon as the holding costs exceed the ordering cost. As with the Silver-Meal heuristic, the next batch is determined in the same way with period $n + 1$ regarded as the first period.

In general this heuristic does not perform as well as the Silver-Meal heuristic. The relative error cannot be arbitrarily large, though. It is bounded by 100 percent, see Axsäter (1982, 1985).

Example 4.8 We consider again the same problem setting as in Examples 4.5 and 4.7, i.e., $A = \$300$ and $h = \$1$ per unit and period. The demands are given in Table 4.3.

Starting with period 2 we compare the accumulated holding costs to the ordering cost.

2 periods	$60 \leq 300$,	
3 periods	$60 + 2 \cdot 90 = 240 \leq 300$,	
4 periods	$60 + 2 \cdot 90 + 3 \cdot 70 = 450 > 300$, which means a new delivery in period 4.	

The same procedure is now applied with period 4 as the first period.

2 periods	$30 \leq 300$,	
3 periods	$30 + 2 \cdot 100 = 230 \leq 300$,	
4 periods	$30 + 2 \cdot 100 + 3 \cdot 60 = 410 > 300$, which means a new delivery in period 7.	

Starting with period 7 as the first period we obtain

2 periods	$40 \leq 300$,
3 periods	$40 + 2 \cdot 80 = 200 \leq 300$,
4 periods	$40 + 2 \cdot 80 + 3 \cdot 20 = 260 \leq 300$.

The solution is given in Table 4.5. It is easy to verify that we get the same costs as with the Silver-Meal heuristic, i.e., 1630. Of course, this might not always be the case.

Table 4.5   Solution with the heuristic.

Period t	1	2	3	4	5	6	7	8	9	10
Quantity	200			200			200			

Other lot sizing heuristics are suggested in e.g., Groff (1979) and Axsäter (1980).

4.9 Exact or approximate solution

We have considered three methods for solving the classical dynamic lot size problem. The Wagner-Whitin algorithm gives the optimal solution while the other techniques are approximate. An exact solution is by no means computationally infeasible. Still it is much more common in practice to use an approximate method. One reason is that the approximate methods are easy to understand. It is also easy to check the computations manually. However, there is also another, probably even more important reason. As we have discussed in Section 4.6, the considered lot sizing techniques are usually applied in a rolling horizon environment. In each period we want to determine lot sizes for the first period. The real infinite horizon is then replaced by a finite horizon. If this finite horizon were sufficiently long we would often still get the optimal solution for the first period. But in practice it is common to use a relatively short horizon. This means that the solution is also approximate when using an exact method like the Wagner-Whitin algorithm. The approximate methods are, in general, not similarly sensitive with respect to the length of the horizon. In such situations a good approximate method may therefore even outperform an exact method (see Blackburn and Millen, 1980).

References

Axsäter, S. 1980. Economic Lot Sizes and Vehicle Scheduling, *European Journal of Operational Research*, 4, 395-398.
Axsäter, S. 1982. Worst Case Performance of Lot Sizing Heuristics, *European Journal of Operational Research*, 9, 339-343.
Axsäter, S. 1985. Performance Bounds for Lot Sizing Heuristics, *Management Science*, 31, 634-640.
Axsäter, S. 1996. Using the Deterministic EOQ Formula in Stochastic Inventory Control, *Management Science*, 42, 830-834.

Baker, K. R. 1989. Lot-Sizing Procedures and a Standard Data Set: A Reconciliation of the Literature, *Journal of Manufacturing and Operations Management*, 2, 199-221.

Benton, W. C., and S. Park. 1996. A Classification of Literature on Determining the Lot Size Under Quantity Discounts, *European Journal of Operational Research*, 92, 219-238.

Bhaskaran, S. and S. P. Sethi. 1988. The Dynamic Lot Size Model with Stochastic Demands: A Decision Horizon Study, *INFOR*, 26, 213-224.

Blackburn, J. D., and R. A. Millen. 1980. Heuristic Lot-Sizing Performance in a Rolling Schedule Environment, *Decision Sciences*, 11, 691-701.

De Matteis, J. J., and A. G. Mendoza. 1968. An Economic Lot Sizing Technique, *IBM Systems Journal*, 7, 30-46.

Federgruen, A., and M. Tzur. 1991. A Simple Forward Algorithm to Solve General Dynamic Lot Sizing Models With n Periods in $O(n \log n)$ or $O(n)$ time, *Management Science*, 37, 909-925.

Federgruen, A., and M. Tzur. 1994. Minimal Forecast Horizons and a New Planning Procedure for the General Dynamic Lot Sizing Model: Nervousness Revisited, *Operations Research*, 42, 456-469.

Federgruen, A., and M. Tzur. 1995. Fast Solution and Detection of Minimal Forecast Horizons in Dynamic Programs with a Single Indicator of the Future: Applications to Dynamic Lot-Sizing Models, *Management Science*, 41, 874-893.

Groff, G. 1979. A Lot-Sizing Rule for Time-Phased Component Demand, *Production and Inventory Management*, 20, 47-53.

Harris, F. W. 1913. How Many Parts to Make at Once, *Factory, The Magazine of Management*, 10, 135-136, 152.

Kunreuther, H., and T. Morton. 1973. Planning Horizons for Production Smoothing with Deterministic Demands, *Management Science*, 20, 110-125.

Lee, H. L., and S. Nahmias. 1993. Single-Product, Single-Location Models, in S. C. Graves et al. Eds., *Handbooks in OR & MS Vol.4*, North Holland Amsterdam, 133-173.

Lundin, R., and T. Morton. 1975. Planning Horizons for the Dynamic Lot Size Model: Zabel vs Protective Procedures and Computational Results, *Operations Research*, 23, 711-734.

Silver, E. A., and H. C. Meal. 1973. A Heuristic for Selecting Lot Size Requirements for the Case of a Deterministic Time-Varying Demand Rate and Discrete Opportunities for Replenishment, *Production and Inventory Management*, 14, 64-74.

Silver, E. A., D. F. Pyke, and R. Peterson. 1998. *Inventory Management and Production Planning and Scheduling*, 3rd edition, Wiley, New York.

Wagelmans, A., S. Van Hoesel, and A. Kolen. 1992. Economic Lot Sizing: An $O(n \log n)$ Algorithm That Runs in Linear Time in the Wagner-Whitin Case, *Operations Research*, 40, S145-S156.

Wagner, H. M., and T. M. Whitin. 1958. Dynamic Version of the Economic Lot Size Model, *Management Science*, 5, 89-96.

Wilson, R. H. 1934. A Scientific Routine for Stock Control, *Harvard Business Review*, 13, 116-128.

Zangwill, W. I. 1966. A Deterministic Multi-Period Production Scheduling Model with Backlogging, *Management Science*, 13, 105-119.

Zheng, Y. S. 1992. On Properties of Stochastic Inventory Systems, *Management Science*, 38, 87-103.

Zipkin, P. H. 2000. *Foundations of Inventory Management*, McGraw-Hill, Singapore.

Problems

4.1* Consider the following data for two products:

Product	A	B
Demand per year, units	48000	72000
Unit cost, $	50	85
Ordering cost, $	125	125

The carrying charge is 24 percent.

a) Use the classical economic lot size model to determine batch quantities.
b) Assume that the real ordering cost is $200 instead of $125. How large are the additional costs incurred by the bad estimate of the ordering cost?

4.2* We consider a company using the classical economic lot size model. How high is the relative real cost increase if the holding cost in the model is only 60 percent of the real holding cost?

4.3 Consider the classical economic order quantity model in Section 4.1. Let I be the inventory on hand. Assume that the linear holding cost hI is only valid for $I \le I'$. For $I > I'$ the holding costs are obtained as $hI' + h'(I-I')$, where $h' > h$. An interpretation is that we need to rent additional and more expensive storage space when the inventory on hand is higher than I'. Determine the costs per unit of time as a function of the batch quantity Q.

4.4 Assume that the classical economic order quantity model in Section 4.1 suggests $Q^* = 25$ for a certain item. However, for some practical reasons the order quantity has to be a multiple of 10. What is then the best order quantity? Explain why.

4.5 Consider the model with finite production rate in Section 4.2. Let $p \to d$ from above. What will happen with Q^* and C^*? Explain.

---

* Answer and/or hint in Appendix 1.

SINGLE-ECHELON – LOT SIZING

4.6* Consider again the model with finite production rate in Section 4.2. Change the assumptions so that the units of a batch can meet demand first when the whole batch is complete. Determine expressions for Q^* and C^*.

4.7 Use the data in Example 4.2 in Section 4.3, except for $v' < v$, which is considered as a problem parameter. Determine for which values of this parameter the optimal solution is obtained from Q', Q'', and Q_0, respectively.

4.8 Consider again Example 4.2 in Section 4.3. Assume instead incremental discounts, i.e., the lower price v' is only for those units that exceed Q_0. Note that this also affects the holding costs. Determine the optimal solution.

4.9 In the model in Section 4.4, replace the backorder cost per unit and time unit, b_1, by a backorder cost per unit, b_2. This means that there is no difference between long and short delays.

a) Show that for a given $0 \leq x \leq 1$ the optimal costs can be expressed as

$$C = \sqrt{2Adh(1-x)} + b_2 dx .$$

What is then the decision rule for x? Explain.

b) Consider the data in Example 4.3, i.e., the constant demand is $d = 300$ units per year. The ordering cost is $A = \$200$ and the holding cost $h = \$20$ per unit and year. What is the optimal solution for different values of b_2?

4.10 Consider again the model in Section 4.4 but assume that demands that can not be met from stock are lost. For each unit that is lost there is a lost sales cost b_2 per unit. Determine the optimal solution.

4.11* We consider the model in Section 4.4 with the data $A = 6$, $d = 1$, $h = 1$, $b_1 = 3$.

a) Determine the optimal solution.
b) Show that the average waiting time for a customer is 1/8.
c) Consider now the problem without the backorder cost b_1 per unit and time unit but with 1/8 as an upper bound for the average waiting time for a customer. Show that the policy in a) is optimal.

4.12 Consider again the model in Section 4.4 without the backorder cost b_1 per unit and time unit. Assume now that there is an upper limit W' for the maximum waiting time for a customer. Determine the optimal policy. Compare with the optimal policy for the original problem.

4.13 Consider again the model in Section 4.4 without the backorder cost b_1 per unit and time unit. Assume now that there is a lower limit for the fill rate, 1-x, i.e., the fraction of demand that can be satisfied immediately from stock on hand. Let the constraint be $x \leq x'$. Determine the optimal policy. Compare with the optimal policy for the original problem.

4.14 Consider the classical dynamic lot size problem. Assume that the period demand takes place in the middle of the period, while the deliveries are still in the beginning of a period. Show how the costs are affected, and that the optimal solution is still the same.

4.15 Consider a more general version of the classical dynamic lot size problem where the ordering and holding costs A_i, $h_i > 0$ may vary with i. Provide a proof that Property 1, the zero-inventory property, is still valid.

4.16* The demand for a product over five periods is 25, 40, 40, 40, and 90 units. The ordering cost is $100 and the holding cost $1 per unit and period. Determine batch quantities by using

a) The Wagner-Whitin algorithm,   *mixed linear*
b) The Silver-Meal algorithm,   *mixed nonlinear*
c) The heuristic in Section 4.8.

4.17* The demand for a product over six periods is 10, 40, 95, 70, 120, and 50 units. The ordering cost is $100 and the holding cost $0.50 per unit and period. Determine batch quantities by using

a) The Wagner-Whitin algorithm,
b) The Silver-Meal algorithm.

4.18* The demand for a product over seven periods is 10, 24, 12, 7, 5, 4, and 3 units. The ordering cost is $100 and the holding cost $4 per unit and period. Determine batch quantities by using

a) The Wagner-Whitin algorithm,
b) The Silver-Meal algorithm.

4.19 In Section 4.6 we use Dynamic Programming with *Forward Recursion*, i.e., we start with the solution for period 1. Given this solution we determine the solution for periods 1 and 2, etc. It is also possible to use *Backward Recursion*, i.e., start with the solution for period T. Given this solution, determine the solution for periods $T - 1$ and T, etc. Formulate such an alternative Dynamic Program and solve Example 4.5 in this way.

4.20 The demand for an item over 7 periods is given.

Period	1	2	3	4	5	6	7
Demand	20	30	20	10	30	50	20

The holding cost is 2 per unit and period and the ordering cost is 100.

a) Determine all optimal solutions by the Wagner-Whitin algorithm.
b) Determine the optimal solutions if deliveries in period 5 are not allowed.

5 SINGLE-ECHELON SYSTEMS: REORDER POINTS

5.1 Discrete stochastic demand

We shall now consider different techniques for determining reorder points, or equivalently safety stocks, when the demand is stochastic. To do this we first of all need a suitable demand model. In practice the demand during a certain time is nearly always a nonnegative integer, i.e., it is a discrete stochastic variable. (Exceptions may occur when we deal with products like oil.) Provided that the demand is reasonably low, it is then natural to use a discrete demand model, which resembles the real demand. However, if the demand is relatively large, it is more practical to use a continuous demand model as an approximation. See Section 5.2.

5.1.1 Compound Poisson demand

A common assumption in stochastic inventory models is that the cumulative demand can be modeled by a nondecreasing stochastic process with stationary and mutually independent increments. Such a process may always be represented as a limit of an appropriate sequence of compound Poisson processes. See, for example, Feller (1966). We shall therefore often make the assumption that the demand is a compound Poisson process. This means that the customers arrive according to a Poisson process with given intensity λ. The size of a customer demand is also a stochastic variable.

In a small time interval Δt the probability for exactly one customer is $\lambda \Delta t$, while the probability for more than one customer can be disregarded, or in other words it is o(Δt). The number of customers in a time interval of length t then has a Poisson distribution and the probability for k customers is

$$P(k) = \frac{(\lambda t)^k}{k!} e^{-\lambda t}, \quad k = 0, 1, 2, \ldots \quad (5.1)$$

Both the average and the variance of the number of customers are equal to λt.

Compound Poisson demand means that the size of a customer demand is also a stochastic variable that is independent of other customer demands and of the distribution of the customer arrivals. The distribution of the demand size is denoted the *compounding* distribution. It is convenient to assume that each customer demands an integral number of units. Let

f_j = probability of demand size j ($j = 1, 2, \ldots$).

If $f_1 = 1$, the demand process degenerates to pure Poisson demand. The demand is then equal to the number of customers and the demand distribution is given by (5.1).

Let us now consider the general case where the demand sizes are varying. Note that we assume that there are no demands of size zero. This is no lack of generality. If $f_0 > 0$, we can replace the demand process by an equivalent process characterized by λ' and f_j' where $\lambda' = \lambda(1 - f_0)$, $f_0' = 0$, and $f_j' = f_j/(1 - f_0)$ for $j > 0$. We shall also assume that not all demands are multiples of some integer larger than one. This is, for example, always the case when $f_1 > 0$. Assume that this condition is not satisfied and that the demand size is, for example, always a multiple of 5. It is then natural to switch to a new unit that consists of 5 old units. With the new unit the condition is satisfied.

Let

f_j^k = probability that k customers give the total demand j,
$D(t)$ = stochastic demand in the time interval t.

We can determine the distribution of $D(t)$ as follows. Note first that $f_0^0 = 1$, and $f_j^1 = f_j$. Given f_j^1 we can obtain the j-fold convolution of f_j, f_j^k, recursively as

SINGLE-ECHELON – REORDER POINTS

$$f_j^k = \sum_{i=k-1}^{j-1} f_i^{k-1} f_{j-i}, \quad k = 2, 3, 4, \ldots \tag{5.2}$$

Using (5.1) we then have

$$P(D(t) = j) = \sum_{k=0}^{\infty} \frac{(\lambda t)^k}{k!} e^{-\lambda t} f_j^k. \tag{5.3}$$

Let us now determine the average and the standard deviation of the demand during one unit of time.

μ = average demand per unit of time,
σ = standard deviation of the demand per unit of time.

Consider one time unit. Let K be the stochastic number of customers during this time unit, J the stochastic demand size of a single customer, and Z the stochastic demand during the time unit considered. Recall that $E(K) = Var(K) = \lambda$, and that K and J are independent. For a given K, Z is simply the sum of K independent customer demands and we have

$$\mu = E(Z) = \underset{K}{E}\{E(Z|K)\} = \underset{K}{E}\{K\,E(J)\} = E(K)E(J) = \lambda \sum_{j=1}^{\infty} jf_j.$$

$$= \lambda E(J) \tag{5.4}$$

To determine σ we first obtain

$$E(Z^2) = \underset{K}{E}\{E(Z^2|K)\} = \underset{K}{E}\{Var(Z|K) + (E(Z|K))^2\}$$
$$= \underset{K}{E}\{K Var(J) + K^2(E(J))^2\}. \tag{5.5}$$

Using $E(K^2) = Var(K) + (E(K))^2 = \lambda + \lambda^2$, we can now determine $\sigma^2 = Var(Z)$ as

$$\sigma^2 = E(Z^2) - \mu^2 = \underset{K}{E}\{K Var(J) + K^2(E(J))^2\} - \mu^2$$

$$= E(K) Var(J) + E(K^2)(E(J))^2 - \mu^2$$

$$= \lambda Var(J) + (\lambda + \lambda^2)(E(J))^2 - \mu^2 = \lambda\left[Var(J) + (E(J))^2\right] \tag{5.6}$$

$$= \lambda E(J^2) = \lambda \sum_{j=1}^{\infty} j^2 f_j.$$

Equations (5.4) and (5.6) give the mean and standard deviation of the demand during one unit of time. Let μ' and σ' be the mean and standard deviation of the demand during the time t. We then simply have $\mu' = \mu t$ and $(\sigma')^2 = \sigma^2 t$.

From (5.4) and (5.6) we note that $\sigma^2/\mu = E(J^2)/E(J) \geq 1$ with equality only for pure Poisson demand. Consequently, it is not possible to model demand processes with $\sigma^2/\mu < 1$ as compound Poisson demand.

We shall use the compound Poisson process when deriving various theoretical results. For items with relatively low demand, it may also be advantageous to use this demand model in practice. An advantage with compound Poisson demand is that it is easy to represent the demand in a simulation model.

In connection with inventory control we usually need the distribution of the demand over a certain time t (e.g., the lead-time). In general we have estimates of μ' and σ' from the forecasting system (see Chapter 2). The standard procedure is then to fit the selected demand model to these two parameters.

5.1.2 Logarithmic compounding distribution

Let us first assume that the demand size has a *logarithmic distribution*. This means that

$$f_j = -\frac{\alpha^j}{\ln(1-\alpha)j} \qquad j = 1, 2, 3, \ldots \tag{5.7}$$

where $0 < \alpha < 1$. This distribution has the following mean and variance

$$E(J) = -\frac{\alpha}{(1-\alpha)\ln(1-\alpha)}, \tag{5.8}$$

$$Var(J) = -\frac{\alpha(\ln(1-\alpha)+\alpha)}{(1-\alpha)^2(\ln(1-\alpha))^2}. \qquad (5.9)$$

Note that this gives

$$E(J^2) = Var(J) + (E(J))^2 = -\frac{\alpha}{(1-\alpha)^2 \ln(1-\alpha)}. \qquad (5.10)$$

When $\alpha \to 0$ the demand size J will approach constant unit demand, i.e., $f_1 \to 1$ and all other $f_j \to 0$. When $\alpha \to 1$ $Var(J)/E(J) \to \infty$.

Let now μ' and σ' be given. From (5.4) and (5.8) we get

$$\mu' = \lambda t E(J) = -\frac{\lambda t \alpha}{(1-\alpha)\ln(1-\alpha)}, \qquad (5.11)$$

and from (5.6) and (5.10)

$$(\sigma')^2 = \lambda t E(J^2) = -\frac{\lambda t \alpha}{(1-\alpha)^2 \ln(1-\alpha)}. \qquad (5.12)$$

It follows immediately from (5.11) and (5.12) that, for given μ' and σ', we can determine α and λ as

$$\alpha = 1 - \frac{\mu'}{(\sigma')^2} = 1 - \frac{\mu}{\sigma^2}, \qquad (5.13)$$

$$\lambda = -\mu'\frac{(1-\alpha)\ln(1-\alpha)}{t\alpha} = -\mu\frac{(1-\alpha)\ln(1-\alpha)}{\alpha}. \qquad (5.14)$$

Using (5.2) and (5.3) we can then determine the distribution of the demand $D(t)$ during the time t. However, when the compounding distribution is logarithmic there is a much simpler way. It is possible to show that in this case $D(t)$ has a negative binomial distribution, i.e., $P(D(t) = 0) = (1-p)^r$, and

$$P(D(t) = k) = \frac{r(r+1)\ldots(r+k-1)}{k!}(1-p)^r p^k, \quad k = 1, 2, \ldots \qquad (5.15)$$

where the parameter r can be any positive number and $0 < p < 1$. In the special case when r is an integer, we can interpret the distribution as the sum of r independent random variables each having a geometric distribution (see below). The negative binomial distribution has $E(D(t)) = rp/(1-p)$ and $Var(D(t)) = rp/(1-p)^2$. So given μ' and σ', we can directly determine p and r as

$$p = 1 - \frac{\mu'}{(\sigma')^2} = \alpha, \qquad (5.16)$$

$$r = \mu' \frac{(1-p)}{p}. \qquad (5.17)$$

It is then easy to apply (5.15).

5.1.3 Geometric compounding distribution

Assume now that the compounding distribution is instead a *delayed* geometric distribution, i.e.,

$$f_j = (1-\beta)\beta^{j-1}, \qquad j = 1,2,3,... \qquad (5.18)$$

where $0 < \beta < 1$. The resulting process is sometimes denoted a *stuttering* Poisson process. The geometric distribution has the following mean and variance

$$E(J) = \frac{1}{1-\beta}, \qquad (5.19)$$

$$Var(J) = \frac{\beta}{(1-\beta)^2}. \qquad (5.20)$$

Given μ' and σ', we can use (5.4) and (5.6) to determine β and λ as

$$\beta = 1 - \frac{2}{1 + (\sigma)^2/\mu}, \qquad (5.21)$$

$$\lambda = \mu(1-\beta). \qquad (5.22)$$

Finally we can apply (5.2) and (5.3) to get the distribution of the demand over the time t. It is, in general, easier to use a logarithmic compounding distribution and apply (5.15)-(5.17).

Example 5.1 Assume that $\mu' = 5$, $(\sigma')^2 = 15$ and that the considered time is $t = 20$. Let us first use a logarithmic compounding distribution. Using (5.16) and (5.17) we get $p = 0.667$ and $r = 2.5$. The corresponding distribution of the demand during t is then easily obtained from (5.15). See Figure 5.1. We can also determine $\alpha = p = 0.667$, and $\lambda = 0.137$ from (5.14). The average demand size is $E(J) = 1.82$ from (5.8).

Consider now instead a geometric compounding distribution. Using (5.21) and (5.22) we get $\beta = 0.5$ and $\lambda = 0.125$. The average demand size is $E(J) = 2$ from (5.19). Finally we apply (5.2) and (5.3) to get the distribution of the demand during t, see Figure 5.1.

It is obvious that the two obtained distributions are very similar. In practice the shape of the distribution is normally unknown. Furthermore, μ' and σ' are uncertain estimates. Given this, it seems better to use a logarithmic compounding distribution to simplify the computations.

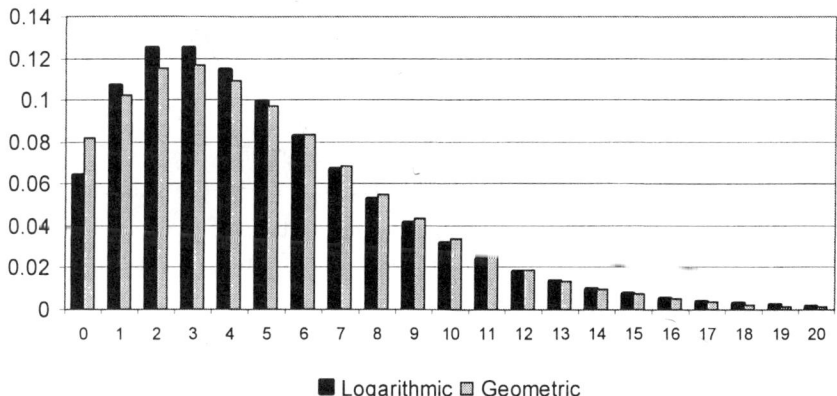

Figure 5.1 Comparison of logarithmic and geometric compounding distribution for the data in Example 5.1.

5.1.4 Smooth demand

If we use compound Poisson demand as our demand model, the total demand during a certain time must have a variance-to-mean ratio larger or equal to one, see Section 5.1.1. In some situations we want to model a smoother demand. Note, however, that this means that the cumulative demand can no longer be interpreted as a nondecreasing stochastic process with stationary and mutually independent increments. Furthermore, as long as we are dealing with a discrete random variable, for a given mean there is always a lower bound for the possible variance. See Adan et al. (1995) for details. They suggest the following technique that can be used for variance-to-mean ratios that are smaller than one.

Let $BIN(k, p)$ be a binomially distributed random variable, where the integer k is the number of trials and p is the success probability, i.e.,

$$P(BIN(k, p) = j) = \binom{k}{j} p^j (1-p)^{k-j}. \tag{5.23}$$

The binomial distribution has mean kp and variance $kp(1-p)$.

Assume now as before that μ' and σ' are given. Determine

$$a = \left(\frac{\sigma'}{\mu'}\right)^2 - \frac{1}{\mu'}. \tag{5.24}$$

Assume also that $-1 \leq a < 0$, otherwise the considered technique cannot be used. Note that $(\sigma')^2 / \mu' < 1$ implies $a < 0$.

In the special case when $a = -1$, use $BIN(1, \mu')$ as the demand distribution. (It can be shown that $\mu' < 1$ in this case.) If $a < -1$, it may also be reasonable to use this distribution as an approximation.

Otherwise determine k such that $-1/k < a \leq 1/(k+1)$, i.e., $k \geq 1$. The demand distribution is then obtained as the mixture of two random variables, $BIN(k, p)$ with probability q and $BIN(k+1, p)$ with probability $1-q$. The probabilities q and p are obtained as

$$q = \frac{1 + a(1+k) + \sqrt{-ak(1+k) - k}}{1+a}, \tag{5.25}$$

$$p = \frac{\mu'}{k+1-q}. \tag{5.26}$$

It can be shown that q and p are always between 0 and 1. We note that $k \to \infty$ as $a \to 0$ and this means that the computational effort will increase.

5.1.5 Fitting discrete demand distributions in practice

In practice it is very convenient and computationally efficient to use the Poisson distribution (5.1), so this is a reasonable choice as long as $(\sigma')^2/\mu' \approx 1$. A suitable decision rule is to use the Poisson distribution for e.g., $0.9 \leq (\sigma')^2/\mu' \leq 1.1$. For $(\sigma')^2/\mu' > 1.1$ it is efficient to use the negative binomial distribution, or in other words, compound Poisson demand with a logarithmic compounding distribution, see Section 5.1.2. For $(\sigma')^2/\mu' < 0.9$ it is quite common to use Poisson demand, even though this means that the variance is overestimated, but an alternative is to use a mixture of binomial distributions as described in Section 5.1.4.

5.2 Continuous stochastic demand

5.2.1 Normally distributed demand

For items with higher demand, it is usually more convenient and efficient to model the demand over a time period by a continuous distribution. For many reasons it is most common to use the normal distribution. We know from the *central limit theorem* that, under very general conditions, a sum of many independent random variables will have a distribution that is approximately normal. In many situations the demand comes from several independent customers, so it reasonable to let the demand be represented by a normal distribution. Furthermore, if the time period considered is long enough, the discrete demand from a compound Poisson process will become approximately normally distributed. The normal distribution has also for a long time been common in practice and is easy to deal with.

A problem with the normal distribution is that there is always at least a small probability for negative demand. Some results which are exact for compound Poisson demand are therefore only approximately true for normal demand. See Browne and Zipkin (1991) for a detailed discussion of various assumptions concerning demand distributions.

Given the mean μ' and standard deviation σ' of the demand during a time period t, we can always fit a unique normal distribution to these parameters. The so-called standardized normal distribution with mean $\mu' = 0$ and standard deviation $\sigma' = 1$ has the density

$$\varphi(x) = \frac{1}{\sqrt{2\pi}} e^{-\frac{x^2}{2}}, \quad -\infty < x < \infty. \tag{5.27}$$

Both the density and the distribution function,

$$\Phi(x) = \int_{-\infty}^{x} \frac{1}{\sqrt{2\pi}} e^{-\frac{u^2}{2}} du, \tag{5.28}$$

are tabulated in Appendix 2. For other values of μ' and σ', we obtain the density as $(1/\sigma')\varphi((x-\mu')/\sigma')$ and the distribution function as $\Phi((x-\mu')/\sigma')$. There is no closed form for the distribution function, but it is still easily available. One possibility is to use tabulated values and interpolation. Another alternative is to use very accurate approximations, see e.g., Abramowitz and Stegun (1964).

5.2.2 Gamma distributed demand

If the ratio σ'/μ' is not considerably less than 1, there is a relatively high probability for negative demand when using the normal distribution. Although the normal distribution usually works quite well also in such cases, it may be more attractive to use some other distribution for which the demand is always nonnegative. An alternative is to use the so-called gamma distribution.

The gamma distribution has the density

$$g(x) = \frac{\lambda(\lambda x)^{r-1} e^{-\lambda x}}{\Gamma(r)}, \quad x \geq 0. \tag{5.29}$$

The two parameters r and λ are both positive, and $\Gamma(r)$ is the *gamma function*

$$\Gamma(r) = \int_0^\infty x^{r-1} e^{-x} dx. \tag{5.30}$$

In the special case when r is an integer $\Gamma(r) = (r-1)!$, and we get a so-called *Erlang distribution* i.e., the sum of r independent exponentially distributed random variables with parameter λ. For any $r > 1$, $\Gamma(r) = (r-1)\Gamma(r-1)$.

The gamma distribution has mean r/λ and variance r/λ^2. Given μ' and σ' it is therefore always easy to determine the corresponding unique parameters r and λ as

$$r = (\mu'/\sigma')^2, \quad (5.31)$$

$$\lambda = \mu'/(\sigma')^2. \quad (5.32)$$

The distribution function cannot be expressed in closed form, but is together with the density available in various software packages, e.g., in Excel.

Example 5.2 Let $\mu' = 25$ and $\sigma' = 10$. From (5.31) and (5.32) we get $r = 6.25$ and $\lambda = 0.25$. In Figure 5.2 gamma distributed demand is compared with normal demand.

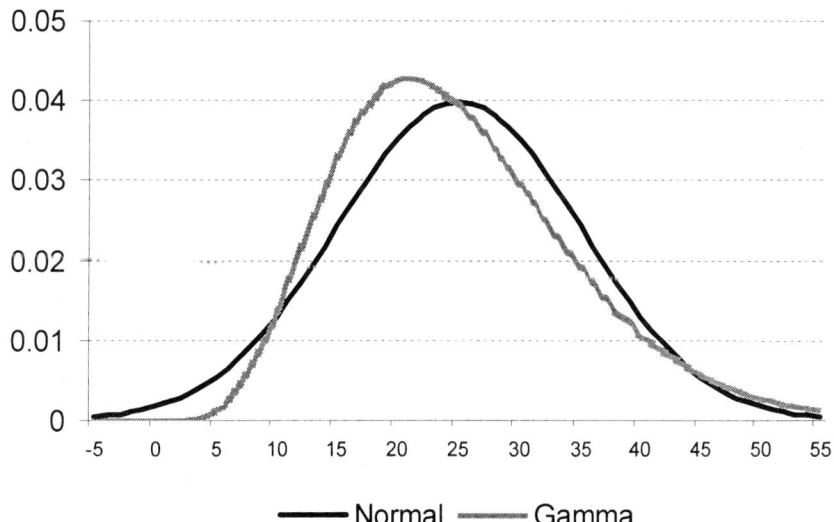

Figure 5.2 Comparison of normal and gamma distributed demand for the data in Example 5.2.

Using the gamma distribution, the demand is always positive. Another difference is that the probability for very high demand is larger with the gamma distribution.

Other possible continuous demand distributions are, for example, the lognormal distribution, the Weibull distribution, and mixtures of Erlang distributions. All these distributions can be fitted to positive random variables with given mean and standard deviation. See Tijms (1994) for details.

5.3 Continuous review *(R, Q)* policy - inventory level distribution

5.3.1 Distribution of the inventory position

Consider now an inventory system with discrete compound Poisson demand, which is controlled by a continuous review (R, Q) policy. We shall derive some important results for such a system. Let

$IP =$ inventory position.

When using an (R, Q) policy, an order is triggered as soon as the integer IP is at or below the reorder point R. The order is for Q units if this is enough to bring IP above R. If not, the order is for the smallest number of batches needed to make the inventory position larger than R. Since an order is triggered as soon as the inventory position is less than or equal to R, we always have $IP \geq R + 1$. Initially the inventory position may be arbitrarily large, but as soon as we have ordered at least once we must have $IP \leq R + Q$. This is so because we always order the smallest number of batches needed to make $IP \geq R + 1$. In steady state we must therefore have $R + 1 \leq IP \leq R + Q$. We shall now prove the following interesting proposition.

Proposition 5.1 In steady state the inventory position is uniformly distributed on the integers $R + 1, R + 2, ..., R + Q$.

Proof Recall that we assume that not all demands are multiples of some integer larger than one. Therefore it is evident that, in the long run, we must reach all the considered inventory positions.

Assume that the inventory position is $R + i$ at some time. Each time a customer arrives, IP will jump to some other value (or possibly to the same value if the size of the demand is a multiple of Q). Let us denote the probability for a jump from $R + i$ to $R + j$ by p_{ij}. It is evident that these probabilities can be determined from the distribution of the demand size. It is also evident that we can see the jumps as a Markov chain. Since all states can be reached, the chain is *irreducible*. In the special case of pure Poisson demand the chain is *periodic*, otherwise *ergodic*. We shall consider the ergodic case. (It is more or less directly evident that the Proposition is also true for Poisson demand.) Because the chain is irreducible and ergodic, it has a unique steady-state distribution. Consequently, we only need to prove that the uniform distribution is a steady-state distribution, i.e., that if the distribution is uniform at some time it is also uniform after the next (unknown) demand. So we need to show that

$$\sum_{i=1}^{Q} \frac{1}{Q} p_{i,j} = \frac{1}{Q}, \quad j = 1, 2, ..., Q, \quad (5.33)$$

or equivalently that $\sum_{i=1}^{Q} p_{i,j} = 1$ for all values of j. When this equality is satisfied the chain is said to be *doubly stochastic*. (Recall that the equality $\sum_{j=1}^{Q} p_{i,j} = 1$ is always true for a Markov chain.)

Let the state be i and consider a certain demand size k. Given k, the next state is known i.e., $p_{i,j}(k)$ is equal to one for a certain j and zero otherwise. Furthermore, it is similarly evident that for a given j, $p_{i,j}(k)$ is one for exactly one value of i. If we know the new state and the demand size, we also know the preceding state. Consequently, $\sum_{i=1}^{Q} p_{i,j}(k) = 1$. The probability $p_{i,j}$ can be expressed as the average of $p_{i,j}(k)$ over k, i.e., $p_{i,j} = E_k p_{i,j}(k)$. We now have

$$\sum_{i=1}^{Q} p_{i,j} = \sum_{i=1}^{Q} E_k \{p_{i,j}(k)\} = E_k \left\{ \sum_{i=1}^{Q} p_{i,j}(k) \right\} = E_k \{1\} = 1. \quad (5.34)$$

This completes the proof.

When dealing with normally distributed demand, we similarly assume that the continuous inventory position is uniformly distributed on the interval $[R, R + Q]$. This is a very accurate approximation provided that we can disregard the possibilities for negative demand.

5.3.2 An important relationship

We shall assume that the replenishment lead-time is constant. Let us introduce the notation:

L = lead-time,
IL = inventory level,
$D(t, t + \tau) = D(\tau)$ = stochastic demand in the interval $(t, t + \tau]$.

Consider an arbitrary time t when the system is in steady state. Let $IP(t)$ be the inventory position at that time. Consider then the time $t + L$. At that time everything that was on order at time t has been delivered. Orders that have been triggered in the interval $(t, t + L]$ have not reached the inventory due to the lead-time. Consequently we have the simple relationship

$$IL(t + L) = IP(t) - D(t, t + L). \qquad (5.35)$$

The demand after t is obviously independent of the inventory position at time t. Furthermore, we know from Proposition 5.1 that the distribution of $IP(t)$ is uniform. We can therefore relatively easily determine the distribution of the inventory level at time $t + L$. Since t is an arbitrary time this is also the case for $t + L$. Consequently we obtain the steady-state distribution.

5.3.3 Compound Poisson demand

For discrete compound Poisson demand we have from (5.35)

$$P(IL = j) = \frac{1}{Q} \sum_{k=\max\{R+1,j\}}^{R+Q} P(D(L) = k - j) \quad j \leq R+Q, \qquad (5.36)$$

where we condition on the inventory position and use the fact that the inventory level at time $t + L$ can never exceed the inventory position at time t, i.e., $k \geq j$.

It is interesting to note that if we have determined these probabilities for one value of R, for example $R = 0$, we can obtain the probabilities for other values of R by a simple conversion. Although this is intuitively clear, we show this formally. Let $P(IL = j | R = r)$ be the probability for inventory level j when $R = r$. We then have from (5.36)

$$P(IL = j \mid R = r) = \frac{1}{Q} \sum_{k=\max\{r+1,j\}}^{r+Q} P(D(L) = k - j) \tag{5.37}$$

$$= \frac{1}{Q} \sum_{k=\max\{1,j-r\}}^{Q} P(D(L) = k - (j-r)) = P(IL = j - r \mid R = 0).$$

In the special case of an S policy we have $R = S - 1$ and $Q = 1$, and the inventory position is always S. This means that (5.36) can be simplified to

$$P(IL = j) = P(D(L) = S - j) \quad j \leq S. \tag{5.38}$$

5.3.4 Normally distributed demand

Let us now derive the corresponding relationship for normally distributed demand. Recall that we assume that the continuous inventory position is uniformly distributed on the interval $[R, R + Q]$. Let $\mu' = \mu L$ and $\sigma' = \sigma L^{1/2}$ denote the mean and the standard deviation of the lead-time demand. Furthermore, let $f(x)$ and $F(x)$ denote the density and the distribution function of the inventory level in steady state. In analogy with (5.36) we then have

$$F(x) = P(IL \leq x) = \frac{1}{Q} \int_{R}^{R+Q} \left[1 - \Phi\left(\frac{u - x - \mu'}{\sigma'}\right)\right] du. \tag{5.39}$$

Given the inventory position u at time t, the inventory level at time $t + L$ is less or equal to x if the lead-time demand is at least $u - x$, see (5.35).

We shall now introduce the *loss function* $G(x)$, tabulated in Appendix 2 and defined as

$$G(x) = \int_{x}^{\infty} (v - x)\varphi(v)dv = \varphi(x) - x(1 - \Phi(x)). \tag{5.40}$$

Note that

$$G'(x) = \Phi(x) - 1, \tag{5.41}$$

which means that $G'(x)$ is negative and increasing. Consequently, $G(x)$ is decreasing and convex. See Figure 5.3.

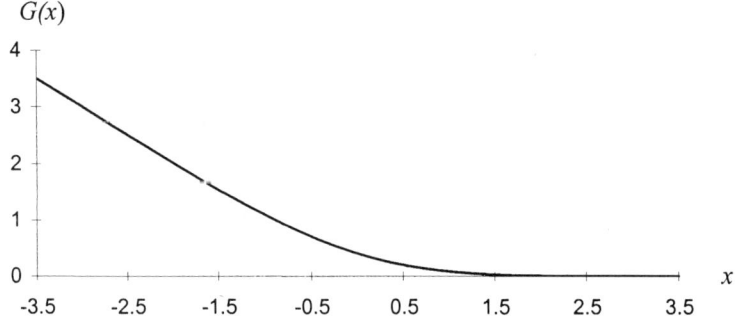

Figure 5.3 The loss function $G(x)$.

Using (5.41) we can reformulate (5.39) as

$$F(x) = \frac{1}{Q}\int_{R}^{R+Q}\left[-G'\left(\frac{u-x-\mu'}{\sigma'}\right)\right]du$$

$$= \frac{\sigma'}{Q}\left[G\left(\frac{R-x-\mu'}{\sigma'}\right) - G\left(\frac{R+Q-x-\mu'}{\sigma'}\right)\right].$$

(5.42)

From (5.42) we obtain the density as

$$f(x) = \frac{1}{Q}\int_{R}^{R+Q}\frac{1}{\sigma'}\varphi\left(\frac{u-x-\mu'}{\sigma'}\right)du = \frac{1}{Q}\left[\Phi\left(\frac{R+Q-x-\mu'}{\sigma'}\right) - \Phi\left(\frac{R-x-\mu'}{\sigma'}\right)\right]$$

(5.43)

Let $Q \to 0$ in (5.42) and (5.43). We get

$$F(x) \to -G'\left(\frac{R-x-\mu'}{\sigma'}\right) = 1 - \Phi\left(\frac{R-x-\mu'}{\sigma'}\right) = \Phi\left(\frac{x-(R-\mu')}{\sigma'}\right),$$

(5.44)

and

$$f(x) = \frac{1}{\sigma'}\varphi\left(\frac{x-(R-\mu')}{\sigma'}\right). \tag{5.45}$$

When $Q \to 0$ this means simply that we are applying an S policy with $S = R$, i.e., the inventory position is kept at R all the time. The distribution of the inventory level is then normally distributed. This is also easy to see from (5.35).

Example 5.3 Consider pure Poisson demand with $\lambda = 2$. Let the lead-time $L = 5$ and consider a continuous review (R, Q) policy with $R = 9$ and $Q = 5$. The lead-time demand is obviously Poisson distributed with mean 10. Applying (5.36) we obtain

$$P(IL = j) = \frac{1}{5}\sum_{k=\max\{10,j\}}^{14} \frac{10^{k-j}}{(k-j)!}e^{-10}.$$

For $j = 1$, for example, we get

$$P(IL = 1) = \frac{1}{5}\sum_{k=10}^{14} \frac{10^{k-1}}{(k-1)!}e^{-10} = 0.106$$

If we instead want to use the normal approximation we have $\mu' = 10$ and $\sigma' = 10^{1/2}$. The distribution function is obtained according to (5.42) as

$$F(x) = \frac{\sqrt{10}}{5}\left[G\left(\frac{-x-1}{\sqrt{10}}\right) - G\left(\frac{-x+4}{\sqrt{10}}\right)\right].$$

It is reasonable to compare $P(IL = j)$ in the Poisson case by $F(j + 0.5) - F(j - 0.5)$, i.e., to the probability that the continuous inventory level is between $j - 0.5$ and $j + 0.5$. For $j = 1$ we obtain the probability of an inventory level $0.5 \leq IL \leq 1.5$ as $F(1.5) - F(0.5) = 0.113$. In Figure 5.4 the distributions are compared for other values of j. We can see that the normal distribution provides a very accurate approximation.

In case of continuous demand and continuous review, an order is triggered exactly when the inventory position reaches the reorder point R. The order will arrive L time units later. The average stock on hand just before the order arrives is denoted the *safety stock*, SS, and is obtained as

$$SS = R - \mu'. \tag{5.46}$$

The safety stock can be interpreted as an additional stock that is used as protection against demand variations. Note that we can replace (5.42) by the equivalent expression

$$F(x) = \frac{\sigma'}{Q}\left[G\left(\frac{SS-x}{\sigma'}\right) - G\left(\frac{SS+Q-x}{\sigma'}\right)\right]. \tag{5.47}$$

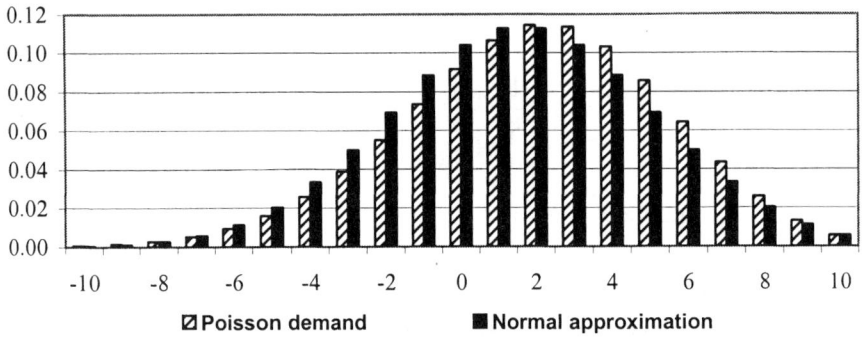

Figure 5.4 Probability distributions in Example 5.3.

5.4 Service levels

When determining a suitable reorder point, or safety stock, we can base the reorder point either on a prescribed service constraint or on a certain shortage or backorder cost. In practice it is often regarded to be easier to specify a service level.

We shall consider three service level definitions:

S_1 = probability of no stockout per order cycle,

S_2 = "fill rate"- fraction of demand that can be satisfied immediately from stock on hand,

S_3 = "ready rate"- fraction of time with positive stock on hand.

The first definition can be seen as the probability that an order arrives in time, i.e., before the stock on hand is finished. This service definition is very easy to use, but it also has some important disadvantages. The problem is that S_1 does not take the batch size into account. If the batch size is large and covers the demand during a long time, it does not matter much if S_1 is low. Most of the time there is still plenty of stock on hand due to the large batch size. If the batch quantity is small, the "real" service can similarly be very low even if S_1 is high. As a consequence S_1 cannot be recommended for inventory control in practice. It is considered in this book because it is still used in many practical applications.

The fill rate and the ready rate make the determination of the corresponding reorder points a bit more complex, but on the other hand, will give a much better picture of the customer service. In case of continuous demand or Poisson demand, S_2 and S_3 are equivalent. This is not the case, though, if a customer may order several units at the same time. Even if the stock on hand is positive, the stock may not be sufficient to cover a large customer order. If the stock contains a small number of units on hand most of the time, the ready rate can be high. Still the fill rate may be very low if there are some large customers who order huge quantities.

Service measures can also be defined in many other ways. For example, it is not necessary to define service by a probability. In some situations it can be more adequate to require that the average customer waiting time does not exceed a certain number of days.

From a practical point of view it is usually most important that the service level is clearly defined and interpreted in the same way throughout the company. It is obviously an advantage if it is possible to follow up the real customer service according to the established definition. In general, it is not suitable to have the same service level for all different items. On the other hand, it may be impractical to use individual service levels for all items. A common solution is to group the items in some way and specify service levels for each group. See also Sections 6.3 and 11.2.

The choice of service level should be based on customer expectations, or in other words, the underlying shortage costs, and on the costs for giving good service. In case of long lead-times and large demand variations, it may become very expensive to maintain high service levels.

5.5 Shortage costs

The alternative to a service level is to use a shortage cost. We shall consider two types of shortage costs:

b_1 = shortage cost per unit and time unit,
b_2 = shortage cost per unit.

The choice of shortage cost definition should, of course, reflect the real costs that are incurred by shortages.

A shortage cost of type b_1, for example, is relevant for a spare part when a shortage implies that a machine has to stop until the spare part is available again. The costs are then proportional to the customer waiting time. Recall that we have already used b_1 in Section 4.4 in connection with deterministic demand. The shortage cost b_1 can also be interpreted as a cost per average number of backorders, as was explained in Section 4.4.

If shortages are covered by overtime production to a higher cost, a shortage cost of type b_2 may be suitable.

One problem with shortage costs is that practitioners usually find it difficult to determine how high they should be. It is, on the other hand, an advantage that a given shortage cost makes it possible to balance shortage and holding costs and find the optimal customer service.

Independent of whether we use a service level or a shortage cost, the optimal reorder point will increase with the service level or the shortage cost used. For a certain item it is, in principle, always possible to determine the service level which gives the same safety stock as a certain shortage cost, and the other way around. Such relationships between service measures and shortage costs will, in general, depend on data for the individual item, like its batch quantity, the mean and standard deviation of lead-time demand, etc. An interesting exception is a relationship between the service levels S_2 and S_3 and b_1, which we shall derive in Section 5.9.2.

5.6 Determining the safety stock for given S_1

The service measure S_1, i.e., the probability of no stockout per order cycle, is usually used in connection with continuous review and continuous demand models. This means that we order a batch Q exactly when the inventory position is R. The batch size is assumed to be given. Our problem is to determine the reorder point R, such that there is a certain specified probability S_1 for the lead-time demand to be lower than R. We will assume that the lead-time demand is normally distributed with average μ' and standard deviation σ'. We obtain

SINGLE-ECHELON – REORDER POINTS

$$P(D(L) \leq R) = S_1 = \Phi\left(\frac{R-\mu'}{\sigma'}\right) = \Phi\left(\frac{SS}{\sigma'}\right), \quad (5.48)$$

since $SS = R - \mu'$. For a given value of S_1 we have to choose the ratio $SS/\sigma' = k$ sufficiently large. The required ratio k (easy to determine from the table in Appendix 2) is often denoted the *safety factor*. After determining the safety factor we obtain the safety stock simply as

$$SS = k\sigma'. \quad (5.49)$$

Finally we get the reorder point as $R = SS + \mu'$. Table 5.1 illustrates how the safety factor grows with service level S_1.

Table 5.1   Safety factors for different values of service level S_1.

Service level S_1	0.75	0.80	0.85	0.90	0.95	0.99
Safety factor k	0.67	0.84	1.04	1.28	1.64	2.33

Note that the safety factor is increasing rapidly for large service levels.

5.7 Fill rate and ready rate constraints

Consider again a continuous review (R, Q) policy with a given batch quantity Q. We will now see how we can determine the reorder point R to satisfy a given fill rate S_2 or ready rate S_3.

5.7.1 Compound Poisson demand

Consider first compound Poisson demand and recall that we have derived the distribution of the inventory level $P(IL = j)$ in (5.36). The ready rate is the probability for positive inventory level, i.e.,

$$S_3 = P(IL > 0). \quad (5.50)$$

It is a little more complicated to determine the fill rate because the demand size can vary. Recall that f_k is the probability for demand size k. We consider a customer demand and get the fill rate as the ratio between the expected satisfied quantity and the expected total demand quantity.

$$S_2 = \frac{\sum_{k=1}^{\infty}\sum_{j=1}^{\infty} \min(j,k) \cdot f_k \cdot P(IL=j)}{\sum_{k=1}^{\infty} k f_k}. \qquad (5.51)$$

If the positive demand size is k and the positive inventory level is j the delivered quantity is min (j, k). In the special case of Poisson demand, we have $f_1 = 1$ and the expression for S_2 in (5.51) degenerates to the expression for S_3 in (5.50), i.e., $S_2 = S_3$ in that case.

It is evident that both S_2 and S_3 will increase with the reorder point R. Furthermore $R \leq -Q$ means that the stock is never positive, implying that both S_2 and S_3 are zero. Consequently, for a given positive S_2 (or S_3) we know that we must have $R > -Q$. Poisson or compound Poisson demand models are normally only used when the demand is relatively low. Therefore, when determining the reorder point we can simply increase R by one unit at a time starting from $R = -Q$ until S_2 obtained from (5.51) (or S_3 from (5.50)) exceeds the required service level. For each value of R we need to recalculate the probabilities $P(IL = j)$. It is then convenient to use (5.37).

5.7.2 Normally distributed demand

For continuous normally distributed demand, it is evident that $S_2 = S_3$ and the service level is the probability of positive stock, which is obtained from (5.42)

$$S_2 = S_3 = 1 - F(0) = 1 - \frac{\sigma'}{Q}\left[G\left(\frac{R-\mu'}{\sigma'}\right) - G\left(\frac{R+Q-\mu'}{\sigma'}\right)\right]. \qquad (5.52)$$

Using $G'(v) = \Phi(v) - 1$ it is easy to show that $dS_2/dR = dS_3/dR > 0$, i.e., $S_2 = S_3$ increases with R. For a given service level we can use a simple *bisection* search to find the smallest R giving the required service level. We start with a lower bound for R, e.g., $\underline{R} = -Q$, and an upper bound $\overline{R}$, which gives a service level that exceeds the required service level. Next we consider $R = (\underline{R} + \overline{R})/2$. If the obtained service is too low, R can replace $\underline{R}$, and otherwise it can replace $\overline{R}$. We continue like this until the gap between $\overline{R}$ and $\underline{R}$ is sufficiently small.

It is common to approximate (5.52) by

$$S_2 = S_3 \approx 1 - \frac{\sigma'}{Q} G\left(\frac{R-\mu'}{\sigma'}\right). \quad (5.53)$$

From the table in Appendix 2 it is obvious that this approximation generally works well for large values of Q. It can be of interest to use the approximation in manual calculations, but it is not needed in a computerized inventory control system. Note that (5.53) underestimates the service levels.

Example 5.4 Consider again the data in Example 5.3, i.e., pure Poisson demand with $\lambda = 2$, $L = 5$, and a continuous review (R, Q) policy with $R = 9$ and $Q = 5$. We want to determine the fill rate S_2 and the ready rate S_3, which in this case are equal because we have pure Poisson demand. Applying (5.50) we obtain

$$S_2 = S_3 = P(IL > 0) = \sum_{j=1}^{14} \frac{1}{5} \sum_{k=\max\{10,j\}}^{14} \frac{10^{k-j}}{(k-j)!} e^{-10} = 0.679.$$

For the normal approximation we can use (5.52) to get

$$S_2 = S_3 = 1 - F(0) = 1 - \frac{\sqrt{10}}{5}\left[G\left(\frac{-1}{\sqrt{10}}\right) - G\left(\frac{4}{\sqrt{10}}\right)\right] = 0.666.$$

The corresponding value of S_1 turns out to be much lower, see (5.48),

$$S_1 = \Phi\left(\frac{R-\mu'}{\sigma'}\right) = \Phi\left(\frac{-1}{\sqrt{10}}\right) = 0.376.$$

5.8 Fill rate - a different approach

So far our approach has been to start by deriving the distribution of the inventory level from (5.35). This must be regarded as the standard technique in inventory theory. There is also another useful approach, which means that we focus instead on what happens to an individual batch after being ordered. To illustrate this other approach we shall once more derive the expression (5.52) for the fill rate, or equivalently, the ready rate.[3]

We are studying a continuous review (R, Q) policy and the lead-time demand is assumed to be continuous and normally distributed with mean $\mu' =$

INVENTORY CONTROL

μL and standard deviation $\sigma' = \sigma L^{1/2}$. We focus on a batch Q that is ordered when the inventory position is R. This batch will later be consumed by customer demand. The reorder point R will cover the demand for R units after the ordering time. So the considered batch will be consumed by the demand for the Q units following after these first R units. When the batch arrives in stock, a part of this demand may have already occurred, i.e., there are backorders waiting for the batch. Let

B = backordered quantity that will be covered by the batch.

Note that we are only considering the backorders that are covered by the batch. Therefore, we must have $0 \leq B \leq Q$. If the backorders exceed Q when the batch arrives in stock, the quantity exceeding Q is covered by future batches. If we know $E(B)$ we can obtain the fill rate (and the ready rate) as

$$S_2 = S_3 = 1 - E(B)/Q. \tag{5.54}$$

It remains to determine $E(B)$. Let u be the normally distributed demand during the lead-time. We need to consider three cases:

$u \leq R$ means $B = 0$,

$R < u \leq R + Q$ means $B = u - R$,

$R + Q < u$ means $B = Q$.

Consequently, $E(B)$ can be determined by averaging over u

$$E(B) = \int_R^{R+Q} (u-R) \frac{1}{\sigma'} \varphi\left(\frac{u-\mu'}{\sigma'}\right) du + \int_{R+Q}^{\infty} Q \frac{1}{\sigma'} \varphi\left(\frac{u-\mu'}{\sigma'}\right) du$$

$$= \int_R^{\infty} (u-R) \frac{1}{\sigma'} \varphi\left(\frac{u-\mu'}{\sigma'}\right) du - \int_{R+Q}^{\infty} (u-R-Q) \frac{1}{\sigma'} \varphi\left(\frac{u-\mu'}{\sigma'}\right) du$$

$$= \sigma' G\left(\frac{R-\mu'}{\sigma'}\right) - \sigma' G\left(\frac{R+Q-\mu'}{\sigma'}\right), \tag{5.55}$$

where we have used the definition of $G(x)$ in (5.40). Combining (5.54) and (5.55) we obtain (5.52).

Note that this derivation also covers the case when R is negative. This simply means that a part of the demand that is covered by Q has already occurred when the order is triggered.

5.9 Shortage cost per unit and time unit

We will now instead consider a shortage (or backorder) cost b_1 per unit and time unit. (Recall that we have dealt with this backorder cost in Section 4.4.) As before, we consider a continuous review (R, Q) policy with given batch quantity Q, and we let μ' and σ' represent mean and standard deviation respectively of the lead-time demand. The holding cost per unit and time unit is denoted h. We optimize the reorder point by minimizing the sum of expected holding and backorder costs. This means that we are balancing backorder costs against holding costs. It is convenient to use the notation

$(x)^+ = \max(x, 0),$

$(x)^- = \max(-x, 0).$

Given the inventory level IL, the holding cost rate is $h(IL)^+$ and the backorder cost rate is $b_1(IL)^-$. Using $x^+ - x^- = x$, we can express the total cost rate in different ways

$$h(IL)^+ + b_1(IL^-) = -b_1 IL + (h+b_1)(IL)^+ = h\,IL + (h+b_1)(IL)^-. \tag{5.56}$$

It is usually most convenient to use the second or third alternative in (5.56).

$E(IL)^-$ is the expected amount of backorders. Let μ be the average demand per unit of time. Then the expected *waiting time* to satisfy a demand is $E(IL)^-/\mu$. This is an application of the well-known *Little's formula* from queuing theory. We can interpret $E(IL)^-$ as the average queue length and μ as the average arrival rate to the queue. Recall the related discussion for a deterministic setting in Section 4.4.

5.9.1 Compound Poisson demand

Consider now compound Poisson demand. Recall that we have derived the distribution of the inventory level in (5.36). When expressing the expected costs per unit of time, C, it is in general, best to use the second alternative in (5.56) for discrete demand. The reason is that we only need to consider a finite number of inventory levels, because the inventory level cannot exceed the maximum inventory position $R + Q$. We get

$$C = -b_1 E(IL) + (h + b_1) E(IL)^+ = -b_1 (R + \frac{Q+1}{2} - \mu') + (h + b_1) \sum_{j=1}^{R+Q} jP(IL = j),$$

(5.57)

where $E(IL)$ is easily obtained as the average inventory position minus the average lead-time demand.

Let us now regard the cost difference when using the reorder point $R + 1$ instead of R.

$$C(R+1) - C(R) = -b_1 + (h + b_1) \left(\sum_{j=1}^{R+1+Q} jP(IL = j | R + 1) - \sum_{j=1}^{R+Q} jP(IL = j | R) \right).$$

(5.58)

From (5.37) we know that $P(IL = j | R) = P(IL = j + 1 | R + 1)$. Using this we have

$$\sum_{j=1}^{R+Q} jP(IL = j | R) = \sum_{j=1}^{R+Q} jP(IL = j + 1 | R + 1) = \sum_{j=2}^{R+1+Q} (j-1)P(IL = j | R + 1).$$

(5.59)

Inserting (5.59) in (5.58) we have

$$C(R+1) - C(R) = -b_1 + (h + b_1) \sum_{j=1}^{R+1+Q} P(IL = j | R + 1) = -b_1 + (h + b_1) S_3(R+1).$$

(5.60)

Because the ready rate increases with the reorder point, it follows that $C(R + 1) - C(R)$ is increasing with R, i.e., the cost is convex in the reorder point.

Since $S_3 = 0$ for $R \leq -Q$ it follows directly from (5.60) that we do not need to consider $R < -Q$. Consequently, to find the optimal R we can start with $R = -Q$ and increase R by one unit at a time until the costs are increasing.

It is also interesting to note from (5.60) that there is always a special relationship between the cost parameters and the ready rate in the optimal solution. The optimal reorder point can be characterized as the largest reorder point giving a ready rate not higher than $b_1/(h + b_1)$. In other words, if R^* is the optimal reorder point we have

$$S_3(R^*) \leq \frac{b_1}{h + b_1} < S_3(R^* + 1). \tag{5.61}$$

In case of pure Poisson demand this is also true for the fill rate.

5.9.2 Normally distributed demand

For continuous normally distributed demand, the relationship between the cost parameters and the service levels fill rate and ready rate is, as we shall see below, even more striking. Recall that $S_2 = S_3$ in this case.

We start by evaluating the expected costs per unit of time. Let us now use the third alternative in (5.56). As before we let $F(x)$ be the distribution function and $f(x)$ the density of the inventory level.

$$C = hE(\text{IL}) + (h + b_1)E(\text{IL})^- = h(R + Q/2 - \mu') + (h + b_1)\int_{-\infty}^{0} F(x)dx$$

$$= h(R + Q/2 - \mu') + (h + b_1)\int_{-\infty}^{0} \frac{1}{Q} \int_{R}^{R+Q}\left[-G'\left(\frac{u - x - \mu'}{\sigma'}\right)\right]dudx$$

$$= h(R + Q/2 - \mu') + (h + b_1)\frac{\sigma'}{Q} \int_{R}^{R+Q} G\left(\frac{u - \mu'}{\sigma'}\right)du. \tag{5.62}$$

In the first line of (5.62) we use that the average inventory position is $R + Q/2$ in the continuous case. Furthermore,

$$E(IL)^- = -\int_{-\infty}^{0} u f(u) du = \int_{-\infty}^{0} f(u) \left[\int_{u}^{0} dx\right] du = \int_{-\infty}^{0} \left[\int_{-\infty}^{x} f(u) du\right] dx = \int_{-\infty}^{0} F(x) dx.$$

(5.63)

In the second line of (5.62) we use (5.42), and when going from the second to the third line we change the order of integration. Recall that $G(\infty)=0$.

At this stage it is useful to define the function $H(x)$ which is tabulated in Appendix 2. (In the literature it is common to use $\overline{H}(x) = 2H(x)$ instead of $H(x)$).

$$H(x) = \int_{x}^{\infty} G(v) dv = \frac{1}{2}\left[(x^2 + 1)(1 - \Phi(x)) - x\varphi(x)\right].$$

(5.64)

Since $H'(x) = -G(x)$, $H(x)$ is decreasing and convex. Recall that $G(x)$ is positive and decreasing. The function $H(x)$ is illustrated in Figure 5.5.

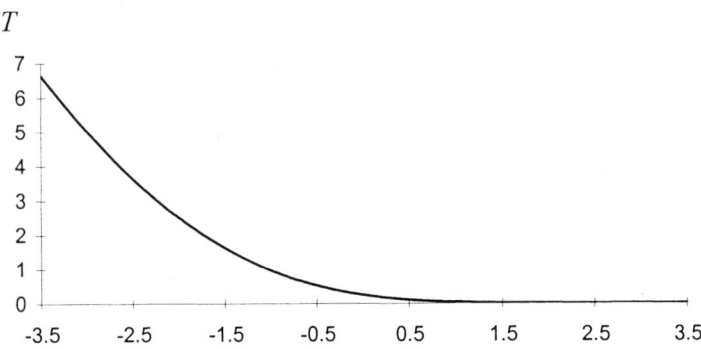

Figure 5.5 The function $H(x)$.

Using $H(x)$ we can express the costs in (5.62) as

$$C = h(R + Q/2 - \mu') + (h + b_1)\frac{\sigma'^2}{Q}\left[H\left(\frac{R - \mu'}{\sigma'}\right) - H\left(\frac{R + Q - \mu'}{\sigma'}\right)\right].$$

(5.65)

SINGLE-ECHELON – REORDER POINTS

Let us now return to the relationship between the cost parameters and the service levels fill rate and ready rate. From (5.65), (5.52), and the definition of $H(x)$ we have

$$\frac{dC}{dR} = h + (h+b_1)\frac{\sigma'}{Q}\left[G\left(\frac{R+Q-\mu'}{\sigma'}\right) - G\left(\frac{R-\mu'}{\sigma'}\right)\right] = h + (h+b_1)(S_2 - 1)$$

$$= -b_1 + (h+b_1)S_2 = -b_1 + (h+b_1)S_3 .$$

(5.66)

Since dC/dR is increasing with R, C is a convex function of R. The optimal R is obtained for $dC/dR = 0$, and in the optimal solution we have

$$S_2 = S_3 = \frac{b_1}{h+b_1} .$$

(5.67)

Given a certain shortage cost b_1, we can easily determine from (5.67) the service level giving exactly the same reorder point. If we reformulate (5.67) as

$$b_1 = \frac{hS_2}{1-S_2} = \frac{hS_3}{1-S_3} ,$$

(5.68)

we can also go in the other direction, i.e., given a service level we can determine an implicit shortage cost. The relations (5.67) and (5.68) (as well as (5.61)) are very useful when choosing service levels or shortage costs in practice. Consider, for example, a company that works with service level constraints because it seems difficult to determine shortage costs. Still it may very well be possible to evaluate whether a certain shortage cost is reasonable or not. Using (5.68), the shortage costs corresponding to chosen service levels can be determined and discussed. Such discussions are often very fruitful when trying to improve inventory control.

Note, however, that the considered "equivalence" between a certain fill rate or ready rate, and a corresponding shortage cost per unit and time unit, is only valid when the order quantity Q is given and the optimization only concerns R. It is not valid if we add ordering costs and carry out a joint optimization of both R and Q.

5.10 Shortage cost per unit

We shall now instead consider a shortage cost b_2 per unit. As before, we consider a continuous review (R, Q) policy. The demand is assumed to be continuous and normally distributed. Except for the different type of shortage cost, the model is the same as in (5.62). We assume that each unit which is demanded when the stock level is zero or negative incurs the cost b_2. Using (5.62) and (5.42), the costs per time unit C can now be expressed as

$$C = h(R + Q/2 - \mu') + h \int_{-\infty}^{0} F(x)dx + b_2 \mu F(0) =$$

$$= h(R + Q/2 - \mu') + h \frac{\sigma'}{Q} \int_{R}^{R+Q} G\left(\frac{u - \mu'}{\sigma'}\right) du \qquad (5.69)$$

$$+ b_2 \mu \frac{\sigma'}{Q} \left[G\left(\frac{R - \mu'}{\sigma'}\right) - G\left(\frac{R + Q - \mu'}{\sigma'}\right) \right].$$

Note that in (5.69) we are using both the average demand per time unit μ and the average lead-time demand μ'. If we use the function $H(x)$ according to (5.64), we can alternatively express (5.69) as

$$C = h(R + Q/2 - \mu') + h \frac{\sigma'^2}{Q} \left[H\left(\frac{R - \mu'}{\sigma'}\right) - H\left(\frac{R + Q - \mu'}{\sigma'}\right) \right]$$

$$+ b_2 \mu \frac{\sigma'}{Q} \left[G\left(\frac{R - \mu'}{\sigma'}\right) - G\left(\frac{R + Q - \mu'}{\sigma'}\right) \right]. \qquad (5.70)$$

It can be shown that the costs in (5.69) and (5.70), although not convex, have a unique local minimum. See Rosling (2002a), and Chen and Zheng (1993).

It is common to replace the exact costs by a simpler approximate cost expression. If there are relatively few backorders we can disregard the second term in (5.69) and (5.70). Furthermore, if Q is relatively large we can also disregard the second term in the last parenthesis. This gives the following approximate costs:

$$\tilde{C} = h(R+Q/2-\mu') + b_2\mu\frac{\sigma'}{Q}G\left(\frac{R-\mu'}{\sigma'}\right). \quad (5.71)$$

It is easy to verify that $\tilde{C}$ is convex in R, and we obtain the optimal reorder point R from

$$\frac{d\tilde{C}}{dR} = h + b_2\mu\frac{1}{Q}\left(\Phi\left(\frac{R-\mu'}{\sigma'}\right)-1\right) = 0, \quad (5.72)$$

or, equivalently,

$$S_1 = \Phi\left(\frac{R-\mu'}{\sigma'}\right) = 1 - \frac{Qh}{b_2\mu}. \quad (5.73)$$

where S_1 is the probability of no stockout per order cycle. Note that the right hand side of (5.73) can become negative for certain parameters. But S_1 is a service level and cannot be negative. The reason that this can happen is our approximation. If this occurs in practice, we have to replace the right hand side in (5.73) by some given lowest service level.

5.11 Continuous review (s, S) policy

We shall now consider an (s, S) policy instead of an (R, Q) policy. We are still assuming continuous review. Recall that an (s, S) policy means that an order is triggered as soon as the inventory position declines to or below the reorder point s. The order size is chosen so that the inventory position increases to S. (See Figure 3.2.)

First of all, we note that an (s, S) policy is equivalent to an (R, Q) policy with $s = R$ and $Q = S - s$ as long as we know that the inventory position is exactly s when the order is triggered. In case of continuous review this equivalence is true both for continuous demand and for pure Poisson demand. It is not true, however, for compound Poisson demand with demand sizes larger than one. Here we will only consider such demand since we have already dealt with (R, Q) policies in Section 5.3.

In Proposition 5.1 in Section 5.3 we proved that the steady-state distribution of the inventory position in case of an (R, Q) policy is uniform on the integers $[R + 1, R + Q]$. With an (s, S) policy the inventory position is obviously always in the corresponding interval $[s + 1, S]$, but it is in general not uniformly distributed on these values. We therefore need to determine the distribution.

Note first that the transitions of the inventory position *IP* are a renewal process. Each time an order is triggered and a new order cycle begins, the inventory position is raised to S and we have a regeneration. We shall now consider the number of customers and the transitions of the inventory position between two orders (regenerations). For each customer arrival we consider the inventory position just after the arrival when both the demand and a possible replenishment order have taken place. Let

m_j = probability to reach $IP = j$ during an order cycle $(s + 1 \leq j \leq S)$.

As before, f_k is the probability of demand size k (recall that we assume that $f_0 = 0$).

Obviously $m_S = 1$ since we start each cycle with $IP = S$. No inventory position can be visited more than once during an order cycle. We can come to $IP = j$ from some $IP = k > j$ provided the next demand is $k - j$. When determining m_j we consider all $IP > j$. We get

$$m_j = \sum_{k=j+1}^{S} m_k f_{k-j}, \quad j = s+1, s+2, ..., S-1. \qquad (5.74)$$

From (5.74) it is easy to determine the probabilities m_j recursively for $j = S - 1, S - 2, ..., s + 1$. Note that m_j can also be interpreted as the expected number of visits to $IP = j$ during an order cycle. Each transition of the inventory position is triggered by a customer. The average total number of customers during an order cycle is therefore $\sum_{j=s+1}^{S} m_j$. Since the customers arrive according to a Poisson process, the average time between two arrivals is always the same. This means that we can obtain the steady-state distribution of the inventory position as

$$P(IP = k) = m_k / \sum_{j=s+1}^{S} m_j, \quad k = s+1, s+2, ..., S. \qquad (5.75)$$

Note that these probabilities only depend on $S - s$, i.e., if, for example, we increase both s and S by the same number, we get the same probabilities. Furthermore, if we have determined the probabilities m_j for $s = s'$ and $S = S'$

SINGLE-ECHELON – REORDER POINTS

and want to consider $s = s' - 1$ and $S = S'$, the probabilities $m_{s'+1}, \ldots, m_{S'}$ are unchanged and we only need to determine $m_{s'}$ from (5.74).

Example 5.5 Let the probabilities for demand sizes 1 and 2 be $f_1 = f_2 = 1/2$. Consider an (s, S) policy with $s = 5$ and $S = 8$. We obtain $m_8 = 1$ and according to (5.74)

$$m_7 = m_8 / 2 = 1/2,$$
$$m_6 = m_7 / 2 + m_8 / 2 = 3/4.$$

The average number of customers in an order cycle is $m_6 + m_7 + m_8 = 9/4$. We finally get the distribution of the inventory position from (5.75) as $P(IP = 6) = 3/9$, $P(IP = 7) = 2/9$, and $P(IP = 8) = 4/9$. Obviously this distribution is far from uniform.

After obtaining the distribution of the inventory position from (5.75) we can derive the distribution of the inventory level in complete analogy with (5.36).

$$P(IL = j) = \sum_{k=\max\{s+1,j\}}^{S} P(IP = k) P(D(L) = k - j), \quad j \leq S. \quad (5.76)$$

The determination of various performance measures for a given inventory level distribution is straightforward and parallels essentially corresponding expressions for (R, Q) policies.

5.12 Periodic review - fill rate

Let us now consider an inventory system that is controlled by a periodic review policy instead of a continuous review policy. Periodic review means that the inventory position is inspected at the beginning of each period and that all replenishments are triggered at these reviews. A period may, for example, be a day or a week. First of all it is obvious that if the review period is short enough there is not much difference between periodic and continuous review. In such a case, continuous review results can be used as an approximation in periodic review systems and the other way around. Quite often, though, the review period cannot be disregarded.

5.12.1 Basic assumptions

As before, we assume a constant lead-time L. Furthermore we let T be the review period. It is common to assume that the lead-time is an integral number of periods, but we shall not introduce that restriction.

It is important to understand that the assumptions regarding when in the period the demand takes place, and when in the period various performance measures are evaluated, may affect the results and the optimal policy. Let us, for example, consider a situation with $L = 0$ and no ordering costs. If all demand takes place at the beginning of a period and it is possible to order immediately after that, we do not need to keep any stock. We just have to order the period demand in each period, and because the lead-time is zero, the delivery will take place instantaneously. We shall assume that the demand in a period is spread out over the whole period as we have assumed in the previously considered continuous review models. This is the most natural assumption in most applications. Evidently this means that we need some stock also if $L = 0$.

Furthermore, we shall focus on evaluation of the fill rate. Assume, for example, that the inventory is controlled by an (R, Q) policy and that the batch quantity Q is given. We can use the results in this section to find the smallest R that satisfies a given fill rate.

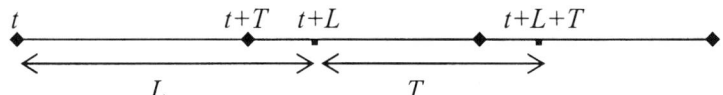

Figure 5.6 Considered time epochs.

Consider Figure 5.6 and let t be an arbitrary review time. The orders can take place at times ... , t, $t + T$, ... A possible order at time t is delivered at time $t + L$. The next possible delivery time is evidently $t + L + T$. When determining the fill rate we consider the interval between these two times. It is easy to determine the expected demand in the considered interval. Furthermore, it is clear that the demand that can be delivered from stock on hand in this interval depends only on the inventory level after the possible delivery at time $t + L$ and the total stochastic demand in the interval.

Note that the fill rate is also the same for most other demand assumptions, e.g., if the whole demand takes place just after the delivery at time $t + L$ or just before the next possible delivery at time $t + L + T$.

In this section we shall first deal with discrete compound Poisson demand both for (R, Q) policy and (s, S) policy. Thereafter we consider an (R, Q)

SINGLE-ECHELON – REORDER POINTS   111

policy and continuous normally distributed demand. We refer to Sobel (2004) and Tijms (1994) for more details and other assumptions.

5.12.2 Compound Poisson demand - (R, Q) policy

We shall first consider discrete compound Poisson demand. Recall that such demand models are mainly of interest in case of relatively low demand. Consider again Figure 5.6. We shall first determine the distribution of the inventory level at the two times $t + L$ and $t + L + T$, i.e., the start and the end of the considered interval. At time $t + L$ we consider the inventory level after a possible delivery, but at time $t + L + T$, a possible delivery has not taken place.

Consider an (R, Q) policy and the transitions of the inventory position at the review times after a possible order. Recall that we in the continuous review case instead considered the times when customer demands occurred. The associated Markov chains are essentially the same if we let the period demand here correspond to the demand per customer in the continuous review case. Under some mild assumptions (see Section 5.3.1), the steady-state distribution of the inventory position will be uniform. For a discrete demand distribution it is uniform on the integers $[R + 1, R + Q]$.

The inventory level IL' at time $t + L$ after a possible delivery can be obtained as

$$IL' = IL(t + L) = IP(t) - D(L), \qquad (5.77)$$

because orders after time t have not been delivered at time $t + L$. Recall that we can determine the distribution of the demand over a given time according to (5.3), or more simply according to (5.15) in case of a logarithmic compounding distribution. The distribution of IL' can consequently be obtained as in (5.36), i.e.,

$$P(IL' = j) = \frac{1}{Q} \sum_{k=\max\{R+1, j\}}^{R+Q} P(D(L) = k - j) \quad j \leq R + Q. \qquad (5.78)$$

The distribution of the inventory level IL'' at time $t + L + T$ before a possible delivery can be determined in exactly the same way. It is just to replace L by $L + T$ in the computations.

Consider now the interval between the two times $t + L$ and $t + L + T$. The expected demand during one unit of time is $\mu = \lambda \sum_{j=1}^{\infty} j f_j$, where λ is the intensity of the customer arrivals and f_j is the probability of demand size j. The expected demand in the considered interval of length T is obviously μT.

Consider then the part of the demand in the considered interval between $t + L$ and $t + L + T$ that cannot be met from stock on hand. This demand is backordered so it can be determined as $E(IL'')^- - E(IL')^-$. This means that we can determine the fill rate as

$$S_2 = 1 - \frac{E(IL'')^- - E(IL')^-}{\mu T}. \tag{5.79}$$

When determining $E(IL'')^-$ and $E(IL')^-$ it is usually most practical to use that $E(IL)^- = E(IL)^+ - E(IL)$, because $IL^+ \leq R + Q$ is bounded. $E(IL')$ and $E(IL'')$ are easily obtained as $R + (Q + 1)/2 - \mu L$ and $R + (Q + 1)/2 - \mu(L + T)$, respectively. Recall that the inventory position immediately after the order at time t is uniform on $[R + 1, R + Q]$.

Example 5.6. Consider pure Poisson demand with $\lambda = 0.1$, the lead-time $L = 5$ and the review period $T = 1.5$. Furthermore, $R = 1$ and $Q = 2$. We get $E(IL'')^- = 0.01925$ and $E(IL')^- = 0.00913$. Using (5.79) we get $S_2 = 0.933$.

5.12.3 Compound Poisson demand - (s, S) policy

We can also handle (s, S) policies essentially as in Section 5.11. Let

g_j = probability that the period demand, i.e., the demand $D(T)$ during a period of length T, is j ($j = 0, 1, ...$),
m_j = average number of periods with $IP = j$ during an order cycle.

Recall that the period demand essentially corresponds to the demand by a single customer in the continuous review case. In Section 5.11 we assumed that the size of a customer demand is at least one unit. We shall now also consider the possibility of zero period demand, which requires a minor modification of the technique in Section 5.11 (see also Zheng and Federgruen, 1991). If the period demand is zero the inventory position will not change. Consequently we have $m_S = 1/(1-g_0)$, and in analogy with (5.74)

$$m_j = \sum_{k=j}^{S} m_k g_{k-j} = m_j g_0 + \sum_{k=j+1}^{S} m_k g_{k-j}, \quad j = s+1, s+2, ..., S-1,$$
$$\tag{5.80}$$

SINGLE-ECHELON – REORDER POINTS

or

$$m_j = \frac{1}{(1-g_0)} \sum_{k=j+1}^{S} m_k g_{k-j}, \quad j = s+1, s+2, ..., S-1. \quad (5.81)$$

From (5.81) we can determine the values of m_j recursively for $j = S-1, S-2, ..., s+1$. We can then get the distribution of the inventory position as in (5.75)

$$P(IP = k) = m_k / \sum_{j=s+1}^{S} m_j, \quad k = s+1, s+2, ..., S. \quad (5.82)$$

Given the distribution of the inventory position we can determine the distributions of IL' (and IL'') essentially as in (5.78)

$$P(IL' = j) = \sum_{k=\max\{s+1,j\}}^{S} P(IP = k)P(D(L) = k - j) \quad j \leq S. \quad (5.83)$$

We can then apply (5.79) just as in Section 5.12.2.

5.12.4 Normally distributed demand - (R, Q) policy

Let us now consider an (R, Q) policy and continuous normally distributed demand. As before, we denote the mean and standard deviation of the demand per unit of time by μ and σ. Furthermore, the mean and standard deviation of the demand during the lead-time L are denoted $\mu' = L\mu$ and $\sigma' = L^{1/2}\sigma$. We shall also consider the demand during the time $L + T$ and the corresponding mean and standard deviation $\mu'' = (L + T)\mu$ and $\sigma'' = (L + T)^{1/2}\sigma$.

Consider again Figure 5.6. The inventory position IP just after the arbitrary review time t is now uniform on the continuous interval $(R, R + Q]$. We can again apply (5.77), and in complete analogy with (5.39) and (5.42) we obtain the inventory level distribution at time $t + L$ after a possible delivery as

$IL' = IL(t+L) = IP(t) - D(L)$

$$F(x) = P(IL' \leq x) = \frac{\sigma'}{Q}\left[G\left(\frac{R-x-\mu'}{\sigma'}\right) - G\left(\frac{R+Q-x-\mu'}{\sigma'}\right)\right].$$

$$(5.84)$$

We then use the same approach as in Sections 5.12.2 and 5.12.3. First we note that the expected demand in the interval between $t + L$ and $t + L + T$ is μT. We then evaluate the expected demand that cannot be met from stock on hand during this interval. This demand is backordered so it can be determined as $E(IL'')^- - E(IL')^-$. We can then get the fill rate as

$$S_2 = 1 - \frac{E(IL'')^- - E(IL')^-}{\mu T}. \qquad (5.85)$$

It therefore only remains to determine $E(IL'')^-$ and $E(IL')^-$. Using (5.63) and (5.84) we have

$$E(IL')^- = \int_{-\infty}^{0} F(x)dx = \int_{-\infty}^{0} \frac{\sigma'}{Q}\left[G\left(\frac{R-x-\mu'}{\sigma'}\right) - G\left(\frac{R+Q-x-\mu'}{\sigma'}\right)\right]dx$$

$$= \frac{(\sigma')^2}{Q}\left[H\left(\frac{R-\mu'}{\sigma'}\right) - H\left(\frac{R+Q-\mu'}{\sigma'}\right)\right].$$

(5.86)

We get $E(IL'')^-$ in exactly the same way. We just have to replace μ' and σ' by μ'' and σ'' in (5.86).

Example 5.7. Consider a normal approximation of the data in Example 5.6. We get $\mu T = 0.15$, $\mu' = 0.5$, $\sigma' = 0.5^{1/2} = 0.707$, $\mu'' = 0.65$ and $\sigma'' = 0.65^{1/2} = 0.806$. For $R = 1$ and $Q = 2$ we get from (5.86) $E(IL'')^- = 0.038$ and $E(IL')^- = 0.017$. Using (5.85) we get $S_2 = 0.86$. Because we are dealing with very low demand, the size of the error is not surprising.

5.13 The newsboy model

Let us now consider a classical inventory problem, the so-called newsboy model. There is a single period with stochastic demand. We assume that it is normally distributed with mean μ and standard deviation σ. We have to determine how much to order before the period starts. There are penalty costs associated with ordering both too much and too little compared to the unknown stochastic period demand. The problem is to optimize the size of the order.

SINGLE-ECHELON – REORDER POINTS

Let

x = stochastic period demand,
S = ordered amount,
c_o = overage cost, i.e., the cost per unit for a remaining inventory at the end of the period,
c_u = underage cost, i.e., the cost per unit for unsatisfied period demand.

Using the original interpretation of the problem, we can think of a newsboy who can buy copies of a newspaper for 25 cents early in the morning each day and sell them for 75 cents during the day. He is not paid anything for unsold copies. In that case we have $c_o = 25$, and $c_u = 75 - 25 = 50$. For a given demand x the costs are

$$C(x) = (S - x)c_o \qquad x < S, \tag{5.87}$$

$$C(x) = (x - S)c_u \qquad x > S, \tag{5.88}$$

and we can express the expected costs $C = E\{C(x)\}$ as

$$C = c_o \int_{-\infty}^{S}(S-x)\frac{1}{\sigma}\varphi(\frac{x-\mu}{\sigma})dx + c_u \int_{S}^{\infty}(x-S)\frac{1}{\sigma}\varphi(\frac{x-\mu}{\sigma})dx$$

$$= c_o(S - \mu) + (c_o + c_u)\int_{S}^{\infty}(x-S)\frac{1}{\sigma}\varphi(\frac{x-\mu}{\sigma})dx \tag{5.89}$$

$$= c_o(S - \mu) + (c_o + c_u)\sigma G\left(\frac{S-\mu}{\sigma}\right).$$

To find the optimal solution we can use the necessary condition $dC/dS = 0$, i.e.,

$$\frac{dC}{dS} = c_o + (c_o + c_u)(\Phi(\frac{S-\mu}{\sigma}) - 1) = 0, \tag{5.90}$$

or, equivalently,

$$\Phi(\frac{S-\mu}{\sigma}) = \frac{c_u}{c_o + c_u}. \qquad (5.91)$$

It is easy to see that C is convex, so (5.91) provides the optimal solution.

Example 5.8 Let us return to our example with $c_o = 25$ and $c_u = 50$. Assume that the daily demand for newspapers is normally distributed with mean $\mu = 300$ and standard deviation $\sigma = 60$. Applying (5.91) we obtain the optimal solution from

$$\Phi(\frac{S-300}{60}) = \frac{50}{25+50},$$

implying that (see the table in Appendix 2)

$$\frac{S-300}{60} = 0.431,$$

or, equivalently, $S \approx 326$ copies of the newspaper.

The same problem occurs in many other contexts. Here are a few examples. A garment retailer has to decide on the quantity of various style goods before the short selling season. A supermarket must decide on the quantities of fresh meat to buy each day. A farmer must decide on quantities of different crops to be planted in a certain season.

So far we have assumed that the problem concerns a single period. The simple newsboy solution (5.91) is, however, also common in multiperiod settings as in Section 5.12. Assume first that the lead-time is zero and that we apply an S policy at the beginning of each period. The period demand is normally distributed with mean μ and standard deviation σ. Units that are left over at the end of a period can be used in the next period, but we have a holding cost h per unit and period. Similarly, a possible shortage at the end of a period is backordered and can be filled in the coming period. The backorder cost per unit and time unit is b_1. The problem considered is a newsboy problem with $c_o = h$, and $c_u = b_1$. If we have a lead-time of L periods the problem is still a newsboy problem. We just have to consider the demand during $L + 1$ periods instead of the demand during a single period. This means that the mean μ and standard deviation σ in (5.91) should be replaced by the mean $\mu' = (L + 1)\mu$ and standard deviation $\sigma' = (L +1)^{1/2}\sigma$ corresponding to $L + 1$ periods.

5.14 A model with lost sales

In this section we leave the assumption of complete backordering and assume instead that demand that cannot be met immediately is lost. Otherwise the model is similar to that in Section 5.10. We refer to Rosling (2002b) for a more complete model.

Again we consider a continuous review (R, Q) policy. Note first that the inventory position can never become negative, because in a lost sales model there are no backorders so the inventory level cannot be negative. We must therefore assume that $R \geq 0$. We shall also make the standard simplifying assumption that $Q > R$. Consequently, if one order is outstanding the inventory position is strictly above the reorder point and no more orders can be triggered. Recall that the inventory level is nonnegative. The simplifying assumption implies, therefore, that there can be at most one outstanding order.

The demand is assumed to be continuous and normally distributed. The demand per time unit has mean μ and standard deviation σ. Consequently the demand during the constant lead-time L, $D(L)$, has mean $\mu' = L\mu$ and standard deviation $\sigma' = (L)^{1/2}\sigma$. There is a holding cost per unit and time unit, h, and a shortage cost per unit lost, b_2. (Note that a shortage cost per unit and time unit does not make sense in a lost sales model.)

We shall also assume that $hQ/2 < b_2\mu$, since otherwise it is not profitable to operate the system. To see this, note that a batch Q will at least incur the average holding costs $hQ^2/2\mu$ corresponding to the case when there is no stock on hand when the batch is delivered. The batch will cover the demand Q. This means that we will avoid the shortage costs Qb_2. The system can only be profitable to operate if the expected holding costs are lower than the avoided shortage costs. This means that we must require that $hQ^2/2\mu < Qb_2$, or, equivalently, $hQ/2 < b_2\mu$.

Consider a complete order cycle between two deliveries of a batch. Recall that $Q > R$. Consequently, when a batch has been delivered, there is first a stochastic time until the reorder point is reached and the next batch is ordered. During this time there are no outstanding orders and the inventory level $IL = IP > R \geq 0$, i.e., no lost sales can occur. When the inventory level reaches R, a new batch is ordered. This batch is delivered after the lead-time L. Lost sales will occur if the stochastic lead-time demand exceeds the reorder point. Let us first determine the average lost sales during the order cycle, $E(R - D(L))^-$. We obtain

$$E(R - D(L))^- = \int_R^\infty (u - R) \frac{1}{\sigma'} \varphi\left(\frac{u - \mu'}{\sigma'}\right) du = \sigma' G\left(\frac{R - \mu'}{\sigma'}\right). \quad (5.92)$$

The average length of an order cycle must be $(Q + E(R - D(L))^-)/\mu$, since all demand is either met or lost.

The average inventory on hand just before the delivery at the beginning of the order cycle considered is

$$E(R - D(L))^+ = E(R - D(L)) + E(R - D(L))^- = R - \mu' + \sigma'G\left(\frac{R - \mu'}{\sigma'}\right). \quad (5.93)$$

Note that the inventory on hand at the beginning of the cycle and the demand during the cycle are independent stochastic variables. Therefore, the average time in stock for an item in the batch that is delivered at the beginning of the order cycle can be obtained as $E(R - D(L))^+/\mu + Q/(2\mu)$. (The first term is the expected time before the batch starts being consumed. As usual we disregard the occurrence of negative demand.) The corresponding holding costs are $hQ[E(R - D(L))^+ + Q/2]/\mu$. The total costs per time unit can consequently be expressed as

$$C = \frac{hQ\left[R - \mu' + \sigma'G\left(\frac{R - \mu'}{\sigma'}\right) + \frac{Q}{2}\right] + b_2\mu\sigma'G\left(\frac{R - \mu'}{\sigma'}\right)}{Q + \sigma'G\left(\frac{R - \mu'}{\sigma'}\right)}. \quad (5.94)$$

The costs in (5.94) can be minimized with respect to R. It is possible to show that the costs have a single local optimum. It is therefore easy to find the optimal solution.

Recall that $R \geq 0$. Furthermore, the model is only an approximation for $R > Q$. If the model is used for $R > Q$, the lost sales are underestimated and an optimization will give a reorder point R that is too low. We have assumed that the lead-time is constant, but independent stochastic lead-times can be handled in essentially the same way.

If we compare (5.94) to the corresponding costs with complete backordering, (5.69) or (5.70), we can see that the cost expressions, as expected, approach each other when the average lost sales during an order cycle are small compared to the batch quantity Q. If we disregard the term $\sigma'G((R - \mu')/\sigma')$ in the first term in the numerator and in the nominator, we obtain the approximate cost expression (5.71) with the simple solution (5.73). This solution is often a good approximation also in the lost sales case (Rosling, 2002b).

It is also easy to evaluate the fill rate in the considered model.

$$S_2 = \frac{Q}{Q + \sigma' G(\frac{R - \mu'}{\sigma'})}. \qquad (5.95)$$

In (5.95) we use the fact that the total demand in an order cycle consists of two parts, the demand that is satisfied from stock on hand Q, and the lost sales $\sigma' G((R - \mu')/\sigma')$.

Other inventory models with lost sales are presented in e.g., Hadley and Whitin (1963), Archibald (1981), Buchanen and Love (1985), Beyer (1994), Johansen and Thorstenson (1996), Johansen (1997), and Johansen and Hill (1998).

5.15 Stochastic lead-times

5.15.1 Two types of stochastic lead-times

So far we have only dealt with constant lead-times. It is quite common, though, that the supply process is also stochastic and we shall therefore show how stochastic lead-times can be incorporated in inventory models.

We shall here describe two different types of stochastic lead-times. (Also other types are possible.) In Sections 5.15.2 and 5.15.3 we show how they can be handled, and in Section 5.15.4 we provide a numerical comparison.

1. Sequential deliveries independent of the lead-time demand.

By sequential deliveries we mean that orders cannot cross in time. This type of stochastic lead-times is most common in practice. Consider, for example, a situation where we order from a factory or from an external supplier, and assume that the orders placed with the outside supplier are delivered according to a first-come, first-served policy. The orders can then not pass each other. The stochastic lead-time for a certain order may depend on the previous demand due to earlier orders that have caused congestion in the supply system, but the demand after the order will not affect the lead-time.

Let us also give an example of when the stochastic lead-times are not independent of the lead-time demand and therefore do not belong to type 1. Assume, for example, that we order from a factory and that the factory produces many different items. If the factory gives higher priority to items for which there are many orders, the lead-times can evidently be affected by the lead-time demand.

2. Independent lead-times

The second type of lead-times, i.e., independent stochastic lead-times, may occur if the orders are served by many independent servers. If the lead-time for a certain order is long, it may then happen that later demands will trigger orders that will be delivered earlier than the order considered.

In many situations it is complicated to model lead-times of type 1. Consider for example, orders to a factory and assume that the stochastic lead-time at a certain time depends on the queues in front of some machines, and that these queues are due to previous orders for the item. To model such lead-times we may need to include both the inventory and the queues in an integrated model. See e.g., Song (1994) and Zipkin (1986, 2000) for a further discussion of different types of stochastic lead-times. Similar situations may also occur in connection with multi-echelon inventory systems, where shortages at a higher echelon may cause stochastic delays for lower echelons. See Chapter 10.

In practice it is usually more difficult to evaluate stochastic variations in the lead-times compared to demand variations. Unless the lead-time variations are large, it is then reasonable to disregard them and replace a stochastic lead-time by its mean. As we shall see in Section 5.15.3, this is also exact in some situations when dealing with independent stochastic lead-times.

5.15.2 *Handling sequential deliveries independent of the lead-time demand*

Consider first a continuous review inventory model with stochastic lead-times of the first type. Since the orders cannot pass each other, it is evident that (5.35) is still valid, i.e.,

$$IL(t + L) = IP(t) - D(t, t + L(t)), \qquad (5.96)$$

where $L(t)$ is the stochastic lead-time that occurs at time t.

Assume that a certain (R, Q) policy is used and define

IP = stochastic inventory position in steady state,
IL = stochastic inventory level in steady state,
D = stochastic lead-time demand when averaging over L,

i.e., for discrete demand

$$P(D = j) = \underset{L}{E} P((D(L) = j)). \tag{5.97}$$

Let now t in (5.96) be a completely arbitrary time. Note that this means that $t + L$ is also a completely arbitrary time. Furthermore, $L(t)$ is an arbitrary stochastic lead-time. We get

$$IL = IP - D, \tag{5.98}$$

which can be used in the same way as we have used (5.35) in case of a constant lead-time.

For compound Poisson demand the distribution of the inventory level can be obtained through a slight generalization of (5.36). We simply average over different lead-times

L continuous

$$P(IL = j) = \underset{L}{E}\left\{\frac{1}{Q} \sum_{k=\max\{R+1,j\}}^{R+Q} P(D(L) = k - j)\right\}$$
$$= \frac{1}{Q} \sum_{k=\max\{R+1,j\}}^{R+Q} P(D = k - j), \quad j \leq R + Q. \tag{5.99}$$

For an S policy, or equivalently $(S - 1, S)$ policy, (5.99) simplifies to

$$P(IL = j) = \underset{L}{E}\{P(D(L) = S - j)\} = P(D = S - j), \quad j \leq S. \tag{5.100}$$

Similarly we get the inventory level distribution for continuous normally distributed demand from (5.39) as

$$F(x) = P(IL \leq x) = \underset{L}{E}\left\{\frac{1}{Q} \int_R^{R+Q} \left[1 - \Phi\left(\frac{u - x - \mu L}{\sigma L^{1/2}}\right)\right] du\right\}. \tag{5.101}$$

Lead-times of the first type should, in principle, be handled as in (5.99)-(5.101). It is often practical, though, to use a simple approximation where the distribution of the demand during a stochastic lead-time is replaced by a normal distribution with correct mean and variance. As before, we denote the mean and standard deviation of the demand per time unit by μ and σ. We obtain the mean of D as

$$E(D) = \mu E(L). \tag{5.102}$$

For a given L, $E\{(D(L))^2\} = \sigma^2 L + (\mu L)^2$. The variance of D can then be determined as

$$Var(D) = E_L\{\sigma^2 L + (\mu L)^2\} - \mu^2 (E(L))^2$$

$$= \sigma^2 E(L) + \mu^2 Var(L) + \mu^2 (E(L))^2 - \mu^2 (E(L))^2 \tag{5.103}$$

$$= \sigma^2 E(L) + \mu^2 Var(L).$$

If we set $\mu' = E(D)$ and $\sigma' = (Var(D))^{1/2}$, we can then, as an approximation, use expressions for constant lead-times like (5.39), (5.42), (5.43), etc.

5.15.3 Handling independent lead-times

We shall consider a well-known and interesting result. Assume an S policy and Poisson demand with intensity λ, i.e., each demand triggers an order. This also means that the orders can be described by the same Poisson process as the demands. The stochastic lead-times are independent but otherwise general. We can see the supply system as an (M/G/∞) system. (M stands for Poisson arrivals, G for a general lead-time distribution, and ∞ for an unlimited number of servers.) According to Palm's theorem (Palm, 1938), the total occupancy in an (M/G/∞) system (Poisson arrivals, general service time and unlimited number of servers) is Poisson distributed with mean $\lambda E(L)$. So the outstanding orders have a Poisson distribution that only depends on the expected value of the lead-time. This means that we can replace the stochastic lead-time by its mean or by exponential service times, or whatever we feel is easy to deal with.

It is a little surprising that in the considered situation a stochastic lead-time will not give any disadvantages compared to a constant lead-time. One reason is that when orders cross in time we can always use the earliest deliveries to satisfy customer demands. So, items do not, in general, satisfy customer demands in the same order as they are replenished from the supplier.

In case of compound Poisson demand, the number of outstanding orders (but not the number of items) is still Poisson.

5.15.4 Comparison of the two types of stochastic lead-times

We shall illustrate the differences between the two types of lead-times by an example.

Example 5.9 Consider pure Poisson demand with intensity $\lambda = 2$ and an $(S-1, S)$ policy with $S = 3$.

Assume first that the orders placed with the outside supplier are served by a single server with exponential service time and service rate $\kappa = 2.5$. This means that the supplier is modeled as an (M/M/1) queuing system. The lead-time, i.e., the waiting time including the service time, is evidently of type 1. It is well-known that the waiting time has an exponential distribution with density $(\kappa - \lambda)e^{-(\kappa - \lambda)t} = 0.5\,e^{-0.5\,t}$. Using (5.100) we obtain

$$P(IL = j) = \underset{L}{E}\{P(D(L) = S - j)\} = \int_0^\infty 0.5 e^{-0.5t} \frac{(2t)^{3-j}}{(3-j)!} e^{-2t} dt \quad j \le 3. \tag{5.104}$$

From (5.104) it is easy to obtain $P(IL = 3) = 0.2$, and for $j < 3$ using partial integration

$$P(IL = j) = \int_0^\infty 0.5 \frac{(2t)^{3-j}}{(3-j)!} e^{-2.5t} dt \tag{5.105}$$

$$= 0.8 \int_0^\infty 0.5 \frac{(2t)^{3-j-1}}{(3-j-1)!} e^{-2.5t} dt = 0.8 P(IL = j+1),$$

i.e., the inventory level has a geometric distribution with $P(IL = j) = 0.2 \cdot 0.8^{3-j}$ ($j \le 3$).

Let us now instead assume that the stochastic lead-times are service times in an (M/M/∞) system, i.e., the supplier is modeled as a queuing system with infinitely many servers and exponential service times. Let us also assume that the service rate for each server is $\kappa = 0.5$. This means that the distribution of the lead-time is exactly the same as in the above situation with a single server. The lead-times are, however, now of the second type, i.e., independent. Using the results in Section 5.15.3 we can simply replace the stochastic lead-time by its mean $1/\kappa = 2$. The lead-time demand is Poisson distributed with mean $\lambda / \kappa = 4$. We obtain the probability for inventory level j as the probability that the lead-time demand is $3 - j$,

$$P(IL = j) = \frac{4^{3-j}}{(3-j)!}e^{-4} \quad j \leq 3. \tag{5.106}$$

The two different inventory level distributions are compared in Figure 5.7.

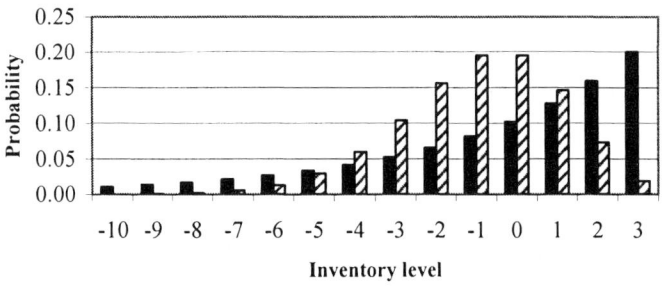

Figure 5.7 Inventory level distributions in Example 5.9.

Recall that the lead-time distribution is the same in both cases. It is obvious that the inventory level varies much less in case of independent lead-times. This is, in general, advantageous since it is easier to avoid large backorder levels.

Note also that a constant lead-time is both of type 1 and of type 2.

References

Abramowitz, M., and I. Stegun. 1964. *Handbook of Mathematical Functions*, National Bureau of Standards, Washington, D. C.

Adan, I., M. van Eenige, and J. Resing. 1995. Fitting Discrete Distributions on the First Two Moments, *Probability in the Engineering and Informational Sciences*, 9, 623-632.

Archibald, B. C. 1981. Continuous Review (s, S) Policies with Lost Sales, *Management Science*, 27, 1171-1177.

Beyer, D. 1994. An Inventory Model with Wiener Demand Process and Positive Lead Time, *Optimization*, 29, 181-193.

Browne, S. and P. H. Zipkin. 1991. Inventory Models with Continuous Stochastic Demands, *The Annals of Applied Probability*, 1, 419-435.

Buchanen, D. J., and R. F. Love. 1985. A (Q, R) Inventory Model with Lost Sales and Erlang-Distributed Leadtimes, *Naval Research Logistics Quarterly*, 32, 605-611.

Chen, F., and Y. S. Zheng. 1993. Inventory Models with General Backorder Costs, *European Journal of Operational Research*, 65, 175-186.

Feller, W. 1966. *An Introduction to Probability Theory and its Applications*, Vol. 2, John Wiley, New York.

Hadley, G., and T. M. Whitin. 1963. *Analysis of Inventory Systems*, Prentice-Hall, Englewood Cliffs, NJ.

Johansen, S. G. 1997, Computing the Optimal (r, Q) Inventory Policy with Poisson Demand and Lost Sales, University of Aarhus.

Johansen, S. G., and A. Thorstenson. 1996. Optimal (r, Q) Inventory Policies with Poisson Demand and Lost Sales: Discounted and Undiscounted Cases, *International Journal of Production Economics*, 46, 359-371.

Johansen, S. G., and R. M. Hill. 1998. The (r, Q) Control of a Periodic-Review Inventory System with Continuous Demand and Lost Sales, University of Aarhus.

Lee, H. L., and S. Nahmias. 1993. Single-Product, Single-Location Models, in S. C. Graves et al. Eds., *Handbooks in OR & MS Vol.4*, North Holland Amsterdam, 133-173.

Palm, C. 1938. Analysis of the Erlang Traffic Formula for Busy Signal Assignment, *Ericsson Technics*, 5, 39-58.

Porteus, E. L. 1990. Stochastic Inventory Theory, in D.P. Heyman, and M. J. Sobel. Eds., *Handbooks in OR & MS Vol.2*, North Holland Amsterdam, 605-652.

Rosling, K. 2002a. Inventory Cost Rate Functions with Nonlinear Shortage Costs, *Operations Research*, 50, 1007-1017.

Rosling, K. 2002b. The (r, Q) Inventory Model with Lost Sales, Växjö University.

Silver, E. A., D. F. Pyke, and R. Peterson. 1998. *Inventory Management and Production Planning and Scheduling*, 3rd edition, Wiley, New York.

Sobel, M. J. 2004. Fill Rates of Single-Stage and Multistage Supply Systems, *Manufacturing & Service Operations Management*, 6, 41-52.

Song, J. S. 1994. The Effect of Leadtime Uncertainty in a Simple Stochastic Inventory Model, *Management Science*, 40, 603-613.

Tijms, H. C. 1994. *Stochastic Models: An Algorithmic Approach*, Wiley, Chichester.

Zheng, Y. S. 1992. On Properties of Stochastic Inventory Systems, *Management Science*, 38, 87-103.

Zheng, Y. S., and A. Federgruen. 1991. Finding Optimal (s, S) Policies Is About as Simple as Evaluating a Single Policy, *Operations Research*, 39, 654-665.

Zipkin, P. H. 1986. Stochastic Leadtimes in Continuous-Time Inventory Models, *Naval Research Logistics Quarterly*, 33, 763-774.

Zipkin, P. H. 2000. *Foundations of Inventory Management*, McGraw-Hill, Singapore.

Problems

5.1* In Section 5.1.1 it is shown how a compound Poisson process with $f_0 > 0$ can be replaced by an equivalent process with $f_0' = 0$. Prove this.

5.2* Consider compound Poisson demand with a logarithmic compounding distribution as in Section 5.1.2. Show that $D(t)$ has a negative binomial distribution.

5.3 Let $\mu' = 3$ and $\sigma' = 1$. Make a computer program that fits a distribution of $D(t)$ to these values according to Section 5.1.4. Verify from the obtained distribution that you get the correct μ' and σ'.

5.4 Show that $G(x) \to -x$ as $x \to -\infty$. Use this to show that $F(x)$ in (5.42) approaches 1 as $x \to \infty$.

5.5 Consider the density of the inventory level $f(x)$ in (5.43). Show that $f(x)$ has its maximum at $x = R - \mu' + Q/2$.

5.6* The lead-time demand for an item is normally distributed with mean 300 and standard deviation 100.

a) Set S_1 to 75, 85, and 95 percent. Determine the corresponding safety stocks.
b) Set the safety factor to 1. Determine S_2 for $Q = 100$, 500, and 1000.
c) Assume that the average demand per week is 25 and 100, respectively. Let the safety stock be determined as the average demand during three weeks. Determine the corresponding values of S_1.

5.7* A company plans to use an (R, Q) policy with continuous review for inventory control. A certain test item is considered. Demand is regarded to be continuous. The lead-time demand has mean 100 and standard deviation 50. The policy $R = 135$ and $Q = 200$ is considered.

a) Determine S_1 and S_2 for normally distributed lead-time demand.
b) Assume instead that the lead-time demand has a uniform distribution with the same mean and standard deviation. Determine S_1 and S_2.

5.8* A company tries to reduce all lead-times by 50 percent. There is a discussion concerning the impact of this change. As an example an item in stock is considered. Demand is continuous and normally distributed. Inventory control is carried out by an (R,Q) policy with continuous review. The demand per week has mean 200 and standard deviation 50. Demands during different weeks are

---

* Answer and/or hint in Appendix 1.

independent. From the beginning the lead-time is four weeks and $R = 850$ and $Q = 600$. Two questions have been raised:

a) Assume that the lead-time is reduced to two weeks, while R and Q are unchanged. How much does S_2 change? Will the stock on hand increase? How much?

b) Assume that the lead-time is reduced to two weeks and that Q is unchanged. How should R be chosen to get the same S_2 as before? How high is then the average stock on hand?

5.9   Consider a continuous review (R, Q) policy with $R = 1$ and $Q = 2$. The demand is Poisson with intensity $\lambda = 0.5$. The lead-time is 4.

a) What is the average inventory level?
b) What is the average inventory on hand?
c) What is the average backorder level?
d) What is the average waiting time for a customer?

5.10   Consider normal lead-time demand with mean 100 and standard deviation 30. An (R, Q) policy is applied with $Q = 50$ given. The holding cost is $h = 2$ per unit and unit time.

a) Determine R so that the average waiting time for a customer is 0.01. Determine the corresponding holding costs.
b) Choose S_1, S_2, and S_3, respectively so that you get the same solution as in a).
c) Choose a backorder cost b_1 per unit and time unit so that you get the same solution as in a).

5.11*   An inventory is controlled by an (R, Q) policy with continuous review. Demand that cannot be met directly is backordered. $R = 1$ and $Q = 3$. The demand per time unit has a Poisson distribution with mean 2. The lead-time is one time unit.

a) What is the probability in steady state to have 2 units in stock?
b) What is the average waiting time for a customer?

5.12*   An item is controlled by an (R, Q) policy with continuous review. The lead-time demand has an exponential distribution with density:

$$f(x) = \frac{1}{m} e^{-\frac{x}{m}} \quad x \geq 0$$

where m is the mean. Determine the fill rate for given R, Q and m. Assume that $R > 0$.

5.13* Consider a continuous review (R, Q) policy with $R = 1$ and $Q = 3$. The demand is Poisson with intensity $\lambda = 0.7$. The lead-time is one time unit.

a) Determine the fill rate S_2 exactly.
b) Determine the fill rate S_2 by using a normal approximation.

5.14 Consider pure Poisson demand and a continuous review (s, S) policy. Use the technique in Section 5.11 to verify that the distribution of the inventory position is uniform.

5.15 Make a computer program to verify the results in Example 5.6.

5.16 a) Consider the periodic review (R, Q) model in Section 5.12.2. Modify the derivation to fit an order-up-to-S policy.
b) Use the results in a) to determine the fill rate under the following assumptions. Poisson demand with mean 0.5 per unit of time, $L = 1$, $T = 0.5$, and $S = 1$.

5.17 Consider Section 5.12.4. Modify the derivation to fit an order-up-to-S policy.

5.18 Consider the newsboy model in Section 5.13 but assume a discrete demand distribution with distribution function $F(x)$. Demonstrate that the optimality condition corresponding to (5.91) is to choose the smallest S satisfying $F(Q) \geq c_u/(c_o + c_u)$.

5.19 Use the result in Problem 5.18 to solve the following newsboy problem. Poisson demand with mean 3, $c_o = 2$, and $c_u = 5$. What are the optimal costs?

5.20 Consider a continuous review (R, Q) policy with $R = 60$ and $Q = 80$. The lead-time demand is normally distributed with mean 100 and standard deviation 20. Determine the fill rate both for complete backordering and for lost sales. Why is the fill rate higher in the lost sales case?

5.21* Consider Example 5.9 and the first case with a stochastic lead-time of type 1. Use (5.102) and (5.103) to find μ' and σ'. Fit then a negative binomial distribution to these values. Using this distribution determine the probabilities for IL equal to 1, 2, 3, and compare to the exact solution.

6 SINGLE-ECHELON SYSTEMS: INTEGRATION - OPTIMALITY

6.1 Joint optimization of order quantity and reorder point

In practice it is most common to determine the batch quantity from a deterministic model. The stochastic demand is then replaced by its mean. In Chapter 4 we have considered different methods for determination of batch quantities under the assumption of deterministic demand. Stochastic variations in the demand, and possibly in the lead-time, are then only taken into account when determining the reorder point. As discussed in Chapter 4 this procedure is, in general, an adequate approximation. In Chapter 5 we have described various techniques for determining the reorder point for a given batch quantity.

It is also possible, though, to optimize the batch quantity and the reorder point jointly in a stochastic model. In this section we shall consider such techniques.

6.1.1 Discrete demand

Assume discrete compound Poisson demand. Each customer demands an integral number of units. As before, we also assume that not all demands are multiples of some integer larger than one. The average demand per unit of time is denoted μ. The lead-time L is constant. The stochastic lead-time demand is denoted $D(L)$ and its mean $\mu' = \mu L$.

Furthermore, we consider a holding cost h per unit and time unit, a shortage cost b_1 per unit and time unit, and an ordering or setup cost A.

6.1.1.1 (R, Q) policy

We shall first deal with a continuous review (R, Q) policy and the joint optimization of the integers R and Q. A periodic review model can be handled in essentially the same way. See also Federgruen and Zheng (1992).

Recall the following standard argument from Section 5.3.2. Let $IP(t)$ be the inventory position at time t. Consider then the time $t + L$. At that time, everything that was on order at time t has been delivered. Orders that have been triggered in the interval $(t, t + L]$ have not reached the inventory due to the lead-time. Consequently we have

$$IL(t + L) = IP(t) - D(L). \qquad (6.1)$$

Let us initially consider the special case of an $(S - 1, S)$ policy with $S = k$, i.e., $R = k - 1$ and $Q = 1$. This means that the inventory position is k at all times. Using (6.1) the inventory level distribution can then be obtained as

$$P(IL = j) = P(D(L) = k - j), \qquad j \le k. \qquad (6.2)$$

Let $g(k)$ be the average holding and shortage costs per time unit. We have from (5.56)

$$g(k) = -b_1 E(IL) + (h + b_1)E(IL^+) = -b_1(k - \mu') + (h + b_1)\sum_{j=1}^{k} jP(IL = j). \qquad (6.3)$$

The results in Section 5.9.1 imply that $g(k)$ is a convex function of the inventory position k. Furthermore $g(k) \to \infty$ as $|k| \to \infty$.

Let us now go back to the (R, Q) policy. Recall that the inventory position is uniform on $[R + 1, R + Q]$. Consequently the total average costs per time unit can be expressed as

$$C(R, Q) = \frac{A\mu}{Q} + \frac{1}{Q}\sum_{k=R+1}^{R+Q} g(k). \qquad (6.4)$$

In (6.4) we obtain the holding and backorder costs by averaging over the inventory position. We assume that each batch incurs an ordering cost A, i.e., if an order for two batches is triggered, the associated ordering cost is $2A$. Our objective is to optimize $C(R, Q)$ with respect to both R and Q.

Let us now define $C(Q)$ as

$$C(Q) = \min_R \{C(R,Q)\}. \qquad (6.5)$$

It is evident that $C(1) = A\mu + \min_k \{g(k)\}$, i.e., if the sum in (6.4) includes a single value of k, we choose a k that gives the minimum cost. We denote an optimal k, i.e., a value of k that minimizes $g(k)$ by k^*. A corresponding optimal reorder point for $Q = 1$ is $R^*(1) = k^* - 1$, i.e., if we use a single value of k, we choose the best one. Consider now $Q = 2$, which means that we use two values of k. Due to the convexity of $g(k)$, the second best k must be either $k^* - 1$ or $k^* + 1$. Clearly, we should use the better of these two values. From (6.4) it is therefore evident that it is optimal to choose $R^*(2) = R^*(1) - 1$ if $g(R^*(1)) \leq g(R^*(1) + 2)$, and $R^*(2) = R^*(1)$ otherwise. We obtain $C(2) = A\mu/2 + [\min\{g(R^*(1)), g(R^*(1) + 2)\} + g(R^*(1) + 1)]/2$, or equivalently, $C(2) = C(1)/2 + \min\{g(R^*(1)), g(R^*(1) + 2)\}/2$. More generally we have

$$R^*(Q+1) = R^*(Q) - 1 \quad \text{if } g(R^*(Q)) \leq g(R^*(Q) + Q + 1),$$

$$R^*(Q+1) = R^*(Q) \quad \text{otherwise,} \qquad (6.6)$$

and

$$C(Q+1) = C(Q)\frac{Q}{Q+1} + \left[\min\{g(R^*(Q)), g(R^*(Q) + Q + 1)\}\right]\frac{1}{Q+1}. \qquad (6.7)$$

It is evident from (6.7) that $C(Q + 1) \geq C(Q)$ if and only if $\min\{g(R^*(Q)), g(R^*(Q) + Q + 1)\} \geq C(Q)$. Furthermore, it is obvious that $\min\{g(R^*(Q)), g(R^*(Q) + Q + 1)\}$ is increasing with Q. Let Q^* be the smallest Q such that $C(Q + 1) \geq C(Q)$. It follows from (6.7) that $C(Q) \geq C(Q^*)$ for any $Q \geq Q^*$. Consequently, Q^* and $R^*(Q^*)$ provide the optimal solution.

To summarize, it is very easy to determine the optimal solution by applying (6.6) and (6.7) until the costs increase.

6.1.1.2 (s, S) policy

Let us now instead consider the optimization of an (s, S) policy. Otherwise the assumptions are exactly the same as above. When using an (s, S) policy the inventory position is no longer uniformly distributed. This makes the optimization more complex. Zheng and Federgruen (1991), however, have developed a very efficient optimization procedure. We shall here only describe the procedure and refer to their paper for proofs and more details.

In Section 5.11 we defined and determined the probabilities

m_j = probability to reach $IP = j$ during an order cycle ($s + 1 \leq j \leq S$).

The average total number of customers during an order cycle is $\sum_{j=s+1}^{S} m_j$. Let λ be the customer arrival rate. The average length of an order cycle is consequently $\sum_{j=s+1}^{S} m_j / \lambda$. The steady state distribution of the inventory position is obtained as

$$P(IP = k) = m_k / \sum_{j=s+1}^{S} m_j, \quad k = s+1, s+2, ..., S. \tag{6.8}$$

See Section 5.11 for more details. Given these probabilities we can determine the average costs per time unit as

$$C(s,S) = \frac{A\lambda}{\sum_{j=s+1}^{S} m_j} + \sum_{k=s+1}^{S} P(IP = k) \cdot g(k). \tag{6.9}$$

We are now ready to describe the optimization procedure.

1. In the first step we set $S^* = k^*$, i.e., a value of k that minimizes $g(k)$. Next, consider $s = S^* - 1$, $s = S^* - 2$, ... , until $C(s, S^*) \leq g(s)$. When this occurs, set $s^* = s$ and the initial best solution as $C^* = C(s^*, S^*)$. Set $S = S^*$.

2. Set $S = S + 1$. If $g(S) > C^*$, s^* and S^* provide the optimal solution with the costs C^*, and the algorithm stops.

3. It is possible to show that the considered S will improve the solution if, and only if, $C(s^*, S) < C^*$. In that case set $S^* = S$. Otherwise go to 2.

4. To find the best s corresponding to S^*, it is only necessary to consider $s = s^*, s^* + 1, \ldots$ The new s^* is obtained as the smallest value of s giving $C(s, S^*) > g(s + 1)$. Update $C^* = C(s^*, S^*)$ and go to 2.

6.1.2 An iterative technique

Now consider normally distributed demand instead, and add ordering costs to the (R, Q) model with backorder cost per unit and unit time that we dealt with in Section 5.9.2. The lead-time demand has mean μ' and standard deviation σ'. The mean per unit of time is μ.

By adding the average ordering costs per time unit to the cost expression in (5.65) we have

$$C(R,Q) = h(R + Q/2 - \mu') + (h + b_1)\frac{\sigma'^2}{Q}\left[H\left(\frac{R - \mu'}{\sigma'}\right) - H\left(\frac{R + Q - \mu'}{\sigma'}\right)\right] + \frac{A\mu}{Q}.$$

(6.10)

Recall the definitions of $H(x)$ and $G(x)$. See Appendix 2.

Our objective is to optimize $C(R, Q)$ with respect to R and Q. As pointed out in Section 5.9.2, we cannot replace the backorder costs by the fill rate (5.67) when carrying out a joint optimization of R and Q.

We shall demonstrate how the optimization can be carried out by a simple iterative procedure. It can be shown that this procedure will always converge to the optimal solution (Rosling, 2002b).

The necessary conditions $\partial C / \partial Q = \partial C / \partial R = 0$ are also sufficient and will guarantee the unique optimal solution.

We obtain $\partial C / \partial R$ as in (5.66)

$$\frac{\partial C}{\partial R} = h + (h + b_1)\frac{\sigma'}{Q}\left[G\left(\frac{R + Q - \mu'}{\sigma'}\right) - G\left(\frac{R - \mu'}{\sigma'}\right)\right]. \quad (6.11)$$

For a given Q we can use (6.11) to determine the corresponding reorder point R giving $\partial C / \partial R = 0$. The resulting R decreases with Q. We can also get $\partial C / \partial Q$ from (6.10) as

$$\frac{\partial C}{\partial Q} = \frac{h}{2} - \frac{A\mu}{Q^2}$$

$$-(h+b_1)\frac{\sigma'^2}{Q^2}\left[H\left(\frac{R-\mu'}{\sigma'}\right) - H\left(\frac{R+Q-\mu'}{\sigma'}\right) - \frac{Q}{\sigma'}G\left(\frac{R+Q-\mu'}{\sigma'}\right)\right].$$

(6.12)

We shall now describe the iterative procedure for finding the optimal solution satisfying $\partial C / \partial Q = \partial C / \partial R = 0$.

We start by determining the batch quantity according to the classical economic order quantity model

$$Q^0 = \sqrt{2A\mu / h}.$$ (6.13)

Next we determine the corresponding reorder point R^0 from (6.11) and the condition $\partial C / \partial R = 0$. In the following step we get a new batch quantity Q^1 from

$$Q^{i+1} = \left[\frac{2A\mu}{h} + \frac{2(h+b_1)}{h}\sigma'^2\left(H\left(\frac{R^i-\mu'}{\sigma'}\right) - H\left(\frac{R^i+Q^i-\mu'}{\sigma'}\right)\right.\right.$$

$$\left.\left. - \frac{Q^i}{\sigma'}G\left(\frac{R^i+Q^i-\mu'}{\sigma'}\right)\right)\right]^{1/2}.$$

(6.14)

After that we determine the reorder point R^1 corresponding to Q^1 from (6.11) and the condition $\partial C / \partial R = 0$. Given Q^1 and R^1, we obtain Q^2 and R^2 in the same way, etc.

It can be shown that the batch quantity increases in each step, $Q^{i+1} \geq Q^i$, while the reorder point decreases, $R^{i+1} \leq R^i$. The costs decrease in each step. Let C^* be the optimal cost. It is possible to show that $C(R^i, Q^{i+1}) - C^* \leq h(Q^{i+1} - Q^i)$, i.e., the remaining gap can be bounded.

Example 6.1 Let $A = 100$, $h = 2$, and $b_1 = 20$. The demand per time unit is normally distributed with $\mu = 50$ and $\sigma = 20$. The lead-time is $L = 4$. We obtain $\mu' = \mu L = 200$ and $\sigma' = \sigma L^{1/2} = 40$.

The results from the iterations when applying the described procedure are shown in Table 6.1. We can see that the costs converge very rapidly. The changes in batch quantity and reorder point as compared to the initial solution are significant. Still the total cost reduction is only about 2.5 percent.

Table 6.1   Results from the iterations for the data in Example 6.1.

Iteration i	0	1	2	3	4	5
Order quantity Q^i	70.71	87.91	93.08	94.59	95.03	95.15
Reorder point R^i	224.76	219.60	218.16	217.75	217.63	217.60
Costs C^i	232.01	226.63	226.24	226.21	226.20	226.20

Similar procedures for models with other types of costs can be designed in the same way. We could, for example, have used a shortage cost per unit as in Section 5.10. A corresponding procedure for a model with lost sales is given in Rosling (2002a).

6.1.3 Fill rate constraint - a simple approach

We shall now consider a different and very simple technique from Axsäter (2004). This technique is especially suitable when optimizing R and Q under a fill rate constraint, because it is in general only necessary to consider relatively few different fill rates. The cost function is the same as in (6.10) except for the backorder costs that are omitted

$$C = h(R + Q/2 - \mu') + h\frac{\sigma'^2}{Q}\left[H\left(\frac{R-\mu'}{\sigma'}\right) - H\left(\frac{R+Q-\mu'}{\sigma'}\right)\right] + \frac{A\mu}{Q}. \quad (6.15)$$

The fill rate is according to (5.52) obtained as

$$S_2 = 1 - \frac{\sigma'}{Q}\left[G\left(\frac{R-\mu'}{\sigma'}\right) - G\left(\frac{R+Q-\mu'}{\sigma'}\right)\right]. \quad (6.16)$$

It is possible to show that the considered problem has a pure optimal strategy, i.e., a single (R, Q) as its optimal solution, see Rosling (2002b).

At a first glance the problem to optimize (6.15) under the constraint (6.16) for a given fill rate S_2 depends on five parameters: h, A, μ, μ', and σ'. However, it is easy to see that the problem does in fact depend on a single parameter only. Define

$$c = C/(h\sigma'),$$
$$q = Q/\sigma',$$
$$r = (R - \mu')/\sigma',$$
$$E = A\mu/(h(\sigma')^2).$$

Substituting in (6.15) and (6.16) we get the equivalent problem to minimize

$$c = r + \frac{q}{2} + \frac{1}{q}[H(r) - H(r+q)] + \frac{E}{q}, \qquad (6.17)$$

under the constraint

$$S_2 = 1 - \frac{1}{q}[G(r) - G(r+q)]. \qquad (6.18)$$

Note that this version of the problem for a certain given S_2 depends on a single problem parameter, $E \geq 0$.

Axsäter (2004) suggests that q^* is determined by linear interpolation of tabulated values (or by using a polynomial approximation). (The tabulated values can, for example, be obtained by using the technique in Section 6.1.2.) The corresponding r^* is then obtained from (6.18) in a second step. A part of the required table is given in Table 6.2.

We are now ready to describe a simple way to solve the original problem for a given fill rate S_2 and any problem parameters. We start by determining E. Next we obtain the solution of the one-parameter problem from the table and (6.18). Given the optimal solution q^* and r^*, we get the solution of the original problem as

$$Q^* = q^* \sigma',$$
$$R^* = r^* \sigma' + \mu'.$$

Note that we can use the same table repeatedly for all items.

SINGLE-ECHELON – INTEGRATION

Example 6.2 Let $S_2 = 0.9$, $A = 100$, $h = 2$, $\mu = 50$, $\mu' = 200$, and $\sigma' = 40$, i.e., except for the backorder cost, the same data as in Example 6.1. We get $E = A\mu/(h(\sigma')^2) = 1.5625$ and $e = \ln(E) = 0.4463$. Using the table values for $e = 0.4$ and $e = 0.5$ and interpolating linearly, we get $q^* = (2.5111(0.5 - 0.4463) + 2.6070(0.4463 - 0.4))/0.1 = 2.5555$. Using (6.18) we get $r^* = 0.3294$. Finally we obtain $Q^* = q^*\sigma' = 2.5555 \cdot 40 = 102.22$ and $R^* = r^*\sigma' + \mu' = 0.3294 \cdot 40 + 200 = 213.18$. The corresponding optimal solution is $Q^* = 102.20$ and $R^* = 213.14$.

Table 6.2 q^* for different fill rates and values of $e = \ln(E)$.

e \ S	60%	70%	80%	85%	90%	95%	99%
-0.2	2.7398	2.4609	2.2323	2.1255	2.0165	1.8926	1.7371
-0.1	2.8408	2.5496	2.3127	2.2025	2.0904	1.9633	1.8047
0.0	2.9462	2.6421	2.3964	2.2828	2.1675	2.0373	1.8756
0.1	3.0562	2.7383	2.4836	2.3664	2.2478	2.1145	1.9498
0.2	3.1712	2.8387	2.5745	2.4536	2.3318	2.1953	2.0276
0.3	3.2914	2.9435	2.6694	2.5446	2.4194	2.2798	2.1092
0.4	3.4172	3.0529	2.7684	2.6397	2.5111	2.3683	2.1949
0.5	3.5490	3.1671	2.8718	2.7391	2.6070	2.4609	2.2848
0.6	3.6872	3.2867	2.9800	2.8430	2.7073	2.5580	2.3791
0.7	3.8322	3.4118	3.0931	2.9518	2.8124	2.6599	2.4783
0.8	3.9846	3.5430	3.2116	3.0658	2.9226	2.7668	2.5824

6.2 Optimality of ordering policies

In all the models that we have considered in this chapter it has been assumed that the policy is either of the (R, Q) type or of the (s, S) type. A natural question to ask is whether other, better policies exist. This is, in general, not the case. In most situations one of these policies is indeed optimal for a single-echelon inventory system with independent items.

In Section 6.2.1 we shall show that an (R, Q) policy is optimal when there are no ordering costs but a given fixed batch quantity Q. After that we comment on the optimality of (s, S) policies in Section 6.2.2.

6.2.1. Optimality of (R, Q) policies when ordering in batches

Consider an inventory system with continuous review. Assume discrete compound Poisson demand and that each customer demands an integral number of units. As before, we assume that not all demands are multiples of some integer larger than one. The average demand per unit of time is denoted μ. The lead-time L is constant. The stochastic lead-time demand is denoted $D(L)$ and its mean $\mu' = \mu L$.

Furthermore, we consider a holding cost h per unit and time unit and a shortage cost b_1 per unit and time unit. There are no ordering costs but all orders must be multiples of a given batch quantity Q. Orders can only be triggered by customer demands. We shall show that an (R, Q) policy is optimal under these assumptions. Our proof follows essentially Chen (2000).

Obviously the inventory position must be an integer at all times. Assume first that the inventory position is k at some arbitrary time t. Using (6.1) the inventory level distribution at time $t + L$ can be obtained as

$$P(IL = j) = P(D(L) = k - j), \qquad j \le k. \tag{6.19}$$

Furthermore, (as in Section 6.1.1.1), let $g(k)$ be the corresponding holding and shortage cost rate at time $t + L$. We have

$$g(k) = -b_1 E(IL) + (h + b_1)E(IL^+) = -b_1(k - \mu') + (h + b_1)\sum_{j=1}^{k} jP(IL = j).$$

$$\tag{6.20}$$

As shown in Section 5.9.1, $g(k)$ is a convex function of the inventory position k. Furthermore $g(k) \to \infty$ as $|k| \to \infty$.

Define now

$$\bar{g}(y) = \sum_{j=1}^{Q} g(y + j),$$

where y is an integer. Clearly $\bar{g}(y)$ is also convex. Denote by R the finite integer y that minimizes $\bar{g}(y)$.

Lemma 6.1 Let x and z be integers. For a given z, $g(z + xQ)$ is convex in x. Let x_z be the unique integer so that $R + 1 \le z + x_z Q \le R + Q$. Then $g(z + xQ)$ is minimized with respect to x for $x = x_z$.

Proof It follows from the convexity of $g(k)$ that $g(z + xQ)$ is convex in x. Note that

$$g(z + (x+1)Q) - g(z + xQ) = \overline{g}(z + xQ) - \overline{g}(z + xQ - 1). \quad (6.21)$$

Consider first any $x < x_z$. This means that $z + xQ \leq R$. Obviously $\overline{g}(z + xQ) - \overline{g}(z + xQ - 1) \leq 0$. Similarly, $x > x_z$ implies $z + xQ > R + Q$ and $\overline{g}(z + xQ) - \overline{g}(z + xQ - 1) \geq 0$. It follows that $g(z + xQ)$ is minimized with respect to x for $x = x_z$.

We are now ready to prove the following proposition.

Proposition 6.1 An (R, Q) policy is optimal.

Proof Consider any feasible policy. Let y_t be the inventory position at time t. The cost rate at time $t + L$ is then $g(y_t)$. From Lemma 6.1 we know that $g(y_t) \geq g(y_t')$ where $y_t' = y_t + nQ$ and n is the unique integer so that $y_t' \in \{R + 1, R + 2, \ldots, R + Q\}$. Consequently, the long-run average cost must be greater than or equal to the long-run average value of $g(y_t')$. To determine these costs consider the stochastic process y_t'. Obviously y_t' is constant between the customer demands. Let D_t be a demand at some time t. Let y_t^- be the inventory position before the demand and y_t^+ the inventory position after the demand. Clearly

$$y_t^+ = y_t^- - D_t + mQ, \quad (6.22)$$

where m is nonnegative. Furthermore, due to our construction we must also have,

$$y_t'^+ = y_t'^- - D_t + m'Q, \quad (6.23)$$

where m' is an integer. Given $y_t'^-$ and D_t, the value of m' is unique because $y_t' \in \{R + 1, R + 2, \ldots, R + Q\}$. The demand sizes are independent, so the different y_t' can be seen as a Markov chain with the finite state space $\{R + 1, R + 2, \ldots, R + Q\}$. The steady state distribution can be shown to be uniform. (This can be done in essentially the same way as the proof of Proposition 5.1 in Section 5.3.1. We omit the details.)

The long-run average value of $g(y_t')$ is therefore $\bar{g}(R)/Q$. This is a lower bound on the long-run cost of any feasible policy. But this lower bound can be achieved by using an (R, Q) policy. Using an (R, Q) policy the inventory position is also uniform on $\{R + 1, R + 2, \ldots, R + Q\}$. (We obtain the costs by setting $A = 0$ in (6.4).) This completes the proof.

Proposition 6.1 can be generalized in different ways, for example to other cost structures and to periodic review. In the special case when $Q = 1$ the (R, Q) policy degenerates to an S policy with $S = R + 1$. This means that Proposition 6.1 also demonstrates the optimality of an S policy in case of no ordering costs and no constraints concerning the batch quantities.

For problems with continuous or Poisson demand, (R, Q) policies and (s, S) policies are equivalent. For such problems (s, S) policies are consequently also optimal.

6.2.2 Optimality of (s, S) policies

If we replace the fixed batch quantity in Section 6.2.1 by an ordering cost the optimal policy is under quite general conditions of the (s, S) type. This is more difficult to show, see e.g., Porteus (2002).

It is interesting to note, however, that (s, S) policies are not necessarily optimal for problems with service constraints. Consider, for example, a problem with discrete integral demand where s and S are integers. It may very well happen that no (s, S) policy provides a certain given service level exactly. The best (s, S) policy that satisfies the service constraint will consequently give a slightly higher service than what is required. In such a situation it may be possible to reduce the costs by varying the policy over time so that the average service level is exactly as prescribed.

Early optimality results were presented by Iglehart (1963) and Veinott (1966). More recent results are provided by Zheng (1991), Rosling (2002b), and Beyer and Sethi (1999).

6.3 Updating order quantities and reorder points in practice

In Chapters 2 - 5 we have presented different techniques for forecasting and determination of batch quantities and reorder points. We shall now illustrate how these techniques can be implemented in an inventory control system. We assume that we are dealing with a single-echelon system and independent items.

SINGLE-ECHELON – INTEGRATION

The forecasts are normally updated with a certain periodicity. In general, it is most practical to also update reorder points and batch quantities at these times, immediately after updating the forecasts. Let

t_F = forecast period.

We can think of the forecast period as, for example, one month. The time unit is not important. We can use one month as the time unit. In that case $t_F = 1$, but we can also express t_F in days ($t_F = 30$), or in years ($t_F = 1/12$). However, to avoid unnecessary errors it is recommended to use the same time unit in all inventory control computations.

Typically the forecasts are updated either by exponential smoothing (Section 2.4), or by exponential smoothing with trend (Section 2.5). It may also be reasonable to use a seasonal method (Section 2.6) for a few items. The forecasting method is, in general, chosen manually for each item. To specify the forecasting method, we also need the smoothing parameters that are part of the different forecasting methods. Usually it is practical to divide the items into a number of inventory control groups and let the forecasting technique as well as various inventory control parameters, like holding cost rate and service level, be identical for all items in the same group. See also Chapter 11.

When using exponential smoothing we update the average demand $\hat{a}_t$ at the end of each forecast period. If instead we use exponential smoothing with trend, we update both the average demand $\hat{a}_t$ and the trend $\hat{b}_t$. In either case we also update some error measure like MAD_t (Section 2.10).

In case of exponential smoothing the average demand per unit of time μ is obtained as

$$\mu = \hat{a}_t / t_F. \tag{6.24}$$

When using exponential smoothing with trend we are, in principle, assuming that the demand is increasing or decreasing linearly with time. Just after the forecast update the estimated demand in the coming period is $\hat{a}_t + \hat{b}_t$. A natural estimate of the demand rate in the middle of this period is then $\hat{a}_t/t_F + \hat{b}_t/t_F$. The trend is $\hat{b}_t/(t_F)^2$. The corresponding estimate of the demand rate in the beginning of the period, i.e., just after the update, is then $\hat{a}_t/t_F + \hat{b}_t/(2t_F)$. The estimated demand rate u time units after the update is $\hat{a}_t/t_F + \hat{b}_t/(2t_F) + u\hat{b}_t/(t_F)^2$.

The standard deviation of the demand per time unit is obtained according to (2.50) and (2.55) as

$$\sigma = \frac{1}{(t_F)^c} \sqrt{\frac{\pi}{2}} MAD_t, \qquad (6.25)$$

where the parameter $c = 1/2$ if we assume that forecast errors in different time periods are independent. This can be regarded as the standard assumption. The parameter c is always in the interval (0.5, 1).

Assume that a continuous review (R, Q) policy is used for inventory control. Since the items are treated independently we shall consider a certain item with lead-time L. The lead-times are in general different for different items. Our first step is to update the batch quantity Q. The most common technique is to use the classical economic order quantity model and let the average demand per time unit μ replace the constant demand per unit of time. As in (4.3) we obtain

$$Q = \sqrt{\frac{2A\mu}{h}}. \qquad (6.26)$$

The ordering cost, A, is in general, the same for items belonging to the same inventory control group. The holding cost, h, is usually determined as a certain percentage of the value of the item. This carrying charge should include capital costs as well as other types of holding costs. Usually the carrying charge is the same for all items in the same inventory control group, but the holding costs vary among the items because of different values of the items. A typical carrying charge could be something like 10 - 15 percent if we use one year as the time unit. The carrying charge is normally higher than the interest rate charged by the bank. See Sections 3.1.1 and 3.1.2.

Although (6.26) is intended for stationary demand, it is often also applied when using exponential smoothing with trend. In that case the average demand rate μ should correspond to the time when the batch is used. Consider a time interval of length τ starting at the time of the update. Let $D(\tau)$ be the stochastic demand during this interval and $g(\tau)$ the expected value of this demand. We have

$$g(\tau) = E\{D(\tau)\} = \int_0^\tau \frac{1}{t_F}(\hat{a}_t + \frac{\hat{b}_t}{2}) + \frac{\hat{b}_t}{t_F^2} u) du = (\hat{a}_t + \frac{\hat{b}_t}{2})\frac{\tau}{t_F} + \frac{\hat{b}_t}{2}\frac{\tau^2}{t_F^2}. \qquad (6.27)$$

SINGLE-ECHELON – INTEGRATION

By setting $g(\tau)$ equal to a certain quantity d and solving for τ, we can estimate the time $\tau(d)$ until a certain quantity d has been demanded. We obtain $\tau(d)$ as the solution of a second order equation

$$\tau(d) = -t_F \left(\frac{\hat{a}_t}{\hat{b}_t} + \frac{1}{2} \right) + t_F \sqrt{\left(\frac{\hat{a}_t}{\hat{b}_t} + \frac{1}{2} \right)^2 + \frac{2d}{\hat{b}_t}}. \qquad (6.28)$$

We can use (6.27) and (6.28) for estimating the demand rate ahead of time in connection with determination of reorder point and batch quantity.

Consider, for example, an order just after the forecast update. Assume that the inventory position is equal to the reorder point R. What is then the average demand to be used in (6.26)? Let Q' be an estimate of the batch quantity, e.g., the previous batch size. We will start to consume the batch around time $\tau(R)$, and the whole batch will be consumed around time $\tau(R + Q')$. About half of the batch has been consumed at time $\tau' = \tau(R + Q'/2)$. A reasonable estimate of the average demand rate during the time when the batch is consumed is then

$$\mu = \frac{1}{t_F}(\hat{a}_t + \frac{\hat{b}_t}{2}) + \frac{\hat{b}_t}{t_F^2}\tau'), \qquad (6.29)$$

and we can use this μ instead of (6.24) in (6.26). Recall also from Section 4.1.2 that the costs are very insensitive to small errors in the batch quantity, so it may also be reasonable to use simpler approximations.

It is a little more complicated to take seasonal demand variations into account when determining batch quantities. In practice it is therefore quite common to disregard the effect of the seasonal variations on the batch quantities.

To be able to determine the reorder point, we next need to determine the distribution of the lead-time demand. Let μ' and σ' be the mean and average of the lead-time demand just after the forecast update. In case of exponential smoothing we have

$$\mu' = \frac{\hat{a}_t}{t_F} L, \qquad (6.30)$$

and in case of exponential smoothing with trend

144   INVENTORY CONTROL

$$\mu' = g(L). \tag{6.31}$$

The standard deviation is obtained as

$$\sigma' = \sigma L^c = \sqrt{\frac{\pi}{2}} MAD_t \left(\frac{L}{t_F}\right)^c. \tag{6.32}$$

It is not common to take stochastic variations in the lead-time into account. One reason is that it is usually difficult to determine the lead-time distribution. If the lead-time variations are known and the deliveries are sequential it is easy to use the approximation based on (5.102) and (5.103).

Given the mean and standard deviation, the most common approach is to assume that the lead-time demand is normally distributed. For items with low demand it may sometimes be more appropriate to use a Poisson distribution or a compound Poisson distribution. If we assume that the normal distribution is used and that there is a given fill rate S_2 (See Section 5.4), it is easy to determine the reorder point R from (5.52). In case of compound Poisson demand we can apply (5.36) and (5.51).

Note that in this section we have used models for stationary demand also when there is a trend in demand. This approximation is usually satisfactory as long as the trend is relatively small compared to the average.

Example 6.3 We shall update forecast, batch quantity, and reorder point for an item controlled by a continuous review (R, Q) policy. The updates take place at the end of each month, and we use one month as our time unit, i.e., $t_F = 1$. We apply exponential smoothing with smoothing constant $\alpha = 0.1$ when updating both the forecast and MAD. At the end of the preceding month we obtained the forecast $\hat{a} = 132$ and $MAD = 42$. We have just received the demand during the last month as 92. The batch quantity is determined according to the classical economic order quantity model. The holding cost is $1.5 per unit and month, and the ordering cost $200. The lead-time is two months. The fill rate is required to be 95 percent. The demand can be regarded as continuous and normally distributed, and forecast errors during different time periods are assumed to be independent.

As our first step we update the forecast and MAD

$$\hat{a} = 0.9 \cdot 132 + 0.1 \cdot 92 = 128,$$

$$MAD = 0.9 \cdot 42 + 0.1 \cdot |132 - 92| = 41.8.$$

Next we determine the batch quantity

$$Q = \sqrt{\frac{2 \cdot 200 \cdot 128}{1.5}} = 184.75 \approx 185.$$

It remains to determine the reorder point R from (5.52). First we obtain μ' and σ' as

$$\mu' = 2 \cdot 128 = 256,$$

$$\sigma' = \sqrt{2} \cdot \sqrt{\frac{\pi}{2}} \cdot 41.8 = 74.09.$$

Finally, using the search procedure described in connection with (5.52) we obtain $R = 313.62 \approx 314$.

We have assumed a continuous review (R, Q) policy. Periodic review can be handled as described in Section 5.12. When using an (s, S) policy instead of an (R, Q) policy it is common in practice to first determine R and Q for an (R, Q) policy and then apply the simple approximation $s = R$ and $S - s = Q$.

References

Axsäter, S. 2004. A Simple Procedure for Determining Order Quantities under a Fill Rate Constraint and Normally Distributed Lead-Time Demand, *European Journal of Operational Research* (to appear).

Beyer, D. and S. P. Sethi. 1999. The Classical Average-Cost Inventory Models of Iglehart (1963) and Veinott and Wagner (1965) Revisited, *Journal of Optimization Theory and Applications*, 101, 523-555.

Chen, F. 2000. Optimal Policies for Multi-Echelon Inventory Problems with Batch Ordering, *Operations Research*, 48, 376-379.

Federgruen. A., and Y. S. Zheng. 1992. An Efficient Algorithm for Computing an Optimal (R, Q) Policy in Continuous Review Stochastic Inventory Systems, *Operations Research*, 40, 808-813.

Hadley, G., and T. M. Whitin. 1963. *Analysis of Inventory Systems*, Prentice-Hall, Englewood Cliffs, NJ.

Iglehart, D. 1963. Optimality of (s, S) Policies in the Infinite-Horizon Dynamic Inventory Problem, *Management Science*, 9, 259-267.

Porteus, E. L. 1990. Stochastic Inventory Theory, in D.P. Heyman, and M. J. Sobel. Eds., *Handbooks in OR & MS Vol.2*, North Holland Amsterdam, 605-652.

Porteus, E. L. 2002. *Stochastic Inventory Theory*, Stanford University Press, Stanford.
Rosling, K. 2002a. The (r, Q) Inventory Model with Lost Sales, Växjö University.
Rosling, K. 2002b. The Square-Root Algorithm for Single-Item Inventory Optimization, Växjö University.
Silver, E. A., D. F. Pyke, and R. Peterson. 1998. *Inventory Management and Production Planning and Scheduling*, 3rd edition, Wiley, New York.
Veinott, A. 1966. On the Optimality of (s, S) Inventory Policies: New Conditions and a New Proof, *SIAM Journal on Applied Mathematics*, 14, 1067-1083.
Zheng, Y. S. 1991. A Simple Proof for Optimality of (s, S) Policies in Infinite-Horizon Inventory Systems, *Journal of Applied Probability*, 28, 802-810.
Zheng, Y. S. 1992. On Properties of Stochastic Inventory Systems, *Management Science*, 38, 87-103.
Zheng, Y. S., and A. Federgruen. 1991. Finding Optimal (s, S) Policies Is About as Simple as Evaluating a Single Policy, *Operations Research*, 39, 654-665.
Zipkin, P. H. 2000. *Foundations of Inventory Management*, McGraw-Hill, Singapore.

Problems

6.1* Consider Example 6.1 and the iterations in Table 6.1. What is the fill rate in the different iteration steps? Why?

6.2 Verify the transformation in Section 6.1.3.

6.3* Consider an item which is controlled by a continuous review (R, Q) policy. The forecast and *MAD* have just been updated by exponential smoothing as $\hat{a} = 100$ and MAD = 40. The forecast period is one month. The lead-time is two months. When adjusting the standard deviation to a different time, the constant c is set to 0.7. The lead-time demand is normally distributed.

a) Determine the reorder point for $S_1 = 90$ percent,
b) For this reorder point determine S_2 for $Q = 25$, $Q = 100$, and $Q = 1200$.

6.4* Simple exponential smoothing is used for updating the forecast each week. The smoothing constant is 0.2. *MAD* is also updated by exponential smoothing with smoothing constant 0.3. The demand during the past five weeks is given in the table. Before week 1 the forecast was 100 and *MAD* was 10.

Week	1	2	3	4	5
Demand	112	96	84	106	110

---

* Answer and/or hint in Appendix 1.

a) Update forecast and *MAD* for weeks 1-5. Determine the expected demand and variance for week 6.
b) Determine (after the update in period 5) batch quantity by the classical economic order quantity model and reorder point under the following assumptions:
Ordering cost: 100
Holding cost: 1 per unit and week
Lead-time: 2 weeks
$S_1 \geq 95\%$
Forecast errors in different periods are assumed to be independent.

6.5* The demand during the past five weeks is given.

Week	16	17	18	19	20
Demand	97	99	100	126	112

Forecasts are determined by both simple exponential smoothing and by exponential smoothing with trend. For simple exponential smoothing, the smoothing constant is 0.2. The same smoothing constant is used when updating the mean with the trend model. The smoothing constant for the trend is 0.4. When updating *MAD* the smoothing constant is 0.2. The forecast errors are assumed to be independent and normally distributed. Before the first week (16) the forecasted demand was 100.0 and the trend was assumed to be zero. *MAD* was 7.

a) Update the forecasts by both methods. Determine mean and standard deviation for week 21 after the update in week 20. Assume stationary stochastic demand. Use the forecast from simple exponential smoothing. Determine batch quantity by the classical economic lot size formula. Determine reorder point such that the fill rate is approximately 95%. Make the following assumptions:
Ordering cost: 2500
Holding cost: 10 per unit and week
Lead-time: 3 weeks
Continuous review
b) Determine S_1 for the chosen reorder point.

6.6 Consider a continuous review (R, Q) policy. The batch quantity $Q = 500$. The lead-time is two weeks. Both S_1 and S_2 must be at least 95%. Before week 1 the forecast was $\hat{x}_{0,1} = 100$ and $MAD_0 = 8$. Use simple exponential smoothing with $\alpha = 0.2$ (for both $\hat{a}$ and *MAD*) to update the forecast for weeks 1 to 5. The demands are:

week	1	2	3	4	5
d	113	101	108	105	95

Use the forecast from week 5 when determining the reorder point. The demand is normally distributed and deviations in different periods are independent.

7 COORDINATED ORDERING

In Chapters 3-6 it was assumed that different items in an inventory could be controlled independently. We shall now leave this assumption and consider situations where there is a need to coordinate orders for different items. In this chapter we shall still, as in Chapters 3-6, assume that the items are stocked at a single location. (Multi-stage inventory systems are dealt with in Chapters 8-10.) We consider traditional inventory costs and constraints, i.e., holding costs, ordering or setup costs, and backorder costs or service constraints.

When coordinating the replenishments for different items, it is common to use cyclic schedules, and especially so-called powers-of-two policies. In Section 7.1 we derive some important results for such policies.

There are two main reasons for coordinating the replenishments of a group of items. One reason, dealt with in Section 7.2, is that we wish to get a sufficiently smooth production load. Assume, for example, that a considered group of items is produced in the same production line. We then want to coordinate the orders for different items so that they are evenly spread over time.

The other main reason for coordinated replenishments, which is treated in Section 7.3, is completely opposite. We want to trigger orders for a group of items at the same time. This can be advantageous in many situations. It may be possible to get a discount if the total order from the same vendor is greater than a certain breakpoint. It may also be possible to reduce the transportation costs, for example, by filling a truckload. Sometimes the setup costs can also be lowered substantially if a group of similar items are produced together in a machine.

7.1 Powers-of-two policies

Both when using mathematical algorithms and when choosing schedules manually, it is very common to use so-called powers-of-two policies in connection with coordinated replenishments. This means that the cycle times are restricted to be powers of two times a certain basic period. If the basic period is, for example, one week, nonnegative powers of two give $2^0 = 1$ week, $2^1 = 2$ weeks, $2^2 = 4$ weeks, $2^3 = 8$ weeks etc. With negative powers of two we also obtain $2^{-1} = 1/2$ week, $2^{-2} = 1/4$ week, etc. A main advantage of such cycle times is that we obtain relatively simple cyclic schedules. Consider, for example, two items that are produced every fourth and every eighth week, respectively. The total cycle time is then eight weeks, since everything is repeated every eighth week. If instead of four and eight weeks we use the similar cycle times five and seven weeks, the total cycle time would be $5 \cdot 7 = 35$ weeks, i.e., more than four times longer. Assume, for example, that both items are produced in week 1. The item with cycle time five weeks is then produced in weeks 1, 6, 11, 16, 21, 26, 31, 36, etc., and the other item, with cycle time seven weeks, in weeks 1, 8, 15, 22, 29, 36, etc. So week 36 is the first time after week 1 when both items are produced.

Consider a number of items with constant continuous demand. Given holding costs and ordering costs we wish to determine suitable batch quantities, or equivalently cycle times. (See Section 4.1.1.) We shall show that a restriction to powers-of-two policies will give a solution which is very close to the optimal solution.

Consider first a single item. Recall the following result from Section 4.1.2.

$$\frac{C}{C^*} = \frac{1}{2}\left(\frac{Q}{Q^*} + \frac{Q^*}{Q}\right), \qquad (7.1)$$

which gives the relative cost increase when deviating from the optimal batch quantity Q^* in the classical economic order quantity model. The expression (7.1) is valid also with a finite production rate (Section 4.2). Furthermore, since we are dealing with constant demand, d, we can just as well express the policy through the cycle time $T = Q/d$ where $T^* = Q^*/d$ is the optimal solution. This means that we can equivalently formulate (7.1) as

$$\frac{C}{C^*} = \frac{1}{2}\left(\frac{T}{T^*} + \frac{T^*}{T}\right). \qquad (7.2)$$

We shall consider cycle times and the representation (7.2) when deriving our results on the approximation errors, but it is important to note that the results are also valid for the batch quantities.

Consider now a powers-of-two solution of a lot sizing problem, i.e., assume that the cycle time T has to be chosen as

$$T = 2^m q, \qquad (7.3)$$

where m can be any integer and where, for the time being, we assume that the basic period q is given. Assume that T^* cannot be expressed according to (7.3). We then have to choose either the next lower or the next higher T satisfying (7.3). Due to (7.3), the ratio between these two values is 2. Note also that the best solution under the constraint (7.3) is not affected if q is multiplied by a power of two. If, for example, q is multiplied by 2, we can reduce m by 1 to get the same result.

What is the worst possible relative cost increase caused by restricting the solution with the constraint (7.3)? Because of the convexity, the worst possible error must occur when two consecutive values of m, say $m = k$ and $m = k + 1$ give the same error. Let $T < T^*$ correspond to $m = k$, and $2T > T^*$ to $m = k + 1$. We obtain

$$\frac{C}{C^*} = \frac{1}{2}\left(\frac{T}{T^*} + \frac{T^*}{T}\right) = \frac{1}{2}\left(\frac{2T}{T^*} + \frac{T^*}{2T}\right). \qquad (7.4)$$

It is easy to see that (7.4) implies that

$$\frac{T^*}{T} = \frac{2T}{T^*} = \sqrt{2}, \qquad (7.5)$$

and

$$\frac{C}{C^*} = \frac{1}{2}\left(\frac{1}{\sqrt{2}} + \sqrt{2}\right) \approx 1.06. \qquad (7.6)$$

We formulate this result as a proposition.

Proposition 7.1 For a given basic period q, the maximum relative cost increase of a powers-of-two policy is 6 percent.

We have only discussed a single item, but Proposition 7.1 is evidently also true if there are several items, because the worst case occurs when all items incur the maximum error of 6 percent.

Let us now assume that it is possible to change q. For a single item we will then get the optimal solution simply by choosing q equal to a power of two times T^*. If, however, we have N items (items $i = 1, 2, \ldots, N$), we can, in general, not fit q perfectly to all cycle times T_i^*, which depend on the problem data for different items. The relative cost increase can be expressed as

$$\frac{C}{C^*} = \frac{\sum_{i=1}^{N} C_i}{\sum_{i=1}^{N} C_i^*} = \frac{\sum_{i=1}^{N} C_i^*(C_i/C_i^*)}{\sum_{i=1}^{N} C_i^*}. \tag{7.7}$$

We know from (7.2) and (7.5) that for a given q, each C_i/C_i^* can be expressed as

$$\frac{C_i}{C_i^*} = e(x_i) = \frac{1}{2}(2^{x_i} + 2^{-x_i}), \quad -1/2 \leq x_i \leq 1/2, \tag{7.8}$$

i.e., $T_i/T_i^* = 2^{x_i}$ $(-1/2 \leq x_i \leq 1/2)$. The end points $x_i = -1/2$ and $x_i = 1/2$ correspond to the worst case (7.6). Let us now interpret the weights for the different values of x_i in (7.7), $C_i^*/\sum_{i=1}^{N} C_i^*$, as probabilities. Denote the corresponding distribution function on $[-1/2, 1/2]$ by $F(x)$, i.e., $F(-1/2) = 0$ and $F(1/2) = 1$. We can see C/C^* in (7.7) as an expected value of $e(x)$ and reformulate (7.7) as

$$\frac{C}{C^*} = \int_{-1/2}^{1/2} e(x) dF(x). \tag{7.9}$$

Assume now that we change q by multiplying by 2^y, where $0 \leq y \leq 1$. (Recall that multiplying q by a power of two does not affect the solution, so we are considering the most general change.) If we do not change x this is equivalent to replacing x by $x + y$. However, if $x + y > 1/2$ it is advantageous to change x to $x - 1$. This means that a certain x is replaced by $x + y$ for $x + y \leq 1/2$, and by $x + y - 1$ for $x + y > 1/2$. Consequently we have

$$\frac{C}{C^*}(y) = \int_{-1/2}^{1/2-y} e(x+y)dF(x) + \int_{1/2-y}^{1/2} e(x+y-1)dF(x)$$

(7.10)

$$= \int_{y-1/2}^{1/2} e(u)dF(u-y) + \int_{-1/2}^{y-1/2} e(u)dF(u-y+1).$$

For a given distribution $F(x)$ the minimum cost increase is obtained by minimizing (7.10) with respect to $0 \le y \le 1$.

We are now ready to prove Proposition 7.2, which shows that the maximum relative error is surprisingly low.

Proposition 7.2 *If we can change the basic period q, the maximum relative cost increase of a powers-of-two policy is 2 percent.*

Proof The average cost increase for $0 \le y \le 1$ must be at least as large as the minimum and we have from (7.10), by changing the order of integration,

$$\min_{0 \le y \le 1} \frac{C}{C^*}(y) \le \int_0^1 \left(\int_{y-1/2}^{1/2} e(u)dF(u-y) + \int_{-1/2}^{y-1/2} e(u)dF(u-y+1) \right) dy$$

$$= \int_{-1/2}^{1/2} e(u) \left(\int_0^{u+1/2} dF(u-y) + \int_{u+1/2}^{1} dF(u-y+1) \right) du$$

$$= \int_{-1/2}^{1/2} e(u)\left(-F(-1/2) + F(u) - F(u) + F(1/2)\right) du = \int_{-1/2}^{1/2} e(u) du = \frac{1}{\sqrt{2} \ln 2} \approx 1.02.$$

The worst case will occur when the distribution $F(x)$ is uniform on $-1/2 \le x \le 1/2$, see (7.9). A change of q will then not make any difference. This completes the proof.

Powers-of-two policies are important ingredients of Roundy's so-called 98 percent approximation. See Sections 7.3.1.2 and 9.2.2.

7.2 Production smoothing

In general, it is a very complicated problem to control the stocks of different items in such a way that we get both low inventory costs and smooth capacity utilization. To simplify the problem it is common to first disregard stochastic demand variations and solve the remaining deterministic problem. Safety stocks are then usually determined in a second step using techniques for independent items, see Chapter 5. However, the deterministic problem can also be very difficult, especially if there are many items with time-varying demand and several capacity constraints.

When the demands for different items are relatively stable over an extended period, it is often advantageous to use cyclic schedules as a means to obtain a smooth production load. This means that each item is ordered periodically, and that the ordering periods for different items are chosen such that the load becomes as smooth as possible.

Consider, for example, a machine which is used for producing four items (items 1 - 4) all with about the same demand. It can then be a good idea to decide that item 1 is produced in weeks 1, 5, 9 ... , item 2 in weeks 2, 6, 10 ... , item 3 in weeks 3, 7, 11 ... , and item 4 in weeks 4, 8, 12 ... . This can be organized by applying periodic review order-up-to-S policies. Each item has a review period equal to four weeks, but the reviews take place in different weeks. Since the machine is producing exactly one of the items each week, the load will be very smooth. As we have discussed in Chapter 3, we need, for a given lead-time, more safety stock when applying a periodic review policy as compared to a policy with continuous review. But the alternative of using a continuous review policy would in this case probably yield large variations in the production load, which would in turn, result in long and uncertain lead-times. Longer lead-times would also mean that we need more safety stock. Furthermore, if we applied a continuous review policy, the capital tied up as work-in-process would increase and lead to additional holding costs. In a situation like this, the total costs would generally be much lower when using the cyclic schedules obtained through the periodic review policy.

In a general case with many items and several production facilities, it can be extremely difficult to find suitable cyclic schedules. In Section 7.2.1 we will deal with different mathematical approaches for solving the problem in the special case when there is only a single machine. In practice it is most common to choose cyclic schedules manually without using mathematical techniques.

When the demand is time-varying, one possible approach is to formulate the problem as a mixed integer program (MIP) and apply mathematical programming techniques to obtain a solution. A typical formulation involves a number of items with given demands over a planning horizon, and holding and setup costs as in the dynamic lot size problem (Section 4.5). To produce an item, we need to use certain production resources with limited capacities. We shall consider such a model in Section 7.2.2.

Situations where it is, for various practical reasons, very difficult to coordinate the replenishments for different items are quite common. It can then still be possible to get a reasonably smooth flow of orders simply by adjusting the order quantities. We discuss such adjustments in Section 7.2.3.

It is also common in practice to smooth production outside the inventory control system. Orders obtained from the inventory control system are then not automatically released to production. Release times and order quantities are instead adjusted with respect to the present production load. One possible planning rule can, for example, be to not release more orders than there is capacity to produce within some fixed time frame. Adjustments of the release times mean essentially that the safety stocks are allowed to vary over time in order to smooth the load. We know from the models in Chapters 4 and 5 that small changes in batch quantities and safety stocks will, in general, not affect the costs significantly, while larger changes can have a substantial impact. When adjusting the orders obtained from the inventory control system, it is therefore important to avoid large changes for individual items. In this context it is also interesting to note that ordering systems of KANBAN type will automatically limit the number of outstanding orders. See Sections 3.2.3.1 and 8.2.1.

7.2.1 The Economic Lot Scheduling Problem (ELSP)

7.2.1.1 Problem formulation

We shall now consider a problem that has been dealt with extensively in the inventory literature, the classical Economic Lot Scheduling Problem. This problem concerns the determination of cyclic schedules for a number of items with constant demands. Backorders are not allowed. The production rate is finite and we wish to minimize standard holding and ordering costs. Everything is as in the elementary model in Section 4.2 with one exception; the items are all produced in a single production facility, i.e., we have a capacity constraint. It turns out that this additional constraint transforms a simple model into a very challenging problem.

We assume that batch quantities and, equivalently, cycle times are kept constant over time. Let us introduce the following notation:

N = number of items,
h_i = holding cost per unit and time unit for item i,

A_i = ordering or setup cost for item i,
d_i = demand per time unit,
p_i = production rate ($p_i > d_i$),
s_i = setup time in the production facility for item i, independent of the sequence of the items,
T_i = cycle time for item i (the batch quantity $Q_i = T_i d_i$).

We shall also for simplicity define:

$\rho_i = d_i/p_i$,
$\tau_i = \rho_i T_i =$ production time per batch for item i excluding setup time,
$\sigma_i = s_i + \tau_i =$ total production time per batch for item i.

Table 7.1 shows data for $N = 10$ items. This sample problem was first presented by Bomberger (1966) and has since then been used extensively in the literature on economic lot scheduling.

Table 7.1 Bomberger's problem (time unit = one day).

Item	1	2	3	4	5	6	7	8	9	10
$h_i \cdot 10^5$	0.2708	7.396	5.313	4.167	116.0	11.15	62.50	245.8	37.50	1.667
A_i	15	20	30	10	110	50	310	130	200	5
d_i	400	400	800	1600	80	80	24	340	340	400
p_i	30000	8000	9500	7500	2000	6000	2400	1300	2000	15000
s_i	0.125	0.125	0.25	0.125	0.50	0.25	1	0.5	0.75	0.125

7.2.1.2 The independent solution

Expressing the costs for item i, C_i, as a function of the cycle time, T_i, we have

$$C_i = \frac{A_i}{T_i} + h_i d_i (1 - \rho_i) \frac{T_i}{2}. \qquad (7.11)$$

The problem is to minimize $\sum_{i=1}^{N} C_i$ subject to the constraint that all items should be produced in the common production facility.

A first approach to solving the problem could be to disregard the capacity constraint and simply optimize each item separately as in Section 4.2. The optimal cycle time when disregarding the capacity constraint is

$$T_i = \sqrt{\frac{2A_i}{h_i d_i (1-\rho_i)}}, \qquad (7.12)$$

and the corresponding cost

$$C_i = \sqrt{2A_i h_i d_i (1-\rho_i)}. \qquad (7.13)$$

Note that (7.11) and (7.12) are equivalent to (4.6) and (4.7) if we replace T_i by Q_i/d_i.

Table 7.2 shows this "independent" solution of Bomberger's problem. The last row gives the corresponding production times per lot for each of the 10 items.

Table 7.2 Independent solution of Bomberger's problem.

Item	1	2	3	4	5	6	7	8	9	10
T_i	167.5	37.7	39.3	19.5	49.7	106.6	204.3	20.5	61.5	39.3
C_i	0.179	1.060	1.528	1.024	4.428	0.938	3.034	12.668	6.506	0.255
σ_i	2.36	2.01	3.56	4.29	2.49	1.67	3.04	5.87	11.20	1.17

The sum of the costs, $\underline{C} = \sum_{i=1}^{10} C_i = 31.62$, is evidently a lower bound for the total costs, since we have disregarded the capacity constraint.

Is the solution feasible? No, it is relatively easy to see that the solution in Table 7.2 cannot be implemented. Consider, for example, items 4, 8, and 9. Assume that the production of item 9 starts at some time t. The production of the following batch will then start at time $t + 61.5$ etc. Since $\sigma_9 = 11.20$, the production facility is occupied by item 9 during the interval $(t, t + 11.20)$. Items 4 and 8 have cycle times 19.5 and 20.5. This means that we must be able to produce one batch of item 4 and one batch of item 8 in the interval $(t, t + 20.5)$. Given the production schedule for item 9, this implies that both items must be produced in $(t + 11.20, t + 20.5)$. The length of this interval is only 9.30, while $\sigma_4 + \sigma_8 = 10.16$. The independent solution can consequently not be implemented.

7.2.1.3 Common cycle time

Does a feasible solution exist? Note first that if at least one setup time is positive, an obvious *necessary* condition for a feasible solution to exist is that

$$\sum_{i=1}^{N} \rho_i < 1. \tag{7.14}$$

The left-hand side of (7.14) is the ratio of the time the production facility must be busy (excluding setup time). Since we also need some time for the setups, we can see that (7.14) is necessary. (If all setup times are zero, (7.14) should be replaced by $\sum_{i=1}^{N} \rho_i \leq 1$.) It is easy to verify that (7.14) is satisfied for Bomberger's problem.

It turns out though, that the condition (7.14) is also *sufficient* for feasibility. If it is satisfied, it is always possible to find a feasible solution where all items have a *common cycle time*. Denote the common cycle time by T. During the cycle time we produce all items, each time in the same order. The production quantity for an item is the demand during the cycle time. By choosing T sufficiently large, we can reduce the ratio of the time needed for setups as much as we need. This explains why (7.14) is sufficient.

Given the assumption of a common cycle time, the problem now is to minimize

$$C = \sum_{i=1}^{N} \left(\frac{A_i}{T} + h_i d_i (1 - \rho_i) \frac{T}{2} \right), \tag{7.15}$$

with respect to the constraint that the common cycle time must be able to accommodate production lots of all items

$$\sum_{i=1}^{N} \sigma_i = \sum_{i=1}^{N} (s_i + \rho_i T) \leq T. \tag{7.16}$$

Note that the constraint (7.16) can also be expressed as

$$T \geq \frac{\sum_{i=1}^{N} s_i}{1 - \sum_{i=1}^{N} \rho_i} = T_{min}, \tag{7.17}$$

i.e., it simply means a lower bound for the cycle time.

If we disregard (7.17), the optimization of (7.15) is similar to (7.12) and we obtain

$$\hat{T} = \sqrt{\frac{2\sum_{i=1}^{N} A_i}{\sum_{i=1}^{N} h_i d_i (1-\rho_i)}}. \qquad (7.18)$$

Since (7.15) is convex in T, the optimal solution, T_{opt}, is obtained as

$$T_{opt} = \max(\hat{T}, T_{min}), \qquad (7.19)$$

i.e., if $\hat{T} < T_{min}$ the best we can do is to choose $T_{opt} = T_{min}$.

For Bomberger's problem we obtain $\hat{T} = 42.75$ and $T_{min} = 31.86$, and consequently, $T_{opt} = \hat{T} = 42.75$. This gives the costs $\overline{C} = 41.17$, which is an upper bound for the optimal solution since we have enforced the additional constraint that all cycle times have to be equal.

At this stage we know that the optimal costs, C^*, are in the interval $\underline{C} = 31.62 \leq C^* \leq \overline{C} = 41.17$. For Bomberger's problem it is rather evident that the upper bound is not especially tight. If we look at the independent solution in Table 7.2, we can see that some of the cycle times are very different, and we can therefore not expect a good solution with a common cycle time. For problems where the individual cycle times are reasonably similar, we can, on the other hand, expect the common cycle approach to give a very good approximation.

We shall now consider two approaches for deriving better solutions.

7.2.1.4 Bomberger's approach

Bomberger (1966) generated feasible upper bound solutions by a relatively simple dynamic programming model. First it is assumed that each cycle time T_i is an integer multiple of a basic period W, i.e., $T_i = n_i W$ for some positive integer n_i. Bomberger also makes the very restrictive assumption that W should be able to accommodate production of all items. To see that this condition is not necessary, consider two items (item 1 and item 2), which both have $T_1 = T_2 = 2W$. We can then produce item 1 in periods 1, 3, 5 ... , and item 2 in periods 2, 4, 6 ... , and we will never have to produce both items in the same basic period. The condition will, however, obviously guarantee that the obtained solution is feasible. Let

160   INVENTORY CONTROL

$F_i(w)$ = minimum cost of producing items $i + 1, i + 2, \ldots, N$ when the available capacity in the basic period is w, i.e., $W - w$ has been used for items $1, 2, \ldots, i$.

We now have

$$F_{i-1}(w) = \min_{n_i}\{C_i(n_iW) + F_i(w - \sigma_i)\}, \qquad (7.20)$$

where $C_i(n_iW)$ are the costs (7.11) for item i with $T_i = n_iW$, $\sigma_i = s_i + \rho_i n_i W$, and the integer n_i is subject to the constraint

$$1 \leq n_i \leq (w - s_i)/\rho_i W. \qquad (7.21)$$

Note that the upper bound in (7.21) is equivalent to $\sigma_i \leq w$.

It is obvious that $F_N(w) = 0$ for all $w \geq 0$. In the first step we determine $F_{N-1}(w)$ for a suitable grid of values of $w \geq 0$ from (7.20). Given $F_{N-1}(w)$, we can next determine $F_{N-2}(w)$, etc. $F_0(W)$ gives the minimum costs when the basic period is equal to W. In the final step we need to optimize $F_0(W)$ with respect to W. Bomberger's solution of his example gave the costs $C = 36.65$ for $W = 40$, $n_i = 1$ for $i \neq 7$, and $n_7 = 3$. This solution is a considerable improvement as compared to the common cycle approach.

Bomberger's approach is simple and will always provide a feasible solution (if a feasible solution exists). Still, the assumption that W should be able to accommodate production of all items is very restrictive. Later contributions (Elmaghraby, 1978, and Axsäter, 1984, 1987), have improved the approach and provide considerably better solutions.

7.2.1.5 A simple heuristic

We shall now consider a completely different heuristic technique (essentially according to Doll and Whybark, 1973). The procedure means that we successively improve the multipliers n_i and the basic period W according to the following iterative procedure:

1. Determine the independent solution and use the shortest cycle time as the initial basic period W.

2. Given W, choose powers-of-two multipliers, ($n_i = 2^m$, $m \geq 0$), to minimize the item costs (7.11).

3. Given the multipliers n_i, minimize the total costs

$$C = \sum_{i=1}^{N} \left(\frac{A_i / n_i}{W} + h_i d_i (1 - \rho_i) n_i \frac{W}{2} \right),$$

with respect to W. We obtain

$$W = \sqrt{\frac{2 \sum_{i=1}^{N} A_i / n_i}{\sum_{i=1}^{N} h_i d_i (1 - \rho_i) n_i}}.$$

4. Go back to Step 2 unless the procedure has converged. In that case, check whether the obtained solution is feasible. If the solution is infeasible, try to adjust the multipliers and then go back to Step 3.

The main disadvantage of the considered heuristic is that there is no guarantee for even a feasible solution. On the other hand, the computations are very easy.

We shall apply the heuristic to Bomberger's problem. In Step 1 we start with the shortest cycle time in Table 7.2 as our basic period $W = 19.5$. In Step 2 we obtain the powers-of-two multipliers $n_1 = 8$, $n_2 = 2$, $n_3 = 2$, $n_4 = 1$, $n_5 = 2$, $n_6 = 4$, $n_7 = 8$, $n_8 = 1$, $n_9 = 4$, and $n_{10} = 2$. In Step 3 we get $W = 20.30$. At this stage the algorithm has converged and we have to check whether the solution is feasible. We can see that this is not the case by using an argument which is very similar to the argument that we used to show that the independent solution is infeasible. We consider again items 4, 8, and 9, and determine $\sigma_4 = 4.45$, $\sigma_8 = 5.81$, and $\sigma_9 = 14.55$. Consider a basic period when item 9 is produced. Since items 4 and 8 are produced in all basic periods, the production of all three items must take place in the considered basic period, i.e., during the time 20.30. This is obviously impossible. A major problem is the long production time for item 9. To reduce σ_9, we change n_9 from 4 to 2 and start the iterations in Step 3. We get $W = 23.42$ and no more changes of the multipliers, see Table 7.3.

It turns out that this solution is feasible, see Table 7.4. The total production time is below $W = 23.42$ in each basic period. The total costs are $C = 32.07$, i.e., very close to the lower bound. This is also the best known solution of the problem.

Table 7.3 Solution of Bomberger's problem with $W = 23.42$.

Item	1	2	3	4	5	6	7	8	9	10
n_i	8	2	2	1	2	4	8	1	2	2
σ_i	2.62	2.47	4.19	5.12	2.37	1.50	2.87	6.63	8.71	1.37

Table 7.4 Feasible production plan.

Basic period	Items	Production time
1	4, 8, 2, 9	22.93
2	4, 8, 3, 5, 10, 1	22.30
3	4, 8, 2, 9	22.93
4	4, 8, 3, 5, 10, 6	21.18
5	4, 8, 2, 9	22.93
6	4, 8, 3, 5, 10, 7	22.55
7	4, 8, 2, 9	22.93
8	4, 8, 3, 5, 10, 6	21.18

The solution in Table 7.4 is repeated every 8th period. Note that it is easier to check feasibility when using powers-of-two policies since the total cycle time is usually relatively short. There are still, however, a large number of plans that correspond to the solution in Table 7.3. If $n_i = 4$, for example, we can produce item i in periods 1 and 5, or in periods 2 and 6, or in periods 3 and 7, or in periods 4 and 8, i.e., there are 4 possibilities. More generally the number of possibilities for item i is equal to n_i. We can, however, without any lack of generality, always allocate one of the items with maximum n_i, say item 1, in some arbitrary period. If, consequently, we disregard the allocation of item 1, the number of remaining possibilities can be obtained as the product of the multipliers of the remaining items 2 - 10, i.e., $2 \cdot 2 \cdot 1 \cdot 2 \cdot 4 \cdot 8 \cdot 1 \cdot 2 \cdot 2 = 1024$.

Although the classical economic lot scheduling problem involves only a single production facility, it can also be of interest in more general situations. For example, if there are several production facilities with limited capacities, it is quite common that one of the constraints constitutes the real bottleneck. It is then a reasonable approach to first derive a plan that only takes this constraint into account, and then in a second step, try to adapt the plan to other capacity limitations.

A more detailed overview of different approaches to solving the problem is provided in Elmaghraby (1978). Feasibility issues are analyzed by Hsu (1983).

7.2.1.6 Other problem formulations

In Sections 7.2.1.1 - 7.2.1.5 we have considered the classical Economic Lot Scheduling Problem. Several papers have dealt with a variation of this problem. This more general formulation of the problem allows the lot sizes to vary over time. There is still a cycle time T, which is the overall period of the system. The schedule repeats itself every T units of time. Each item is produced during T but some items may be produced more than once. Furthermore, the batch sizes of these runs may be different. See, for example, Dobson (1987), Roundy (1989), and Zipkin (1991).

Gallego and Roundy (1992) allow backorders, while Gallego and Moon (1992) consider a situation where setup times can be exchanged by setup costs.

There are also quite a few papers dealing with stochastic demand. In case of stochastic demand it is necessary to allow backorders and/or capacity variations, for example by using overtime production. See Sox et al. (1999) for a review. They classify the existing research approaches in two categories: cyclic sequencing and dynamic sequencing. The cyclic sequencing category uses a fixed cyclic schedule on the production facility, while the lot sizes are varied to meet demand variations. The cyclic schedule can, for example, be obtained from a deterministic model. Dynamic sequencing means that both the production sequence and the lot sizes are varied. Examples of the cyclic sequencing approach are Gallego (1990), Bowman and Muckstadt (1993, 1995), and Federgruen and Katalan (1996a, b). Papers considering dynamic sequencing are e.g., Graves (1980), and Sox and Muckstadt (1997).

7.2.2 Time-varying demand

7.2.2.1 A generalization of the classical dynamic lot size problem

Recall the classical dynamic lot size problem in Section 4.5. In this section we shall consider a generalization of this problem. Instead of a single item there are N items. Furthermore, these items are produced in the same machine, which has limited capacity. (The generalization to several machines is relatively straightforward.) For simplicity, it is assumed (as in Section 4.5) that all events take place in the beginning of a period. A quantity that is produced in period t can also be delivered to customers in period t. The demands for different items are given but vary over time. No backorders are allowed. Let us introduce the following notation:

N	=	number of items,
T	=	number of periods,
$d_{i,t}$	=	demand for item i in period t,
a_i	=	setup time for item i,
b_i	=	operation time per unit for item i,
q_t	=	available time in the machine in period t,
$M_{i,t}$	=	upper bound for the production of item i in period t,
A_i	=	ordering or setup cost for item i,
h_i	=	holding cost per unit and time unit for item i,
$x_{i,t}$	=	production quantity of item i in period t,
$y_{i,t}$	=	inventory of item i after the demand in period t, $y_{i,0} = 0$,
$\delta_{i,t}$	=	$\begin{cases} 1 \text{ if } x_{i,t} > 0, \\ 0 \text{ otherwise,} \end{cases}$
C	=	total variable costs.

We wish to choose production quantities in different periods so that the sum of the ordering and holding costs C are minimized. The considered problem can be formulated as a Mixed Integer Program (MIP).

$$C = \min \sum_{i=1}^{N} \left(A_i \sum_{t=1}^{T} \delta_{i,t} + h_i \sum_{t=1}^{T} y_{i,t} \right), \quad (7.22)$$

subject to

$$\sum_{i=1}^{N} (a_i \delta_{i,t} + b_i x_{i,t}) \leq q_t, \quad t = 1,2,....,T, \quad (7.23)$$

$$y_{i,t} = y_{i,t-1} + x_{i,t} - d_{i,t}, \quad t = 1,2,....,T, i = 1,2,...,N, \quad (7.24)$$

$$x_{i,t} - M_{i,t} \delta_{i,t} \leq 0, \quad t = 1,2,....,T, i = 1,2,...,N, \quad (7.25)$$

$$x_{i,t} \geq 0, \quad y_{i,t} \geq 0, \quad \delta_{i,t} = 0 \text{ or } 1, \quad t = 1,2,....,T, i = 1,2,...,N. \quad (7.26)$$

We need the constraint (7.25) to enforce that production in a period implies that there is a corresponding setup time and setup cost. If there is no upper bound on $x_{i,t}$, we can let $M_{i,t}$ be very large. In the sequel it is assumed that this is the case.

Although the considered problem can be seen as a minor variation of the classical dynamic lot size problem, the model is quite complex. If, for example $N = 100$ and $T = 12$, there are 1200 integer variables and 2400 nonnegative continuous variables. The number of constraints (7.23) - (7.25) is 2412.

A possible approach is to eliminate the capacity constraints (7.23) by a Lagrangian relaxation. Let $\lambda_t \geq 0$ be the multiplier for period t. We get the Lagrangean:

$$L = -\sum_{t=1}^{T} \lambda_t q_t + \min \sum_{i=1}^{N}\sum_{t=1}^{T}\left((A_i + a_i\lambda_t)\delta_{i,t} + h_i y_{i,t} + b_i x_{i,t}\lambda_t\right). \quad (7.27)$$

It is easy to see that the following proposition is true (Problem 7.7).

Proposition 7.3 For any nonnegative multipliers $L \leq C$.

Consider (7.27) together with the constraints (7.24) - (7.26). Because we have eliminated the capacity constraint (7.23) we can determine L by optimizing each item separately for certain given multipliers. For item i, the problem that we need to solve is a generalized dynamic lot size problem (see Section 4.5) with a time-variable setup cost $A_i + a_i\lambda_t$ and an additional time-variable production cost per unit $b_i\lambda_t$. Also this more general problem is easy to solve, e.g., by dynamic programming.

If we want to get a tight lower bound for C we can solve the *dual* problem.

$$D = \max_{\lambda_1, \lambda_2, \ldots, \lambda_T} L. \quad (7.28)$$

In general, we will get a duality gap i.e., $D < C$ because our MIP is nonconvex. A tight lower bound is very useful, though. We can, for example, check whether a heuristic feasible solution to the original problem gives a cost that is reasonably close to the lower bound. In that case we know that the approximate solution is acceptable. Furthermore, if we want to solve the original problem by a branch-and-bound procedure, we can determine the needed associated lower bounds from the dual problem.

Let us now consider another interesting approach for solving the problem (7.22) - (7.26). Note first that in an optimal solution of the considered problem all items must have $y_{i,T} = 0$. Otherwise we would just get unnecessary holding costs. Let us now focus on one of the items. (For simplicity, we suppress index i.) We shall say that a production plan, $(x_1, x_2, \ldots, x_T)$, is *demand-feasible* if all demands are satisfied without delays and the end stock is zero. It is easy to verify that the set of demand-feasible plans is convex, i.e., if two plans are demand-feasible, a convex combination of these plans is also demand-feasible (Problem 7.8).

In Section 4.5 we showed that the optimal solution of the dynamic lot size problem must satisfy *Property 1*, meaning that *the production in a period must cover the demand in an integer number of consecutive periods*. This is no longer true for our more general problem with a capacity constraint. However, it turns out that the set of such plans constitute the extreme points of the convex set of all demand-feasible plans.

Proposition 7.4 The set of plans satisfying *Property 1* constitute the extreme points of all demand-feasible plans.

Proof Consider first a demand-feasible plan, which does not satisfy *Property 1*. Then there must exist a period j such that the demand in this period is covered partly by the production in some period $k < j$, and partly by the production in period j. We denote this plan by x. Let ε be a small number and consider two other plans x' and x'', which are almost identical to x. The only differences are that in x' we replace x_k by $x_k + \varepsilon$ and x_j by $x_j - \varepsilon$, and in x'' we replace x_k by $x_k - \varepsilon$ and x_j by $x_j + \varepsilon$. Clearly these plans are demand-feasible provided ε is sufficiently small. But $x = (x' + x'')/2$, so x is not an extreme point.

Consider then the set of plans satisfying *Property 1*. Let x be such a plan and assume that it can be expressed as a convex combination of two other such plans y and z. We shall show that this assumption leads to a contradiction. It follows that all periods without production in x must be without production also in y and z. This means that the difference between y (or z) and x is that some of the batches associated with x have been aggregated. (Recall that we by assumption cannot have the same batches.) Any given holding cost $h > 0$ will then give higher holding costs for y and z than for x. It is easy to see that this must then also be true for a convex combination of y and z. This is a contradiction and completes the proof.

Proposition 7.4 implies that any demand-feasible plan can be expressed as a convex combination of plans satisfying *Property 1*. Let us now reintroduce index i for item i. Consider the plans for item i satisfying *Property 1* and assume that these plans are numbered in some way. Let R_i be the total number of plans. ($R_i \leq 2^{T-1}$, see Problem 7.9.) Furthermore, let

$x_{i,t,r}$ = production of item i in period t when using plan r,

CORDINATED ORDERING

$c_{i,r}$ = costs for item i when using plan r,
$\beta_{i,t,r}$ = capacity requirements for item i in period t when using plan r.

Using Proposition 7.4 we can express the production of item i in period t for any demand-feasible plan as

$$x_{i,t} = \sum_{r=1}^{R_i} x_{i,t,r} \theta_{i,r}, \tag{7.29}$$

where

$$\sum_{r=1}^{R_i} \theta_{i,r} = 1, \tag{7.30}$$

and $\theta_{i,r} \geq 0$.

We are now ready to formulate the following linear programming model:

$$\min C = \sum_{i=1}^{N} \sum_{r=1}^{R_i} c_{i,r} \theta_{i,r}, \tag{7.31}$$

$$\sum_{i=1}^{N} \sum_{r=1}^{R_i} \beta_{i,t,r} \theta_{i,r} \leq q_t, \quad t = 1,2,\ldots,T, \tag{7.32}$$

$$\sum_{r=1}^{R_i} \theta_{i,r} = 1, \quad i = 1,2,\ldots,N, \tag{7.33}$$

$$\theta_{i,r} \geq 0, \quad i = 1,2,\ldots,N, r = 1,2,\ldots,R_i. \tag{7.34}$$

Although (7.29) is exact, the considered linear program is an approximation unless all $\theta_{i,r}$ are 0 or 1. Assume, for example, that for item i we have $\theta_{i,1} = \theta_{i,2} = 1/2$, i.e., we are using a *mixed* strategy with weights 1/2 for plans 1 and 2. Assume also that plan 1 has production in period t but not plan 2. According to (7.31) the setup cost in period t is only half of the correct setup cost. In the same way we will underestimate the setup time in (7.32).

Although, the linear program (7.31) - (7.34) is approximate, it turns out that the approximation in many important cases is very good. If we solve the linear program we will get a solution that has at most $T + N$ nonzero variables, i.e., the number of constraints (excluding nonnegativity constraints).

Assume that there are m items that have more than one positive $\theta_{i,r}$. The total number of nonzero variables is then less or equal to $2m + N - m = m + N \leq T + N$ so we get $m \leq T$. If there are many items and few time periods the fraction of mixed strategies will be very small. Let, for example, $N = 500$ and $T = 5$. We then have $m \leq 5$. This means that the fraction of items with mixed strategies is at most 1 percent. The resulting solution may not be optimal and may also be infeasible. Still, from a practical point of view, the solution is probably very good. The real capacity constraint is usually not that rigid. It may, for example, be possible to use some overtime. The resulting costs will be very close to the optimal costs. Note, however, that in case of a small N and a large T the considered model is of less interest. See, e.g., Example 7.1 below.

The first model of this type was formulated by Manne (1958).

Example 7.1 Consider a problem with two items and three periods. See Table 7.5. There are no initial stocks. The available production capacity is 100 time units per period. For both items the setup time is 15 time units and the operation time is one unit of time per unit. The holding cost is 1 per unit and time unit. There is no setup cost.

Table 7.5 Demands in different periods.

Item	Demand, period 1	Demand, period 2	Demand, period 3
1	25	25	75
2	20	50	25

First we determine the extreme points of the set of demand-feasible plans, i.e., the plans satisfying *Property 1*. See Table 7.6.

Table 7.6 Demand-feasible plans.

Item	Plan	Production, period 1	Production, period 2	Production, period 3
1	1	125	0	0
1	2	50	0	75
1	3	25	100	0
1	4	25	25	75
2	1	95	0	0
2	2	70	0	25
2	3	20	75	0
2	4	20	50	25

The holding costs (no setup costs) and the capacity requirements for the demand-feasible plans are:

$c_{1,1} = 175 \quad c_{1,2} = 25 \quad c_{1,3} = 75 \quad c_{1,4} = 0$
$c_{2,1} = 100 \quad c_{2,2} = 50 \quad c_{2,3} = 25 \quad c_{2,4} = 0$

$\beta_{1,1,1} = 140 \quad \beta_{1,1,2} = 65 \quad \beta_{1,1,3} = 40 \quad \beta_{1,1,4} = 40$
$\beta_{2,1,1} = 110 \quad \beta_{2,1,2} = 85 \quad \beta_{2,1,3} = 35 \quad \beta_{2,1,4} = 35$

$\beta_{1,2,1} = 0 \quad \beta_{1,2,2} = 0 \quad \beta_{1,2,3} = 115 \quad \beta_{1,2,4} = 40$
$\beta_{2,2,1} = 0 \quad \beta_{2,2,2} = 0 \quad \beta_{2,2,3} = 90 \quad \beta_{2,2,4} = 65$

$\beta_{1,3,1} = 0 \quad \beta_{1,3,2} = 90 \quad \beta_{1,3,3} = 0 \quad \beta_{1,3,4} = 90$
$\beta_{2,3,1} = 0 \quad \beta_{2,3,2} = 40 \quad \beta_{2,3,3} = 0 \quad \beta_{2,3,4} = 40$

The linear program (7.31) - (7.34) is then obtained as:

$$\min C = 175\theta_{1,1} + 25\theta_{1,2} + 75\theta_{1,3} + 100\theta_{2,1} + 50\theta_{2,2} + 25\theta_{2,3}, \quad (7.35)$$

$$\begin{aligned} 140\theta_{1,1} + 65\theta_{1,2} + 40\theta_{1,3} + 40\theta_{1,4} \\ + 110\theta_{2,1} + 85\theta_{2,2} + 35\theta_{2,3} + 35\theta_{2,4} \leq 100, \end{aligned} \quad (7.36)$$

$$115\theta_{1,3} + 40\theta_{1,4} + 90\theta_{2,3} + 65\theta_{2,4} \leq 100, \quad (7.37)$$

$$90\theta_{1,2} + 90\theta_{1,4} + 40\theta_{2,2} + 40\theta_{2,4} \leq 100, \quad (7.38)$$

$$\theta_{1,1} + \theta_{1,2} + \theta_{1,3} + \theta_{1,4} = 1, \quad (7.39)$$

$$\theta_{2,1} + \theta_{2,2} + \theta_{2,3} + \theta_{2,4} = 1, \quad (7.40)$$

$$\theta_{1,1}, \theta_{1,2}, \theta_{1,3}, \theta_{1,4}, \theta_{2,1}, \theta_{2,2}, \theta_{2,3}, \theta_{2,4} \geq 0. \quad (7.41)$$

Let m be the number of items that have more than one positive $\theta_{i,r}$. Our bound $m \leq T = 3$ is in this case of no interest, because there are only two items.

When solving the linear program (7.35) - (7.41) we get the following nonnegative $\theta_{i,r}$: $\theta_{1,2} = 3/4$, $\theta_{1,4} = 1/4$, and $\theta_{2,3} = 1$. The corresponding production plans are (175/4, 25/4, 75) for item 1 and (20, 75, 0) for item 2. The optimal cost obtained from the linear program is 175/4. It is easy to see that the obtained solution is not feasible (Problem 7.10).

7.2.2.2 Application of mathematical programming approaches

Recently there has been a renewed interest in mathematical programming models for production-inventory planning. One reason is that such models are quite often included as so-called Advanced Planning Systems (APS) in modern Enterprise Resource Planning (ERP) systems. An overview of APS systems is given by Fleischmann and Meyr (2003). ERP systems are discussed in Section 8.2.4.

As illustrated in Section 7.2.2.1, detailed models dealing with individual items are, in general, quite complex due to nonlinear setup times and costs. Still, it is possible to solve relatively large problems of this kind quite efficiently. Shapiro (1993) gives an overview of these types of models. See also Billington et al. (1983) and Eppen and Martin (1987).

One possibility to partly avoid the difficulties associated with setup times and setup costs is to use a *hierarchical planning* procedure. This means that the planning is carried out at two (or more) hierarchical levels. This is in line with standard industrial planning procedures. At an upper level an *aggregate* problem is considered. This problem concerns aggregate entities like product groups and machine groups. At the lower level the aggregate plan is *disaggregated* into a detailed plan for individual items and machines. The disaggregation is usually carried out by a simple heuristic procedure. In a model representing the aggregate level, it is usually necessary and reasonable to disregard the nonlinearities because they are associated with individual items. It is consequently possible to use linear programming. Furthermore, the number of variables and constraints are reduced by considering product groups and machine groups instead of individual items and machines. Therefore the model becomes less complex.

An overview of hierarchical planning models is given by Bitran and Tirupati (1993). See also the discussion of different planning concepts in De Kok and Fransoo (2003).

7.2.3 Production smoothing and batch quantities

In many situations it is not practical to smooth production by active coordination of the replenishments. If, for example, the demands of different items are varying substantially over time, it is difficult to smooth production by using cyclic schedules. It may still be an important goal to avoid excessive queues in production, though. A remaining possibility is to try to adjust the batch quantities in order to obtain a reasonably smooth load.

Many companies have found that smaller order quantities may smooth the production load and reduce the queues in production. There is a simple explanation for this. If the orders arriving to production can be seen as a stochastic process, smaller batches will reduce the variations over time due to the laws of large numbers. The load during a certain time is then built up of a larger number of smaller batches. If the orders are more or less independent, this will clearly smooth the load. On the other hand, we cannot use order quantities that are too small, since this will mean that the setup times will take too much of the available capacity into account and increase the queues. As usual, we have to find a suitable middle way.

The models which we have dealt with in Chapters 4 - 6 do not address these types of questions. In these models it is assumed that the lead-time is constant, or that the lead-time distribution is given but independent of the batch quantity. But from our discussion it is obvious that there are situations when this is not true and the lot sizes strongly affect the lead-time distribution. There are some lot-sizing models that explicitly take the queuing situation in production into account. We shall consider a simple such model essentially according to Karmarkar (1987, 1993). See also Axsäter (1980, 1986), Bertrand (1985), and Zipkin (1986).

Consider a machine in a large multi-center shop. This machine is used for producing a number of similar items having the same batch size. We define

d = average total demand for all items, units per time unit,
p = average processing rate, units per time unit,
Q = batch quantity for an item,
s = setup time for a batch,
T = average time in the system for a batch.

We shall, for simplicity, assume that the batches arrive at the machine as a Poisson process with rate $\lambda = d/Q$. This is a reasonable approximation if there are relatively many items. Let us also, for simplicity, think of the processing time as exponentially distributed. The average processing time for a batch is $1/\kappa = s + Q/p$ and the service factor $\rho = \lambda/\kappa = ds/Q + d/p$. The average time in the considered (M/M/1) queuing system is then

$$T = \frac{1/\kappa}{1-\rho} = \frac{s+Q/p}{1-ds/Q-d/p}. \tag{7.42}$$

Note that we must require $\rho < 1$, or equivalently, $Q > ds/(1 - d/p)$, i.e., the cycle time must on average accommodate both the setup time and the operation time.

By minimizing the average time in the system, T, with respect to Q, it is possible to show that the minimum occurs for

$$Q = \frac{ds}{1 - d/p} + \frac{s\sqrt{dp}}{1 - d/p}. \tag{7.43}$$

The average time in the system is large both when using batch quantities that are too small and when using batch quantities that are too large. When the batch quantities are too small the setup times will require too much capacity. When the batch quantities are too large the production time for a batch is long and there are also more stochastic variations in the production load, which will cause longer queuing times.

If we wish to minimize the sum of holding costs for work-in-process we should, in the considered case, use Q according to (7.43). However, there are also other objectives when choosing the batch size. For example, there are holding costs for stocks of completed items. There may also be ordering costs. The holding costs for completed items are clearly affected by the replenishment lead-time. We can see (7.42) as the lead-time (or part of the lead-time) when modeling the stocks of completed items. (This means that we disregard that the lead-time is stochastic.) Doing so, we can include holding costs for both work-in-process and the stocks of completed items in the same model when determining batch quantities and reorder points.

Note that the average time in the system for a batch is, in steady state, not affected by the reorder points for completed items.

7.3 Joint replenishments

In Section 7.2 we have considered situations where we want to have orders for different items spread out over time in order to get a smooth production load. In this section we shall deal with the opposite problem, i.e., we shall consider a group of items which should be replenished jointly as much as possible. As we have discussed previously, there are several possible reasons for joint replenishments, for example, joint setup costs, quantity discounts, coordinated transports, etc.

The methods for determination of batch quantities and reorder points that we have dealt with in Chapters 4 - 6 are evidently not directly applicable in case of joint replenishments. If there is a joint setup cost, for example, we wish to have a total order size that is sufficiently large while the individual lot sizes may be of less importance. We would also like to choose the batch sizes such that we have reasonable possibilities to coordinate future orders. Furthermore, we can normally not use reorder points that have been determined individually. Assume that the individual reorder points correspond to a certain service level that we would like to maintain. If we make a joint replenishment as soon as one of the items reaches its reorder point, other items will be ordered too early. This means that the service level (and the holding costs) will be higher than what was intended.

7.3.1 A deterministic model

A common way of modeling joint replenishment problems is to assume that there are two types of ordering or setup costs: individual setup costs for each item, and a joint setup cost for the whole group of items. The joint setup cost does not depend on the number of items that are ordered. We shall first consider such a model with constant continuous demand. No backorders are allowed. The cycle times, or equivalently the batch quantities, are constant. The production time can be disregarded. We can also without any lack of generality disregard the lead-times, provided they are the same for all items. Let us introduce the following notation:

N = number of items,
h_i = holding cost per unit and time unit for item i,
A = setup cost for the group,
a_i = setup cost for item i,
d_i = demand per time unit for item i,
T_i = cycle time for item i.

The problem is to determine cycle times in order to minimize the sum of holding and setup costs. Given a cycle time T_i, the corresponding batch quantity Q_i is obtained as $Q_i = T_i d_i$. To simplify the notation we shall replace the holding costs h_i by $\eta_i = h_i d_i$, and set all demands equal to one. It is easy to check that this will not change the problem. (See Problem 7.12.) Furthermore, we assume for simplicity, that the items are ordered so that $a_1/\eta_1 \leq a_2/\eta_2 \leq ... \leq a_N/\eta_N$. We shall consider two approaches to solve the problem. The first approach is a simple iterative technique. The second one is based on Roundy's 98 percent approximation.

7.3.1.1 Approach 1. An iterative technique

If there were no joint setup cost the optimal cycle times could be determined by the classical economic lot size formula as:

$$T_i = \sqrt{\frac{2a_i}{\eta_i}}, \qquad (7.44)$$

i.e., T_1 would be the smallest cycle time. A natural approach to solving the problem is therefore to assume that the cycle times of items 2, 3, ... , N are integer multiples n_i of the cycle time for item 1, or equivalently,

$$T_i = n_i T_1, \quad i = 2, 3, \ldots, N. \qquad (7.45)$$

Our objective is then to minimize the total costs per time unit,

$$C = \frac{A + a_1 + \sum_{i=2}^{N} a_i / n_i}{T_1} + \frac{T_1(\eta_1 + \sum_{i=2}^{N} \eta_i n_i)}{2}, \qquad (7.46)$$

with respect to $T_1, n_2, n_3, \ldots, n_N$.

Given $n_2, n_3, \ldots, n_N$, the optimal T_1 and the corresponding costs are obtained as:

$$T_1^*(n_2, n_3, \ldots, n_N) = \sqrt{\frac{2(A + a_1 + \sum_{i=2}^{N} a_i / n_i)}{\eta_1 + \sum_{i=2}^{N} \eta_i n_i}}, \qquad (7.47)$$

$$C^*(n_2, n_3, \ldots, n_N) = \sqrt{2(A + a_1 + \sum_{i=2}^{N} a_i / n_i)(\eta_1 + \sum_{i=2}^{N} \eta_i n_i)}. \qquad (7.48)$$

Note that T_1^* is not chosen according to (7.44).

If we disregard that $n_2, n_3, \ldots, n_N$ have to be integers and optimize the costs (7.48), it is possible to show (Problem 7.13) that:

CORDINATED ORDERING

$$n_i = \sqrt{\frac{a_i}{\eta_i} \frac{\eta_1}{(A+a_1)}} \cdot \qquad i=2,\ldots,N \qquad (7.49)$$

Inserting (7.49) in (7.48) we get the following lower bound for the costs:

$$\underline{C} = \sqrt{2(A+a_1)\eta_1} + \sum_{i=2}^{N}\sqrt{2a_i\eta_i} \,. \qquad (7.50)$$

The lower bound (7.50) can also be understood in the following way. Assume that there is no joint setup cost but that the individual setup cost for item 1 is $A + a_1$. If we optimize each item separately we obtain the total costs (7.50). But the considered problem is clearly a relaxation of the original problem since we can have setups of items 2, 3, ... , N without any joint setup cost. Consequently (7.50) is a lower bound for the total costs.

We are now ready to formulate a heuristic for the original problem where $n_2, n_3, \ldots, n_N$ are integers.

1. Determine start values of $n_2, n_3, \ldots, n_N$ by rounding (7.49) to the closest positive integers.

2. Determine the corresponding T_1 from (7.47).

3. Given T_1, minimize (7.46) with respect to $n_2, n_3, \ldots, n_N$. This means that we are choosing n_i as the smallest positive integer satisfying:

$$n_i(n_i+1) \geq \frac{2a_i}{\eta_i T_1^2} \,. \qquad (7.51)$$

Return to Step 2 if any multiplier n_i has changed since the last iteration.

Since the procedure gives an improvement in each step it will obviously converge, but not necessarily to the optimal solution. The resulting costs can be compared to the lower bound (7.50).

Example 7.2 Consider $N = 4$ items with a joint setup cost $A = 300$. The individual setup costs are $a_1 = a_2 = a_3 = a_4 = 50$, and the holding costs are $h_1 = h_2 = h_3 = h_4 = 10$. The demands per time unit are $d_1 = 5000$, $d_2 = 1000$, $d_3 = 700$, and $d_4 = 100$. Note that in this case $\eta_i = h_i d_i = 10 d_i$. As requested, a_i/η_i is nondecreasing with i. When applying the heuristic we obtain:

1. $n_2 = \sqrt{\dfrac{50}{1000 \cdot 10} \dfrac{5000 \cdot 10}{350}} \approx 1,\; n_3 \approx 1,\; n_4 \approx 3$.

2. $T_1 = \sqrt{\dfrac{2 \cdot (350 + 50 + 50 + 50/3)}{10 \cdot 5000 + 10 \cdot 1000 + 10 \cdot 700 + 10 \cdot 100 \cdot 3}} = 0.1155$.

3. We again obtain the multipliers $n_2 = 1$, $n_3 = 1$, $n_4 = 3$, i.e., the algorithm has already converged.

We get the resulting costs from (7.48) as $C = 8082.9$, which can be compared to the lower bound according to (7.50) $\underline{C} = 8069.0$.

Related procedures are described in e.g., Goyal and Satir (1989), and Silver et al. (1998). A technique for finding a solution with an arbitrarily small deviation from the optimal value is given by Wildeman et al. (1997).

7.3.1.2 Approach 2. Roundy's 98 percent approximation

We shall now look for a solution where the joint setups have cycle time $T_0 \geq 0$, and all other cycle times are nonnegative powers-of-two times T_0, i.e.,

$$T_i = 2^{k_i} T_0, \qquad i = 1, 2, ..., N, \tag{7.52}$$

where k_i is a nonnegative integer. Using the notation $a_0 = A$, and $\eta_0 = 0$, we can express our objective as:

$$\min_{T_0, T_1, ..., T_N} \sum_{i=0}^{N} \left(\eta_i \dfrac{T_i}{2} + a_i \dfrac{1}{T_i} \right), \tag{7.53}$$

subject to the constraints (7.52). This means that we let the joint setups be represented by a fictive item 0.

Let us now relax the considered problem by replacing (7.52) by

$$T_i \geq T_0, \qquad i = 1, 2, ..., N. \tag{7.54}$$

It is a relaxation because (7.52) implies (7.54) while the opposite is not true. The resulting solution will therefore give a lower bound for the costs.

CORDINATED ORDERING

Consider the relaxed problem, i.e., (7.53), with respect to the constraints (7.54). Since the objective function is convex and the constraints are linear, we can get the solution from the following Lagrangean relaxation:

$$\max_{\lambda_1,\lambda_2,\ldots,\lambda_N} \min_{T_0,T_1,\ldots,T_N} \sum_{i=0}^{N}\left(\eta_i \frac{T_i}{2} + a_i \frac{1}{T_i}\right) + \sum_{i=1}^{N} \lambda_i (T_0 - T_i), \quad (7.55)$$

where the multipliers λ_i are also required to be nonnegative. Let us define:

$$\eta'_0 = \eta_0 + 2\sum_{i=1}^{N} \lambda_i ,$$

$$\eta'_1 = \eta_1 - 2\lambda_1,$$

$$\vdots \quad (7.56)$$

$$\eta'_N = \eta_N - 2\lambda_N.$$

It is easy to see that the optimal solution must have all $\eta'_i > 0$. Otherwise $T_i \to \infty$, which is obviously not optimal.

Given the multipliers, the optimal solution of the relaxed problem can be obtained by solving:

$$\min_{T_0,T_1,\ldots,T_N} \sum_{i=0}^{N}\left(\eta'_i \frac{T_i}{2} + a_i \frac{1}{T_i}\right), \quad (7.57)$$

without any constraints on the cycle times, i.e., we have $N + 1$ independent classical lot sizing problems.

Let us now go back to the original joint replenishment problem but with cost parameters a_i and η'_i. Consider the effect of λ_i on the holding costs. The holding cost of item i is reduced by $2\lambda_i$ and the holding cost of item 0 is increased by $2\lambda_i$. This will mean a cost reduction for any solution of the joint replenishment problem, even if we allow the periods between orders to vary over time. To see this, consider the period between two orders for item $i > 0$. If there are additional joint setups between the two orders the total holding costs during the considered period will decrease, otherwise they will be the same.

But given the cost parameters a_i and η'_i, (7.57) will obviously provide the minimum cost since no constraints are considered. The optimal solution of the relaxed problem will consequently give a lower bound for the costs of any solution. This bound is tighter than (7.50) because of the constraints (7.54).

Assume that we have solved the relaxed problem. We can then adjust this solution by rounding the cycle times so that they can be expressed as:

$$T_i = 2^{m_i} q \qquad (7.58)$$

for some number $q > 0$. We know from Proposition 7.1 in Section 7.1 that if q is given, the maximum cost increase is at most 6 percent, and if we can also adjust q to get a better approximation, the cost increase is at most 2 percent according to Proposition 7.2. Due to (7.54), we know that $m_i \geq m_0$ and the cycle times obtained must consequently satisfy (7.52). We have now obtained Roundy's solution of the problem. This solution has an important quality. The cost increase compared to the optimal solution is at most 2 percent, since it is at most 2 percent compared to the lower bound (7.57).

It is possible to use the considered Lagrangean relaxation for numerical determination of Roundy's solution, but in general, it is much simpler to use the following technique. Since the items are ordered so that a_i/η_i are nondecreasing with i for $i > 0$, it is obvious from (7.53) and (7.54) that the optimal cycle times in the relaxed problem are nondecreasing with i for $i > 0$. It is also clear that we must have $T_1 = T_0$ in the optimal solution of the relaxed problem because $a_0 = A$, and $\eta_0 = 0$. This means that

$$T_i \geq T_{i-1}, \qquad i = 1, 2, \dots, N. \qquad (7.59)$$

Consider the relaxed problem with (7.54) replaced by (7.59). This will not change the optimal solution. Without the constraints (7.59) it would be optimal to use $T_i^* = (2a_i/\eta_i)^{1/2}$ for all i. Consequently, if a_i/η_i is increasing with i, we have found the optimal solution since the resulting batch quantities will satisfy (7.59). Since $\eta_0 = 0$ this is never the case initially. Assume that for some i, $a_i/\eta_i < a_{i-1}/\eta_{i-1}$, or equivalently that $T_i^* < T_{i-1}^*$. Assume furthermore that in the optimal solution $T_i > T_{i-1}$. Because of the convexity this implies that $T_i \leq T_i^*$ since we would otherwise reduce T_i, and similarly that $T_{i-1} \geq T_{i-1}^*$ since we would otherwise increase T_{i-1}. But this means that $T_i < T_{i-1}$ which is a contradiction. Consequently, $T_i = T_{i-1}$ in the optimal solution of the relaxed problem. But this implies that we can aggregate items $i-1$ and

i into a single item with cost parameters $a_{i-1} + a_i$, and $\eta_{i-1} + \eta_i$. Next we consider the resulting reduced problem with one item less. If $a_i/\eta_i < a_{i-1}/\eta_{i-1}$ for some i we can aggregate the two items, otherwise we obtain the optimal solution from the classical economic lot size model, etc.

Since $\eta_0 = 0$ we will always aggregate items 0 and 1. After aggregation we have an item with cost parameters $A + a_1$ and η_1. Next we check whether $a_2/\eta_2 < (A + a_1)/\eta_1$. If this is the case the aggregate item should include also item 2, etc. When no more aggregations are possible, we can optimize the resulting aggregate items individually.

Example 7.3 Consider the same data as in Example 7.2, i.e., $N = 4$, $A = 300$, $a_1 = a_2 = a_3 = a_4 = 50$, $\eta_1 = 50000$, $\eta_2 = 10000$, $\eta_3 = 7000$, and $\eta_4 = 1000$.

To solve the relaxed problem we first aggregate items 0 and 1. The combined item has cost parameters $A + a_1 = 350$, and $\eta_0 + \eta_1 = 50000$. Consider then item 2. Since $a_2/\eta_2 = 50/10000 < 350/50000$, item 2 should also be added to the combined item. The resulting cost parameters are obtained as $A + a_1 + a_2 = 400$, and $\eta_0 + \eta_1 + \eta_2 = 60000$. Consider item 3. Since $50/7000 > 400/60000$, item 3 should not be added. Compare finally item 4 and item 3. We get $50/1000 > 50/7000$, i.e., items 3 and 4 should not be combined. This gives the cycle times $T_0 = T_1 = T_2 = (800/60000)^{1/2} = 0.1155$, $T_3 = (100/7000)^{1/2} = 0.1195$, and $T_4 = (100/1000)^{1/2} = 0.3162$. The resulting lower bound for the costs is $\underline{C} = C_{0+1+2} + C_3 + C_4 = 6928.2 + 836.7 + 316.2 = 8081.1$. Note that this bound is better than the bound obtained in Example 7.2.

Consider then cycle times that can be expressed as powers of two, i.e., let $q = 1$ in (7.58). We obtain $T_0 = T_1 = T_2 = 2^{-3} = 0.125$, $T_3 = 2^{-3} = 0.125$, and $T_4 = 2^{-2} = 0.250$. The resulting costs are $C = C_{0+1+2} + C_3 + C_4 = 6950 + 837.5 + 325 = 8112.5$, i.e., 0.39 percent above the lower bound.

By minimization of (7.10) it is possible to show that it is optimal to have $q = 1.88$ in (7.58). The corresponding cycle times are $T_0 = T_1 = T_2 = 2^{-4}q = 0.1175$, $T_3 = 2^{-4}q = 0.1175$, and $T_4 = 2^{-3}q = 0.235$. The resulting costs are $C = C_{0+1+2} + C_3 + C_4 = 6929.3 + 836.8 + 330.3 = 8096.3$, i.e., 0.19 percent above the lower bound. The solution in Example 7.2, which is not a powers-of-two policy, is still slightly better.

A similar approach is used for multi-stage lot sizing in Section 9.2.2. See also Jackson et al. (1985), Roundy (1985, 1986), and Muckstadt and Roundy (1993).

The two approaches considered assume constant demand, but can also be used in case of stochastic demand. We then replace the stochastic demands by their means when determining cycle times. Given the cycle times we can, using the techniques in Section 5.12, determine appropriate periodic review S policies for each item. The order-up-to inventory positions should include suitable amounts of safety stock. Next, in Section 7.3.2, we will consider a different model that is more directly focused on stochastic demand.

7.3.2 A stochastic model

We shall now instead consider a stochastic model. The demands for the items are independent and stationary stochastic processes. Each customer demand is for an integral number of units. We can, for example, consider Poisson or compound Poisson demand processes. We assume complete backordering. Let us introduce the following notation to describe the problem:

N = number of items,
h_i = holding cost per unit and time unit for item i,
$b_{1,i}$ = shortage cost per unit and time unit for item i,
A = setup cost for the group,
a_i = setup cost for item i,
L = constant lead-time.

Viswanathan (1997) suggests the following technique that, in his numerical tests, outperforms other suggested methods.

In the first step we disregard the joint setup cost and consider the items individually for a suitable grid of review periods T. For each review period we determine the optimal individual (s, S) policies for all items and the corresponding average costs. This can be done very efficiently as explained in Section 6.1.1.2 Let

$C_i(T)$ = average costs per time unit for item i when using the optimal individual (s, S) policy with a review interval of T time units.

In the second step we determine the review period T by minimizing,

$$C(T) = A/T + \sum_{i=1}^{N} C_i(T). \qquad (7.60)$$

Note that the actual costs are lower than the costs according to (7.60), since the major setup cost A is not incurred at reviews where none of the items are ordered.

Atkins and Iyogun (1988) also use periodic review policies but in a different way.

Other policies that have been used frequently in the inventory literature are so-called *can-order policies*. When using such a policy there are two reorder points for each item: a *can-order* level and a lower *must-order* level. An item must be ordered when its inventory position reaches the must-order level. When an item in the group is ordered, other items with inventory positions at or below their respective can-order levels are also ordered. This type of policy was first suggested by Balintfy (1964). Techniques for designing can-order policies have been suggested by Silver (1981) and Federgruen et al. (1984).

Renberg and Planche (1967) suggested a so-called (S_i, Q) policy. According to this policy all items are replenished up to certain levels S_i when the total demand for the whole group since the preceding replenishment has reached Q.

References

Atkins, D. R., and P. Iyogun. 1988. Periodic Versus Can-Order Policies for Coordinated Multi-Item Inventory Systems, *Management Science*, 34, 791-796.

Axsäter, S. 1980. Economic Order Quantities and Variations in Production Load, *International Journal of Production Research*, 18, 359-365.

Axsäter, S. 1984. Lower Bounds for the Economic Lot Scheduling Problem Using Aggregation, *European Journal of Operational Research*, 17, 201-206.

Axsäter, S. 1986. Evaluation of Lot-Sizing Techniques, *International Journal of Production Research*, 24, 51-57.

Axsäter, S. 1987. An Extension of the Extended Basic Period Approach for Economic Lot Scheduling Problems, *Journal of Optimization Theory and Applications*, 52, 179-189.

Balintfy, J. L. 1964. On a Basic Class of Multi-Item Inventory Problems, *Management Science*, 10, 287-297.

Bertrand, J. W. M. 1985. Multiproduct Optimal Batch Sizes with In-Process Inventories and Multiwork Centers, *IIE Transactions*, 17, 157-163.

Billington, P. J., J. O. McClain, and L. J. Thomas. 1983. Capacity-Constrained MRP Systems, *Management Science*, 29, 1126-1141.

Bitran, G. R., and D. Tirupati. 1993. Hierarchical Production Planning., in S. C. Graves et al. Eds. *Handbooks in OR & MS Vol. 4*, North Holland Amsterdam, 523-568.

Bomberger, E. A. 1966. A Dynamic Programming Approach to a Lot Size Scheduling Problem, *Management Science*, 12, 778-784.

Bowman, R. A., and J. A. Muckstadt. 1993. Stochastic Analysis of Cyclic Schedules, *Operations Research*, 41, 947-958.

Bowman, R. A., and J. A. Muckstadt. 1995. Production Control of Cyclic Schedules with Demand and Process Variability, *Production and Operations Management*, 4, 145-162.

De Kok, A. G. , and J.C. Fransoo. 2003. Planning Supply Chain Operations: Definition and Comparison of Planning Concepts, in A. G. de Kok and S. C. Graves. Eds. *Handbooks in OR & MS Vol. 11*, Elsevier Amsterdam, 597-675.

Dobson, G. 1987. The Economic Lot-Scheduling Problem: Achieving Feasibility Using Time-Varying Lot Sizes, *Operations Research*, 35, 764-771.

Doll, C. L., and D. C. Whybark. 1973. An Iterative Procedure for the Single-Machine Multi-Product Lot Scheduling Problem, *Management Science*, 20, 50-55.

Elmaghraby, S. E. 1978. The Economic Lot Scheduling Problem (ELSP): Review and Extensions, *Management Science*, 24, 587-598.

Eppen, G. D., and R. K. Martin. 1987. Solving Multi-Item Capacitated Lot-Sizing Problems Using Variable Reduction, *Operations Research*, 35, 832-848.

Federgruen, A., H. Groenevelt, and H. C. Tijms. 1984. Coordinated Replenishments in a Multi-item Inventory System with Compound Poisson Demands, *Management Science*, 30, 344-357.

Federgruen, A., and Z. Katalan. 1996a. The Stochastic Economic Lot Scheduling Problem: Cyclical Base-Stock Policies with Idle Times, *Management Science*, 42, 783-796.

Federgruen, A., and Z. Katalan. 1996b. The Impact of Setup Times on the Performance of Multi-Class Service and Production Systems, *Operations Research*, 44, 989-1001.

Fleischmann, B., and H. Meyr. 2003. Planning Hierarchy, Modeling and Advanced Planning Systems, in A. G. de Kok and S. C. Graves. Eds. *Handbooks in OR & MS Vol. 11*, Elsevier Amsterdam, 457-523.

Gallego, G. 1990. Scheduling the Production of Several Items with Random Demands in a Single Facility, *Management Science*, 36, 1579-1592.

Gallego, G., and I. Moon. 1992. The Effect of Externalizing Setups in The Economic Lot Scheduling Problem, *Operations Research*, 40, 614-619.

Gallego, G., and R. Roundy. 1992. The Economic Lot Scheduling Problem with Finite Backorder Costs, *Naval Research Logistics*, 39, 729-739.

Goyal, S. K., and A. T. Satir. 1989. Joint Replenishment Inventory Control: Deterministic and Stochastic Models, *European Journal of Operational Research*, 38, 2-13.

Graves, S. C. 1980. The Multi-Product Production Cycling Problem, *AIIE Transactions*, 12, 233-240.

Hsu, W. 1983. On the General Feasibility Test of Scheduling Lot Sizes for Several Products on One Machine, *Management Science*, 29, 93-105.

Jackson, P. L., W. Maxwell, and J. A. Muckstadt. 1985. The Joint Replenishment Problem with a Powers-of-Two Restriction, *IIE Transactions*, 17, 25-32.

Karmarkar, U. S. 1987. Lot Sizes, Lead Times and In-Process Inventories, *Management Science*, 33, 409-418.

Karmarkar, U. S. 1993. Manufacturing Lead Times, Order Release and Capacity Loading, in S. C. Graves et al. Eds. *Handbooks in OR & MS Vol. 4*, North Holland Amsterdam, 287-329.

Manne, A. S. 1958. Programming of Economic Lot Sizes, *Management Science*, 4, 115-135.

Muckstadt, J. A., and R. Roundy. 1993. Analysis of Multistage Production Systems, in S. C. Graves et al. Eds. *Handbooks in OR & MS Vol. 4*, North Holland Amsterdam, 59-131.

Renberg, B., and R. Planche. 1967. Un Modele Pour La Gestion Simultanee Des n Articles D'un Stock, *Revue Francaise d'Informatique et de Recherche Operationelle*, 6, 47-59.

Roundy, R. 1985. 98%-Effective Integer-Ratio Lot-Sizing for One-Warehouse Multi-Retailer Systems, *Management Science*, 31, 1416-1430.

Roundy, R. 1986. 98%-Effective Lot-Sizing Rule for a Multi-Product Multi-Stage Production/Inventory System, *Mathematics of Operations Research*, 11, 699-729.

Roundy, R. 1989. Rounding Off to Powers of Two in Continuous Relaxations of Capacitated Lot Sizing Problems, *Management Science*, 35, 1433-1442.

Shapiro, J. F. 1993. Mathematical Programming Models and Methods for Production Planning and Scheduling, in S. C. Graves et al. Eds. *Handbooks in OR & MS Vol. 4*, North Holland Amsterdam, 371-443.

Silver, E. A. 1981. Establishing Reorder Points in the (S, c, s) Coordinated Control System under Compound Poisson Demand, *International Journal of Production Research*, 9, 743-750.

Silver, E. A., D. F. Pyke, and R. Peterson. 1998. *Inventory Management and Production Planning and Scheduling*, 3rd edition, Wiley, New York.

Sox, C. R., P. L. Jackson, A. Bowman, and J. A. Muckstadt. 1999. A Review of the Stochastic Lot Scheduling Problem, *International Journal of Production Economics*, 62, 181-200.

Sox, C. R., and J. A. Muckstadt. 1997. Optimization-Based Planning for the Stochastic Lot Scheduling Problem, *IIE Transactions*, 29, 349-357.

Viswanathan, S. 1997. Periodic Review (s, S) Policies for Joint Replenishment Inventory Systems, *Management Science*, 43, 1447-1454.

Wildeman, R. E., J. B. G. Frenk, and R. Dekker. 1997. An Efficient Optimal Solution Method for the Joint Replenishment Problem, *European Journal of Operational Research*, 99, 433-444.

Zipkin, P. H. 1986. Models for Design and Control of Stochastic, Multi-Item Batch Production Systems, *Operations Research*, 34, 91-104.

Zipkin, P. H. 1991. Computing Optimal Lot Sizes in the Economic Lot Scheduling Problem, *Operations Research*, 39, 56-63.

Problems

7.1 Show that (7.4) implies (7.5).

7.2 a) Show that $\int_{-1/2}^{1/2} e(u)du = \dfrac{1}{\sqrt{2\ln 2}}$.

b) Show that the worst case will occur in (7.9) if x is uniform on $(-1/2, 1/2)$. What happens if we change q? Why?

7.3* Three products are produced in a single machine. The demands are constant and continuous. No backorders are allowed. The following data are given

Product	Demand, units per day	Holding cost per unit and day	Setup cost	Setup time, days	Production rate, units per day
1	48	0.060	800	0.50	200
2	20	0.040	500	0.25	100
3	32	0.048	1000	1.00	100

a) Determine the independent solution and the corresponding lower bound for the optimal costs.
b) Show that the independent solution is infeasible.
c) Derive the common cycle solution and an upper bound for the total costs.
d) Apply Doll and Whybark's technique. Is the solution feasible?

7.4 Demonstrate that the iterative procedure in Section 7.2.1.5 will converge.

7.5* Three products are produced in the same machine. Various data are given in the table.

Product	I	II	III
Demand per week	100	50	20
Production per week	1000	500	250
Setup time in weeks	0.8	0.4	0.1
Ordering cost	10000	10000	5000
Holding cost per unit and week	10	20	10

a) Use a cyclic schedule with a common cycle. Determine batch quantities.
b) By using the technique by Doll and Whybark the following solution has been found:

$Q_1 = 460$
$Q_2 = 230$
$Q_3 = 184$

---

* Answer and/or hint in Appendix 1.

CORDINATED ORDERING

Does this solution reduce the costs, and in that case how much. Is the solution feasible?

7.6 A company is producing three products in the same machine. There are 250 working days in one year, and 8 working hours per day. The following data are given:

Product	Demand, units per year	Holding cost per unit and year	Setup cost	Setup time, minutes	Operation time, minutes per unit
1	8000	20	120	20	5
2	12000	15	100	10	3
3	5000	30	200	30	8

Determine the best cyclic plan with a common cycle time.

7.7 Prove Proposition 7.3.

7.8 Show that the set of *demand-feasible* production plans is convex.

7.9 Show that the number of plans satisfying *Property 1* is at most 2^{T-1}.

7.10 Consider Example 7.1.

a) Represent the following plan for item 1: (30, 40, 55) as a convex combination of plans satisfying *Property 1*. Is the solution unique?
b) Show that the plan obtained by the linear program is infeasible.
c) Demonstrate that the optimal solution of the problem is to use the plan (50, 0, 75) for item 1 and (20, 75, 0) for item 2.

7.11 Consider (7.42). Show that the average time in the system T is minimized by choosing Q according to (7.43).

7.12 Consider the model in Section 7.3.1. Show that we, without changing the problem, can replace the holding costs h_i by $\eta_i = h_i d_i$, and set all demands equal to one.

7.13 Derive the optimal continuous n_i, (7.49), from (7.48).

7.14 Derive the condition (7.51).

7.15 Start with the solution obtained in Example 7.3 and check whether the technique in Section 7.3.1.1 can improve the solution. This means that we no longer require a powers-of-two policy.

8 MULTI-ECHELON SYSTEMS: STRUCTURES AND ORDERING POLICIES

So far we have considered a single installation. In practice, though, it is also common to see multi-stage, or multi-echelon, inventory systems, where a number of installations are coupled to each other. For example, when distributing products over large geographical areas, many companies use an inventory system with a central warehouse close to the production facility and a number of local stocking points close to the customers in different areas. Similarly, in production, stocks of raw materials, components, and finished products are coupled to each other. To obtain efficient control of such inventory systems it is necessary to use special methods that take the connection between different stocks into account. In Chapters 8 - 10 we will show when and how such methods can be used.

The interest in *Supply Chain Management* in general, and for multi-echelon inventory control in such chains, is growing rapidly. The management of multi-echelon inventory systems is a crucial part of supply chain operations. Furthermore, the possibilities for efficient control of multi-echelon inventory systems have increased substantially during the last two decades. One reason is the progress in research, which has resulted in new techniques that are both more general and more efficient. Another reason is new information technologies which have created a completely different infrastructure and greatly increased the possibilities for efficient supply chain coordination. Examples of such new technologies are EDI (Electronic Data Interchange), satellite communication systems, and RFID (Radio Frequency Identification).

Multi-echelon inventory systems in supply chains are not always part of a single company. Quite often, different companies work together to improve the coordination of the total material flow. An example is the implementation of so-called VMI (Vendor Managed Inventory) systems.

8.1 Inventory systems in distribution and production

Multi-echelon inventory systems are common in both distribution and production. When analyzing such systems it is important to understand that inventory systems which may physically appear to be very different can still be very similar in an inventory control context. Consider for example the two installations in Figure 8.1. The customer demand takes place at installation 1. Installation 1 is replenished from installation 2, which in turn, replenishes from an outside supplier.

In a distribution system, installation 1 can be a local sales stock in a distant country, while installation 2 is a central warehouse close to the factory. The replenishment lead-time at installation 1 is then essentially the transportation time.

In production, installation 1 may be the stock of a final product while installation 2 is the stock of a subassembly which is used when producing the final product. In this case the stocks contain different items, yet the coupling is the same. In both situations installation 1 can be seen as a customer at installation 2. In the production situation the replenishment lead-time at installation 1 is essentially the production time, and the two stocks may very well be situated in the same building.

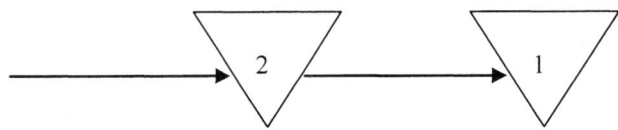

Figure 8.1 An inventory system with two coupled inventories.

8.1.1 Distribution inventory systems

Distribution inventory systems are, in general, divergent, i.e., the number of parallel installations increase along the material flow. In a pure *distribution system*, or *arborescent system*, each stock has at most a single immediate predecessor. Figure 8.2 shows such a system with two levels: a central warehouse, and a number of retailers.

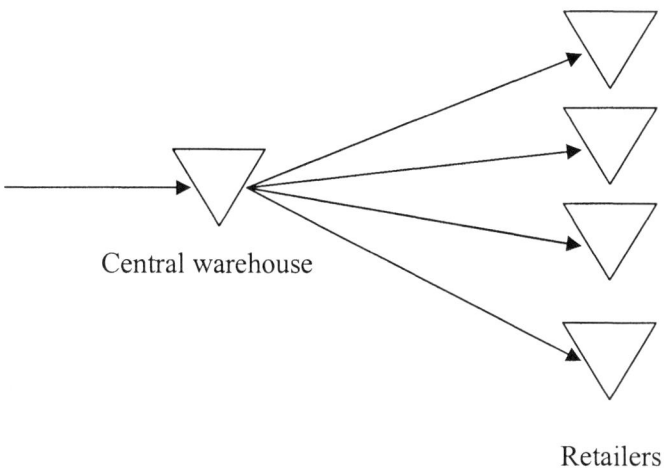

Figure 8.2 Distribution inventory system.

In the special case when each installation has also at most one immediate successor, as in Figure 8.1, we have a *serial system*.

The stocks at the retailers in Figure 8.2 are needed to maintain a high service at different local markets. The central warehouse supports all retailers. More stock at the central warehouse will give shorter and less variable lead-times for the retailers, and this can enable the retailers to reduce their stocks. On the other hand, the holding costs at the central warehouse will increase. The best distribution of the total system stock will depend on the structure of the system, the demand variations, the transportation times, and the unit costs. Situations exist when there should be a relatively large stock at the central warehouse, but it is more common that the optimal solution means a very low warehouse stock, much lower than what most practitioners would expect.

8.1.2 Production inventory systems

Inventory systems in production may also sometimes have a structure like that in Figure 8.2. Such systems are especially common in process industries, but also occur quite frequently at the end of the production chain in mechanical industries. The central warehouse in Figure 8.2 can then correspond to the stock of a subassembly that is used as a basic module when producing a number of different end products. The stocks of these end products correspond to the retailer stocks in Figure 8.2.

In production it is more common, though, to have inventory systems with many parallel stocking locations early in the material flow and successively fewer stocks later in the flow, i.e., a convergent flow. If each installation has at most one immediate successor, as in Figure 8.3, we have an *assembly system*. Note that a serial system is also a special case of an assembly system.

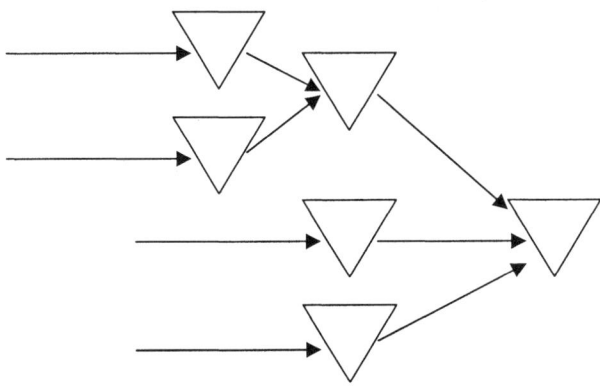

Figure 8.3 An assembly system.

For an assembly system there also exist various reasons for keeping stock early in the material flow. Raw materials and components early in the flow usually have much lower values than subassemblies and final products at the end of the flow. This means that the holding costs are lower. Sometimes high setup costs in early production stages may also make it necessary to use large batch quantities.

It is, in general, considerably easier to deal with serial systems than with other types of multi-echelon systems. Some results for serial systems however, can be generalized to pure distribution systems or assembly systems. In production it is very common, though, to also encounter more general inventory systems which are much more difficult to handle. Figure 8.4 shows such a system. Note that a facility can have several successors as well as several predecessors.

For a certain multi-echelon inventory system in production there exists an equivalent *bill of material*, or *product structure*. In Figure 8.4, stock 1 can be seen as the stock of a final product that is produced from items 2 and 3. Item 2 is produced from items 4 and 5 etc. The complete bill of material corresponding to Figure 8.4 is shown in Figure 8.5.

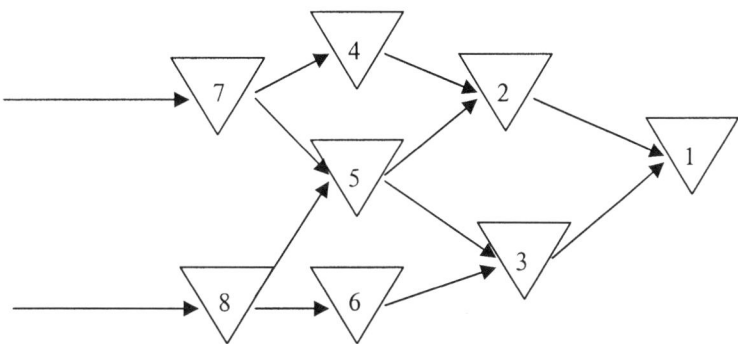

Figure 8.4 A general multi-echelon inventory system.

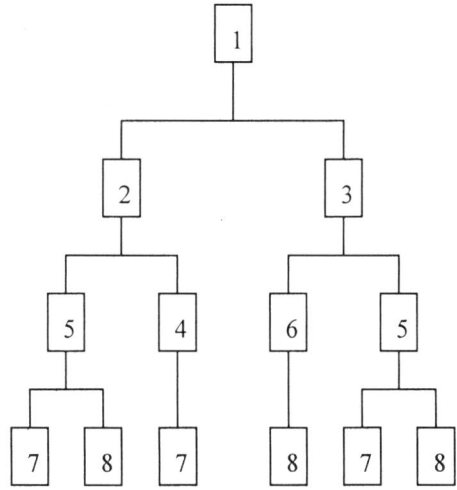

Figure 8.5 Bill of material corresponding to the inventory system in Figure 8.4.

In production, it is also common that we need more than one unit of a component when producing some other item. For example, item 1 may consist of two units of item 2 and three units of item 3. For serial and assembly systems this does not add to the complexity. Since an item can only be part of a single item we can always redefine the unit so that we always need exactly one unit of a lower level item. If, for example, a certain item contains two units of some other item, we can let two units of this item be the new unit. But we cannot handle a general system in this way. To see this, consider again Figure 8.5. It may happen that item 2 contains two units of item 5 while item 3 contains three units of the same item. We cannot circumvent this difficulty by redefining the unit.

8.1.3 Repairable items

In this book we deal with *consumable* items. However, it is interesting to note that *repairable* items can in many situations be handled in essentially the same way. Some of the early important work concerning multi-echelon inventory systems (e.g., Sherbrooke, 1968) was focused on repairable items, which are especially common in military applications.

To illustrate the relationship between consumable and repairable items, consider again Figure 8.2. Assume that the stocks contain repairable spare items which can replace items that fail. Furthermore, assume that all replenishments are one-for-one. When an item fails it is sent to the central site for repair. It is then stocked at the central site. The failed item is, if possible, replaced by another item at the local site. If no item is available, the item is backordered. In any case a new item is ordered from the central site. Consider then a corresponding system with consumable items. The only difference is the warehouse lead-time. In the case of consumable items, this lead-time is given or has a given distribution. In the case of repairable items, the corresponding warehouse lead-time is the transportation time from a local site plus the repair time. By using this equivalence we can use models and techniques for consumable items for repairable items as well.

A review of repairable item inventory control is given by Nahmias (1981).

8.1.4 Lateral transshipments in inventory systems

Consider again the distribution inventory system in Figure 8.2. We can say that the advantage of the central warehouse is that it enables *pooling* of stock. The stock in the warehouse can be used by any of the retailers. But pooling of stock can also be obtained by other inventory systems. One possibility is to use so-called *lateral transshipments* between parallel local stock points. Consider Figure 8.6.

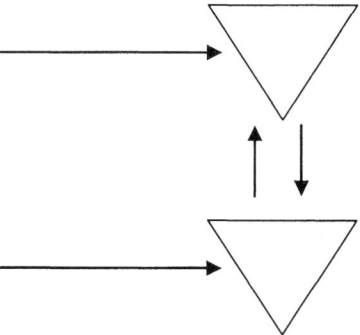

Figure 8.6 Lateral transshipments between parallel stocks.

The local stocks normally replenish from an outside supplier, for example by applying (R, Q) policies. But in emergency situations it is also possible to use lateral transshipments between adjacent retailers. These transshipments are faster but incur additional costs. Such lateral transshipments are common in practice. Models with lateral transshipments are usually more difficult to handle, and the available results are less general.

There are two problems associated with lateral transshipments:

1. What is a suitable decision rule for deciding on the source and size of a lateral transshipment?

2. Given a certain decision rule for lateral transshipments, how can this policy be evaluated and how does it affect the policy for normal replenishments?

Many of the earlier papers dealing with lateral transshipments consider the second problem and assume that some simple decision rule is given. A frequent assumption is that a demand at a local stock point that can not be met by stock on hand is, if possible, supplied by a lateral transshipment from some other stock point with inventory on hand. Examples of such approaches are Lee (1987), Axsäter (1990), Dada (1992), Alfredsson and Verrijdt (1999), Grahovac and Chakravarty (2001), Kukreja et al. (2001), Sherbrooke (1992), Muckstadt (2005), and Vidgren (2005).

A possible approach when trying to optimize decisions concerning lateral transshipments (Problem 1) is to use stochastic dynamic programming. See e.g., Das (1975), Robinson (1990), Archibald et al. (1997), and Olsson (2005). Tagaras and Cohen (1992) evaluate different decision rules by simulation. Given the policy for normal replenishments, the decision rule in Axsäter (2003) provides a performance guarantee; it will always lead to an improvement compared to not using lateral transshipments. Another heuristic decision rule is suggested by Minner et al. (2003). Rudi et al. (2001) consider decentralized decisions concerning transshipments and determine transshipment prices that lead to joint-profit maximization. We shall not discuss lateral transshipments any further in this book.

Problems concerning transshipments are related to problems with substitutions. See e.g., Bassok et al. (1999) and the references therein for a further discussion of such problems.

Another related problem area is emergency deliveries (from the supplier) in an inventory system. See e.g., Moinzadeh and Nahmias (1988), Johansen and Thorstenson (1998), Minner (2003), and Axsäter (2005).

8.1.5 Inventory models with remanufacturing

It has become more and more common to use production-inventory systems where manufacturing and remanufacturing operations occur simultaneously. For example, it may be possible to process used components of certain products so that they satisfy the same quality standards as new components. This means that new components can be replaced by remanufactured used components when available. See Figure 8.7.

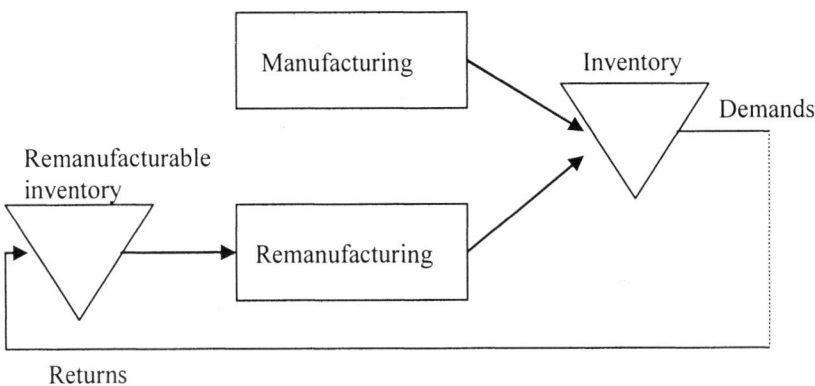

Figure 8.7 Production-inventory system with remanufacturing.

There are several motivations for such manufacturing systems. One important reason is new regulations that force manufacturing companies to collect and handle used products after customer usage. Another reason is that remanufacturing processes tend to become more competitive.

Remanufacturing policies will lead to inventory systems with coupled inventories. The demands as well as the returns are usually stochastic processes. It is obvious that the control of the system in Figure 8.7 involves many complex questions. For example, what does a suitable production strategy look like, and how much safety stock is needed in the two stocks?

The literature dealing with such inventory models is growing rapidly. See e.g., Van der Laan et al. (1998), Van der Laan et al. (1999), Guide et al. (2003), Dekker et al. (2004), and the references therein.

8.2 Different ordering systems

For a single-echelon system we know that an (s, S) policy is optimal under very general conditions. Some optimality results exist for serial systems, see Clark and Scarf (1960), Chen (2000), and Muharremoglu and Tsitsiklis (2003). But for most multi-echelon systems the structure of the optimal policy is unknown. Generally we can expect an optimal control policy for a multi-echelon inventory system to be quite complex. An optimal decision to send a batch from one site to another site may depend on the inventory status at all sites. Such a centralized decision system is usually associated with several disadvantages. A lot of information is needed and although information technology has improved dramatically, the data movement will lead to additional costs. Furthermore, completely general centralized policies are difficult to derive. It is therefore usually more suitable to limit the degree of centralization and use relatively simple local control rules that are similar to policies used in single-echelon systems. Multi-level methods mean then that we try to coordinate such local decision rules in a suitable way. It is common to use simple reorder point policies in multi-echelon inventory systems, for example. Our problem may then be to find reorder points for the different facilities that will give a satisfactory coordinating control of the system as a whole.

The need to decentralize the control is generally larger in a distribution system than in a production system. Within distribution, different facilities are normally situated at locations far from each other, and quite often relatively independent organizational units make the replenishment decisions.

Also for multi-echelon systems we can choose between continuous review and periodic review. In Section 7.2 we have discussed the possibility of using periodic review systems in order to smooth the production load. This is, of course, also a possibility when dealing with multi-echelon inventory systems. Furthermore, there can be other similar reasons for using periodic review in connection with multi-echelon systems. Consider, for example, the distribution system in Figure 8.2. If the four retailers use periodic review systems with a review period of four weeks, and with different ordering weeks, the demand at the central warehouse will be very smooth. This will simplify the inventory control and also lead to a smooth production load in the factory delivering to the warehouse.

In this section we will describe different common ordering systems and show how they are related to each other. The ordering systems are all quite simple, but regardless of which ordering system we choose, it is difficult to determine suitable control parameters like batch quantities and safety stocks. We shall deal with these questions in Chapters 9 and 10.

8.2.1 Installation stock reorder point policies and KANBAN policies

One common way to control a multi-echelon inventory system is to control each facility by an (R, Q) policy or by an S policy. Recall that for discrete integral demand an S policy is a special case of an (R, Q) policy where $R = S - 1$ and $Q = 1$. The reorder points and batch quantities do not, of course, need to be the same for different facilities. As we have described in Section 3.2.3, each facility considers its own inventory position, i.e., stock on hand plus outstanding orders and minus backorders. When the inventory position declines to or below R, a batch of size Q (or possibly a number of batches) is ordered so that the new inventory position is above R and less or equal to $R + Q$.

As we shall describe in Section 8.2.2, it is also common to use a different type of reorder point policy in multi-echelon inventory systems. When using this modified reorder point policy, the inventory position is defined in another way and not based only on the stock at the installation. To avoid misinterpretation the normal (R, Q) policy is therefore often denoted *installation stock* (R, Q) policy in connection with multi-echelon inventory control.

As pointed out above, the optimality of the (s, S) policy for single-echelon systems does not, in general, carry over to multi-echelon systems. Therefore it is more common to let the installations use the somewhat simpler (R, Q) policy in a multi-echelon setting.

It is also common to apply KANBAN policies in multi-echelon systems. As described in Section 3.2.3.1, a KANBAN policy can be interpreted as an installation stock (R, Q) policy where backorders are not subtracted from the inventory position. This means that the number of KANBANs is an upper bound for the number of outstanding orders and will limit work-in-process, which is sometimes an advantage, see Veatch and Wein (1994) and Spearman et al. (1990).

8.2.2 Echelon stock reorder point policies

Consider the inventory system and the corresponding product structures in Figure 8.8.

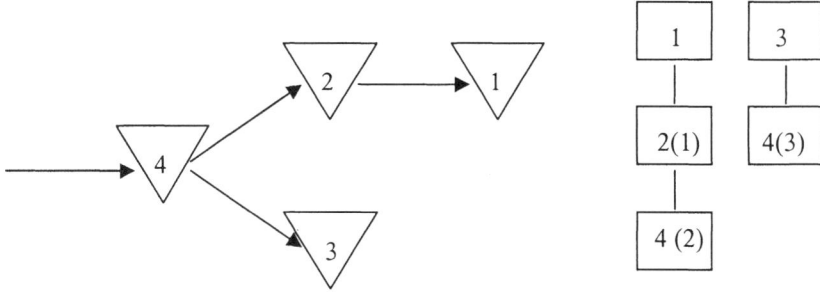

Figure 8.8 Inventory system and corresponding product structures.

We use a number in parenthesis to illustrate the number of units contained in the higher level item. For example, 2 units of item 4 are used as components when producing one unit of item 2. Similarly, 3 units of item 4 are needed when producing one unit of item 3.

Consider the stock of item 4. If the inventory position is low, it is clearly an indication that we might need to replenish. But on the other hand, if the stocks of items 1, 2, and 3 are very high, there is no immediate need for more units of item 4 and it may not be necessary to replenish. The idea of using the *echelon stock* is to take this into account by also including downstream stock of an item. This means that we also add units of item 4 contained in the inventory positions of items 1, 2 and 3 when deciding whether to replenish item 4. This strategy does not always work well, though. Assume that the stock of item 3 is very high, while the stocks of items 1 and 2 are low. The echelon stock of item 4 may then be very high, and no replenishment is triggered even though the stocks of items 1 and 2 need to be replenished. The replenishments of item 2 will cause corresponding demands for item 4. The problem is that these demands cannot be covered by the stock of item 3.

We shall explain the echelon stock concept in detail by an example.

Example 8.1 Consider the inventory system in Figure 8.8 and assume installation inventory positions as in the second column of Table 8.1.

Table 8.1 Installation and echelon stock inventory positions in Figure 8.8.

Item	Installation stock inventory position	Echelon stock inventory position
1	5	5
2	2	7
3	3	3
4	5	28

Items 1 and 3 have no downstream stock. The echelon stock inventory positions are therefore the same as the installation stock inventory positions. Consider item 2. Since each unit of item 1 contains one unit of item 2, the echelon stock inventory position of item 2 is obtained as the installation inventory position plus the stock included in the inventory position of item 1, i.e., $2 + 5 = 7$ units. Item 4 is part of items 1, 2, and 3. Each unit of items 1 and 2 contains two units of item 4 and each unit of item 3 contains three units. Consequently we obtain the echelon stock inventory position of item 4 as $5 + 2 \cdot 5 + 2 \cdot 2 + 3 \cdot 3 = 28$. Alternatively, we can use the echelon stock positions of items 2 and 3 to obtain the same result, i.e., $5 + 2 \cdot 7 + 3 \cdot 3 = 28$.

Since the echelon stock inventory position is normally higher than the installation stock inventory position, an echelon stock reorder point policy will generally use larger reorder points than an installation stock policy to achieve similar control.

Note that if the number of units contained in a higher level item is always one, the echelon stock inventory position is simply obtained as the installation stock inventory position plus the sum of the installation inventory positions of all downstream items. This is always the case in a distribution system where all stocks contain the same item. Recall also that we can always obtain this situation when dealing with serial or assembly systems by a suitable change of units.

8.2.3 *Comparison of installation stock and echelon stock policies*

We shall now consider the serial inventory system in Figure 8.9. Item 1 is the final product facing customer demand. Item N is raw material obtained from an outside supplier. When producing item n ($1 \leq n \leq N - 1$) we need one unit of item $n + 1$.

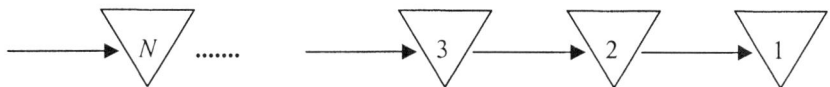

Figure 8.9 Serial inventory system with N installations.

We shall analyze the relationship between installation stock (R, Q) policies and echelon stock (R, Q) policies. The batch quantities are assumed to be given and the same for both policies,

Q_n = batch quantity at installation n.

We make the additional assumption that the batch quantity at installation n is an integer multiple of the batch quantity at installation $n - 1$ (it is convenient to define $Q_0 = 1$), i.e.,

$$Q_n = j_n Q_{n-1}, \qquad (8.1)$$

where j_n is a positive integer. This assumption is natural if the rationing policy is to satisfy all or nothing of an order, since the installation stock at installation n should then always consist of an integer number of the next downstream lot size Q_{n-1}.

Let us now introduce some additional notation:

IP_n^i = installation inventory position at installation n,

IP_n^e = $IP_n^i + IP_{n-1}^i + ... + IP_1^i$ = echelon stock inventory position at installation n,

R_n^i = installation stock reorder point at installation n,

R_n^e = echelon stock reorder point at installation n.

We shall assume that the system starts with initial inventory positions IP_n^{i0} and IP_n^{e0} satisfying:

$$R_n^i < IP_n^{i0} \leq R_n^i + Q_n, \qquad R_n^e < IP_n^{e0} \leq R_n^e + Q_n. \qquad (8.2)$$

Note that these conditions are always satisfied as soon as each installation has ordered at least once.

Let us now make some observations. An installation stock policy is always *nested*, i.e., installation n will not order unless its inventory position has just been reduced by an order from installation $n - 1$. But if installation $n - 1$ orders, this must then also be the case for installation $n - 2$ etc., and consequently for all downstream installations.

The echelon stock inventory position at installation n is not affected by orders at installations $1, 2, \ldots, n - 1$. If, for example, installation $n - 1$ orders a batch from installation n, IP_{n-1}^i is increased by Q_{n-1} and IP_n^i is reduced by Q_{n-1}. This will clearly not affect IP_n^e. The echelon stock inventory position at installation n is only changed by the final demand at installation 1 and by replenishment orders at installation n.

We shall now prove Propositions 8.1 and 8.2 below for the special case of unit demand and continuous review. The propositions are also valid for other types of demand and for periodic review. They can furthermore be extended to assembly systems. See Axsäter and Rosling (1993).

We shall also, without any lack of generality, assume that $IP_n^{i0} - R_n^i$ is an integer multiple of Q_{n-1}. All demands at installation n are for multiples of the batch quantity at installation $n - 1$, Q_{n-1}, and all replenishments are also multiples of Q_{n-1} due to (8.1). Therefore this assumption simply means that we will hit the reorder point exactly when ordering. The reorder point $R_n^i + y$, where $1 \leq y < Q_{n-1}$, will trigger orders at the same times and inventory positions. The only difference is that the inventory position will be y units below the reorder point when ordering because the reorder point is y units higher.

Proposition 8.1 An installation stock reorder point policy can always be replaced by an equivalent echelon stock reorder point policy.

Proof Assume that the installation stock policy is given. Consider installation n. Since the installation stock policy is *nested*, and since we have chosen the reorder points such that all inventory positions will hit the reorder points exactly when orders are triggered, the echelon stock inventory position just after ordering must be:

$$IP_n^e = \sum_{k=1}^{n} (R_k^i + Q_k). \tag{8.3}$$

Because of the unit demand, the echelon stock inventory position is reduced by one unit at a time. Consequently, the orders will be triggered at exactly the same times as with the installation stock policy if we use an echelon stock policy with reorder points:

$$R_n^e = R_n^i + \sum_{k=1}^{n-1}(R_k^i + Q_k). \tag{8.4}$$

This proves the proposition.

If the condition that $IP_n^{i0} - R_n^i$ is a multiple of Q_{n-1} is not satisfied, we first have to change to an equivalent R_n^i satisfying the condition. After that we can apply (8.4) to get the equivalent echelon stock policy. Note that Proposition 1 concerns only the inventory positions. Consequently, the result is not affected by the lead-times.

Under the same assumptions we shall also prove:

Proposition 8.2 An echelon stock reorder point policy which is nested can always be replaced by an equivalent installation stock reorder point policy.

Proof Let the echelon stock policy be given. For installation 1 there is no difference between an echelon stock and an installation stock policy. Consequently, $R_1^i = R_1^e$ will trigger orders at the same times as the echelon stock policy. Consider installation $n > 1$. Due to the unit demand, all installations will always hit their reorder points exactly when ordering. Since the policy is nested we know that if installation n orders, installations 1, 2, ... , $n - 1$ will also order. Immediately after ordering, the installation inventory position at installation n is:

$$IP_n^i = IP_n^e - IP_{n-1}^e = R_n^e + Q_n - R_{n-1}^e - Q_{n-1}. \tag{8.5}$$

Consequently, an installation stock policy with reorder points

$$R_1^i = R_1^e, \quad R_n^i = R_n^e - R_{n-1}^e - Q_{n-1} \quad n > 1, \tag{8.6}$$

will trigger orders at the same times as the echelon stock policy. (Note that (8.6) is equivalent to (8.4).) This completes the proof.

Example 8.2 Consider a serial system with $N = 3$ installations and batch quantities $Q_1 = 5$, $Q_2 = 10$, and $Q_3 = 20$. Assume that the initial inventory positions are: $IP_1^{i0} = 10$, $IP_2^{i0} = 20$, and $IP_3^{i0} = 30$. Assume furthermore that the demand for item 1 is one unit per time unit. Consider first an installation stock policy with reorder points: $R_1^i = 5$, $R_2^i = 10$, and $R_3^i = 10$. Note that our assumptions that Q_n as well as $IP_n^{i0} - R_n^i$ are multiples of Q_{n-1} are satisfied. It is easy to check that installation 1 will order at times 5, 10, 15, ... The demand at installation 2 is consequently 5 units at each of these times. This means that installation 2 will order at times 10, 20, 30, ... Demands for 10 units at installation 3 at these times will trigger orders at times 20, 40, 60, ...

Using (8.4) we obtain the equivalent echelon stock policy as $R_1^e = 5$, $R_2^e = 20$, and $R_3^e = 40$. Recall that when considering the echelon stock, the inventory positions at all installations are reduced by the final demand, i.e., by one unit each time unit, and not by the internal system orders. The initial inventory positions are: $IP_1^{e0} = 10$, $IP_2^{e0} = 30$, and $IP_3^{e0} = 60$. It is easy to verify that the orders will be triggered at the same times as with the installation stock policy.

Assume then that we change the echelon stock reorder point at installation 3 to $R_3^e = 42$. This will not change the orders at installations 1 and 2, but the orders at installation 3 will be triggered 2 time units earlier, i.e., at times 18, 38, 58, ... Recall that the echelon stock is reduced by one unit at a time. The resulting echelon stock policy is not nested and it is impossible to get the same control by an installation stock policy.

Propositions 8.1 and 8.2 show that in any serial inventory system an installation stock policy is simply a special case of an echelon stock policy. This is also true for assembly systems. Recall that these results depend on the assumption (8.1). One advantage of an installation stock policy is that once the reorder points are determined only local information is needed for controlling the replenishments. To implement an echelon stock policy, we need both the installation inventory position and the inventory positions of all downstream installations. An alternative is, at least in principle, to have information of the initial echelon stock inventory position and be able to keep track of the final customer demand. In practice it is usually difficult though, to determine the echelon stock from these data because of various changes in the inventory positions, for example, due to damage and obsolescence.

The higher generality of an echelon stock policy can be an advantage in certain situations. We shall illustrate this by an example.

Example 8.3 Consider a serial system (Figure 8.9) with $N = 2$ installations and batch quantities $Q_1 = 50$ and $Q_2 = 100$. Assume that the final demand at installation 1 is constant and continuous, 50 units per time unit. The lead-time at installation 1 (excluding possible delays at installation 2) is one time unit, and at installation 2 it is 0.5 time units. No shortages are allowed, and the holding costs at installation 1 are higher than at installation 2. The optimal control of the system is illustrated in Figure 8.10.

Consider, for example, the two batches that are delivered to installation 1 at times 2 and 3. These two batches were ordered by installation 1 at times 1 and 2. The delivery of these 100 units to the warehouse took place, just-in-time, at time 1. Immediately after the delivery, 50 units were sent to installation 1. Since the lead-time at installation 2 is 0.5 time units, the corresponding order was triggered at time 0.5. But this can never be achieved by an installation stock (R, Q) policy because it is nested. Installation 2 can only order when installation 1 orders, i.e., at times 0, 1, 2, ... This will lead to additional costs (See Axsäter and Juntti, 1996). The optimal control can be obtained by an echelon stock policy with $R_1^e = 50$ and $R_2^e = 75$. Just before time 0.5 the installation inventory position is zero at installation 2 and 75 at installation 1, since one batch is on order at installation 1.

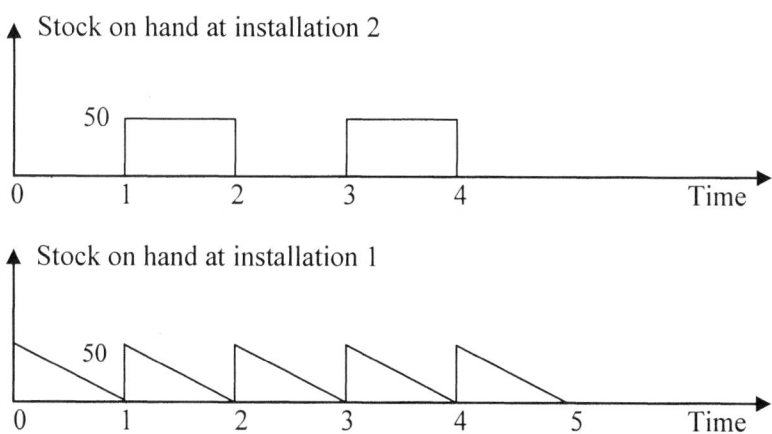

Figure 8.10 Inventory development in the optimal solution.

Let us also state the following result from (Axsäter and Rosling, 1993).

Proposition 8.3 S policies based on the echelon stock are equivalent to S policies based on the installation stock for general systems.

The dominance of echelon stock policies for serial and assembly systems does not carry over to distribution systems. For such systems an installation stock policy may perform better in certain situations, but the opposite can also be true. In general, the differences are relatively small (Axsäter and Juntti, 1996), but in special situations very large differences may occur (Axsäter, 1997).

A KANBAN policy can be interpreted as an installation stock (R, Q) policy with constraints on the number of outstanding orders. By applying the same constraints to echelon stock policies, we obtain a class of policies which, for serial and assembly systems, contains the KANBAN policies as a subset. See Axsäter and Rosling (1999).

8.2.4 Material Requirements Planning

When using reorder point systems, based on the installation stock or on the echelon stock for multi-echelon inventory control, it is generally implicitly assumed that the demands at the different installations are stochastic but still relatively stationary. The reorder points are updated with a certain periodicity but do not change much over time. However, especially in production, it is common to find settings where such reorder point systems do not perform well. Consider, for example, a component for which nearly all demand occurs twice per year when assembling a certain product. Assume that the production lead-time of the component is three weeks. In such a case it is generally suitable to start production of the component a little more than three weeks before the two times per year when large numbers of the component are needed in production. A reorder point system could in principle provide this control, provided that the reorder point is updated continuously with respect to the immediate lead-time demand caused by production of higher level items. In the considered case the expected lead-time demand would then be close to zero unless we are within three weeks of one of the production times. *Material Requirements Planning (MRP)* is essentially such a reorder point system, where the reorder points are updated continuously with respect to known discrete requirements.

To be able to update the lead-time demand it is necessary to keep track of all future production plans for higher level items that may affect the lead-time demand. The time horizon for the planning system should therefore exceed the total system lead-time, i.e., the time from ordering raw materials to deliveries of final products.

MULTI-ECHELON – STRUCTURES AND ORDERING

We shall now describe material requirements planning in detail. The planning procedure is usually applied in a periodic review, rolling horizon setting. The review period can be, for example, one week. The planning procedure is based on the following data:

- A production program for final products, a *Master Production Schedule (MPS)*. In a make-to-order environment the master schedule is usually comprised of plans for final assembly and delivery to customers. The master production schedule must cover the total system lead-time, i.e., the time span between ordering raw materials and delivering end products to customers. Production beyond that horizon cannot affect today's plans for components and raw materials.

- External demands of other items, e.g., demands that do not emanate from production plans for higher level items. An example is the demand for spare parts.

- A bill of material for each item specifying all of its immediate components and their numbers per unit of the parent.

- Inventory status for all items, i.e., stock on hand, stock on order, and backorders.

- Constant lead-times for all items.

- Rules for safety stocks (and/or safety times) and batch quantities.

The procedure starts with end products, i.e., items that only face external demand. The requirements of these items are given by the master production schedule. These items are sometimes denoted Level 0 items. Next we consider all items that are immediate components of Level 0 items. These items, which can also face external demand, are denoted Level 1 items. More generally, Level n items are direct components of Level $n - 1$ items. They can also be components of higher level items and have external demand.

Let us consider the simple product structure in Figure 8.11. We start with the end item, item 1. Table 8.2 illustrates how the net requirements of item 1 are determined.

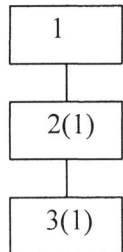

Figure 8.11 Considered product structure.

Table 8.2 Net requirements of item 1.

Item 1	Period	1	2	3	4	5	6	7	8
Lead-time = 1	Gross requirements	10		25	10	20	5		10
Order quantity = 25	Scheduled receipts		25						
Safety stock = 5	Projected inventory	22	12	37	12	2	-18	-23	-23	-33
	Net requirements				3	20	5		10

In the first column the planning parameters, lead-time, order quantity, and safety stock, are given. If the order quantity is one, this is usually termed *lot-for-lot* in connection with material requirements planning. Table 8.2 describes the planning situation at the end of period 0, i.e., just before the start of period 1. The inventory on hand at that time is 22. The given *gross requirements* of item 1 in different periods are obtained from the master production schedule, since item 1 is an end product. Usually the requirements as well as the deliveries are assumed to occur in the beginning of a period. A requirement occurs just after a possible delivery. The *scheduled receipts* in Table 8.2 show planned deliveries of outstanding real orders. The *projected inventory* is the inventory level after the gross requirement and the scheduled receipt. In period 1 the initial inventory is reduced by the gross requirement in period 1 and we obtain the projected inventory 22 - 10 = 12. In period 2 the receipt is added and we obtain 12 + 25 = 37 etc. In period 4 the projected inventory declines to 2 units. The safety stock is 5 units. The most common convention when determining the *net requirement* is to require that backorders plus the safety stock should be covered. This means that the net requirement in period 4 is 3 units. In period 5 the *additional* net requirement is 20, since 3 + 20 - 18 = 5. It is straightforward to complete the table.

We shall in this section follow the mentioned convention that the net requirement should cover the safety stock. Sometimes another rule is used, though. According to the alternative rule, there is no net requirement unless the projected inventory is negative. But when there is a net requirement, the number of units should also cover the safety stock. In Table 8.2 this means that there is no net requirement in period 4, while the net requirement in period 5 is increased by 3 units to 23.

Let us now replace the net requirements in Table 8.2 by *planned orders*. We shall also let the planned orders affect the projected inventory. Note that the order quantity is 25 and that the lead-time is one period. To cover the net requirement in period 4 we need to trigger an order of size 25. This order must be placed in period 3 due to the lead-time. See Table 8.3. Note that a planned order is represented by its starting time, while a scheduled receipt is represented by its delivery time. A planned order is replaced by a scheduled receipt when the order is started. This is because we want to make a difference between real and planned orders. If for example, in period 3, we choose to initiate the planned order in that period, it is removed from planned orders and added as a scheduled receipt in period 4.

Table 8.3 MRP record for item 1 with a planned order in period 3.

Item 1	Period	1	2	3	4	5	6	7	8
Lead-time = 1	Gross requirements	10		25	10	20	5		10
Order quantity = 25	Scheduled receipts		25						
Safety stock = 5	Projected inventory 22	12	37	12	27	7	2	2	-8
	Planned orders			25					

The projected inventory is still below the safety stock in period 6. We therefore need a second order in period 5. This gives the final MRP record in Table 8.4.

Table 8.4 MRP record for item 1 with planned orders in periods 3 and 5.

Item 1	Period	1	2	3	4	5	6	7	8
Lead-time = 1	Gross requirements	10		25	10	20	5		10
Order quantity = 25	Scheduled receipts		25						
Safety stock = 5	Projected inventory 22	12	37	12	27	7	27	27	17
	Planned orders			25		25			

We can interpret the planning of orders as the application of an installation stock reorder point system, where the reorder point is the lead-time demand plus the safety stock minus one (the projected inventory should be larger or equal to the safety stock). Without an order in period 3, the inventory position after the gross requirement is 12. The reorder point in period 3 is 10

+ 5 - 1 = 14. This means that an order is triggered. After the order the inventory position is 37. In period 4 the inventory position is 27 and the reorder point 20 + 5 - 1 = 24, i.e., no order is triggered. In period 5, the inventory position is 7 and the reorder point 5 + 5 - 1 = 9, and the second order is triggered.

Tables 8.2 and 8.3 have mainly been used to explain the MRP logic. From now on we will use the representation in Table 8.4 directly. Recall that Table 8.4 represents the planning situation at the end of period 0. At the end of period 1 we consider a similar table covering periods 2 - 9.

Let us now go back to Figure 8.11. We assume that item 2 is only needed for production of item 1, and that item 3 is only needed for production of item 2. According to Table 8.4 we plan to start production of 25 units in the beginning of periods 3 and 5. This means that we need 25 units of item 2 at the same times, i.e., we have determined the gross requirements of item 2. Given the gross requirements, we can apply the same planning procedure for item 2 and determine the planned orders, which in turn are the gross requirements of item 3. The whole MRP procedure is shown in Figure 8.12.

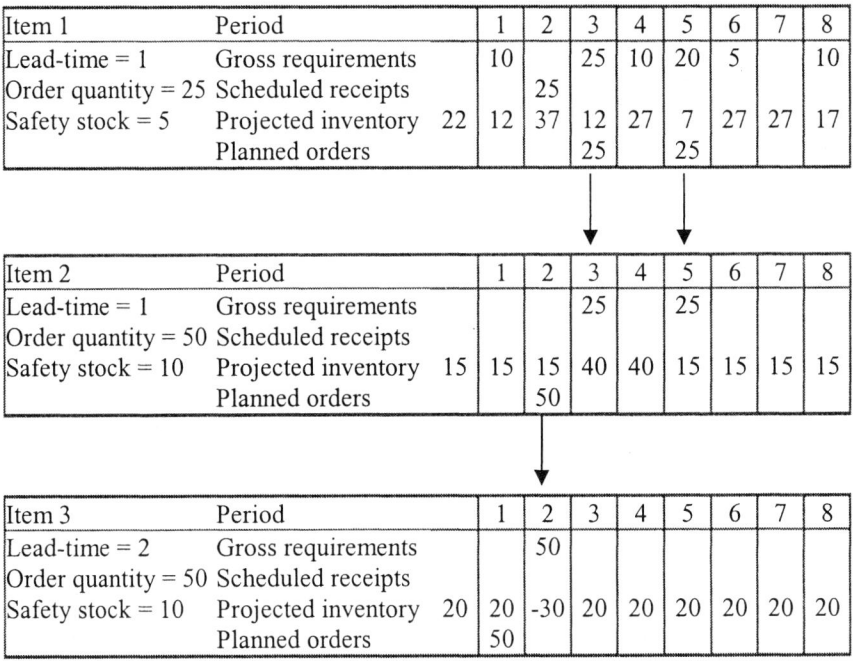

Figure 8.12 Material requirements planning for items 1, 2, and 3 in Figure 8.11.

Note that the planning procedure in Figure 8.12 gives a shortage of item 3 in period 2. To avoid the shortage the order should have been started in period 0, but we are already at the end of this period. The shortage means that the plans for items 1 and 2 cannot be followed. In practice there are usually several ways to overcome the problem. Perhaps it is possible to temporarily reduce the lead-time of item 3 to one period. Another alternative is to adjust the basis for the planning. It may, for example, be possible to change the master production schedule, which determines the gross requirements of item 1, or to adjust order quantities and safety stocks.

So far we have assumed that safety stocks are used to cover variations in demand and lead-times. When applying MRP it is also common to use a *safety time* as an alternative or supplement. This means that we plan the orders so that the deliveries are expected to occur the safety time before the corresponding net requirements. Table 8.5 shows the MRP record if we add a safety time of one period to the planning situation in Table 8.4.

Table 8.5 MRP record for item 1 with a safety time.

Item 1	Period	1	2	3	4	5	6	7	8
Lead-time = 1	Gross requirements	10		25	10	20	5		10
Order quantity = 25	Scheduled receipts		25						
Safety stock = 5	Projected inventory 22	12	37	37	27	32	27	27	17
Safety time = 1	Planned orders		25		25				

In Figure 8.12 it is assumed that item 2 is only part of item 1 and that item 3 is only part of item 2. It is very common, though, that an item is a component of many higher level items. Figures 8.13 and 8.14 illustrate how this is handled in MRP. Recall that when processing a certain item, plans for all items at higher levels are already complete.

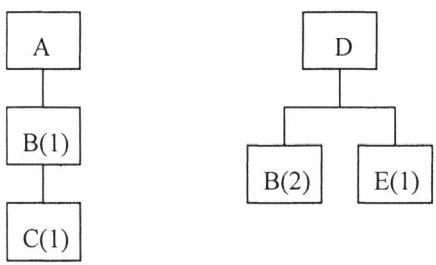

Figure 8.13 Product structures.

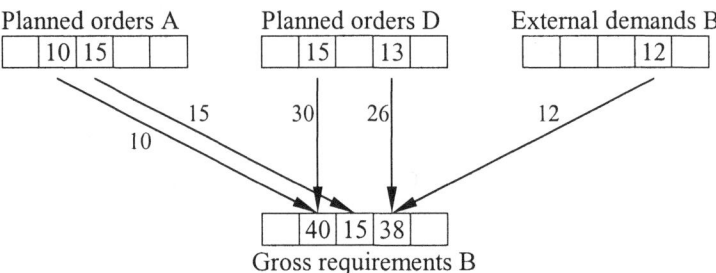

Figure 8.14 Gross requirements from different sources.

Sometimes it is important to keep track of which higher level items have generated different gross requirements. If for example, we obtain an infeasible plan such as for item 3 in Figure 8.12, it is interesting to see if it is possible to avoid the difficulties by changing the plans for one or more of the parent items. To be able to do that we need to *"peg"* generated gross requirements with identifications of the items generating them.

MRP is often referred to as a *push system* since orders are in a sense triggered in anticipation of future needs. Reorder point systems and KANBAN systems are similarly said to be *pull systems* because orders are triggered when downstream installations need them. This notation is unclear and misleading. An MRP system is, in fact, a more general ordering system than an installation stock reorder point system. Recall that the MRP logic can be interpreted as application of a reorder point system, where the reorder point in each period is the lead-time demand plus the safety stock minus one. The lead-time in the MRP system is a planning parameter that does not necessarily need to correspond to an estimate of the real lead-time. If we set the lead-time to zero, the reorder point is constant and equal to the safety stock minus one. The safety stock is also a planning parameter that can be chosen arbitrarily. Consequently, we can state the following proposition:

Proposition 8.4 For a general inventory system, an MRP system can give the same control as any installation stock reorder point system.

In Axsäter and Rosling (1994) it is also shown that:

Proposition 8.5 For serial and assembly systems, MRP can work as any echelon stock reorder point system.

Consequently we can conclude that if MRP is a push system, it is also a pull system.

The MRP logic is simple, yet the computational effort can be very large if there are thousands of items and complex multi-level product structures. In such systems it is also difficult to update all data. As we have discussed above, the major advantage of MRP is that it enables us to keep track of relatively infrequent large requirements. Many low level components often have a large number of parents, and the gross requirements obtained from the MRP calculations will, for statistical reasons, be very smooth. Quite often it is more practical for such items to forecast the demand by exponential smoothing and use a simple reorder point system. These items can then be excluded from the bills of material and the remaining product structures will be simpler.

One problem with MRP can be that the plans for individual items change too frequently over time. Even a very small change in the master production schedule can lead to relatively large changes in the plans for many lower level items. This "instability" of MRP is often called *"nervousness"*. One possibility to reduce the nervousness is to *"freeze"* orders within a certain time frame. This means that orders within this time frame are not changed according to the MRP logic.

It is most common to use MRP in production. But since a product structure corresponds to a multi-echelon inventory system (See e.g., Figure 8.8), we can also use the same ordering technique in a distribution system. In this context the term MRP is sometimes replaced by *DRP, Distribution Requirements Planning*.

Usually the MRP procedures that we have described are the core in a planning system that also contains other planning modules. It is especially common to include modules for capacity planning. In such more general systems MRP is then sometimes interpreted as *Manufacturing Resource Planning*. The term MRP II is also used.

Capacity requirements can be estimated in two ways. One possibility is to forecast the needed production capacity directly from the master production schedule. Such a forecast is usually not very accurate since the inventory status of different items is not taken into account. In general, this is denoted *Rough Cut Capacity Planning (RCCP)*. It is more accurate to estimate the capacity requirements after carrying out the complete MRP calculations. This is usually denoted *Capacity Requirements Planning (CRP)*. Figure 8.15 illustrates how the modules can be coupled to each other.

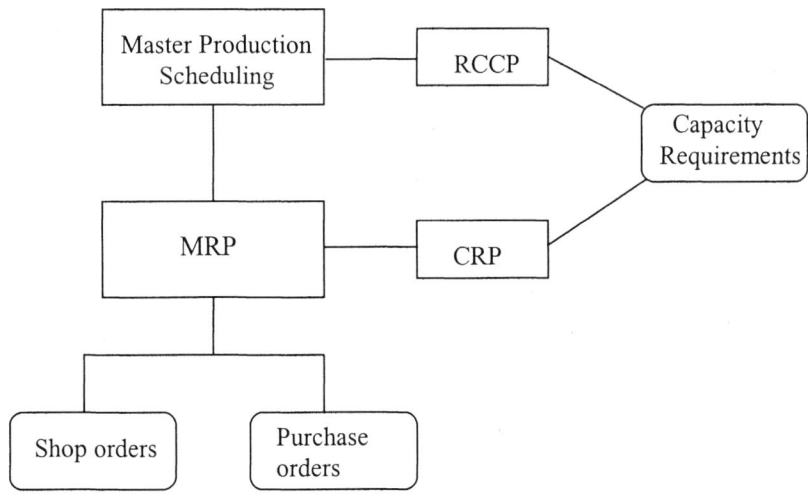

Figure 8.15 Manufacturing Resource Planning.

When choosing the master schedule, the capacity requirements are estimated by Rough Cut Capacity Planning. After choosing the master schedule and going through the MRP calculations, a more accurate estimate is obtained from Capacity Requirements Planning. If there are large deviations between these two estimates of the capacity requirements, and the more accurate estimate is infeasible, it may be necessary to go back and adjust the master schedule.

Although the system in Figure 8.15 contains more than material requirements planning, it is obvious that MRP is the heart of the system. The capacity planning modules essentially give only the estimated capacity consequences of the MRP plan. A planning framework of this type is natural if the main task of the system is materials management. If the main problem is to achieve high and smooth capacity utilization, it can be more suitable to start with the capacity planning and let the results of this planning be a basis for the materials management.

A so-called *Enterprise Resource Planning (ERP)* system can be seen as a further extension of MRP. Such systems usually provide a number of additional functions, many of which are outside the production area, like warehouse management, shipping, financial information and planning, etc. ERP can also be a multilocation system and handle production planning across multiple sites, which is usually very important for large corporations.

We refer to Orlicky (1975), Baker (1993), Silver et al. (1998), and Vollman et al. (1997) for a more detailed treatment of MRP.

8.2.5 Ordering system dynamics

A major advantage of installation stock reorder point policies and KANBAN policies is that these policies are easy to implement. In particular, they only need local information about inventory positions after the control parameters are set. One disadvantage, however, is that the information in the supply chain is delayed. Consider again the material requirements planning in Figure 8.12. Assume that the gross requirements of item 1 will increase due to larger demand from the final customers. This may lead to a change in the planned orders for item 1 so that the order in period 3 will instead be placed in period 2. When using material requirements planning, this will immediately lead to corresponding changes in the plans for items 2 and 3. We are consequently able to let changes in the master production schedule directly affect plans for all items. Echelon stock reorder point policies provide similar possibilities, since the echelon stock is immediately affected by the final customer demand. If the stocks are controlled in a decentralized manner by installation stock reorder point policies, the information concerning the increasing customer demand is not transmitted in the same rapid way, however. With an installation stock reorder point system, the order for item 1 in Figure 8.12 may also be placed in period 2 instead of period 3. The problem is that this information is not available at installation 2 until the order actually occurs in period 2. Nothing happens during period 1. In some situations with large batch quantities and many stages, such information delays can cause major problems.

There is also considerable evidence, theoretical as well as practical, that decentralized installation stock policies in multi-echelon systems can yield very large demand variations early in the material flow, even though the final demand is very stable. Consider the simple serial system in Figure 8.16.

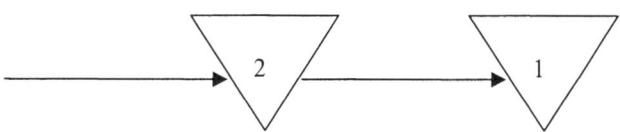

Figure 8.16 A simple serial two-echelon inventory system.

Let us assume that the inventory system is part of a distribution system and that the installations are situated far from each other. We also assume that the lead-times are five weeks for each installation. The two installations are adapting their reorder points with respect to the demand at their site. The demand at installation 2 is the orders from installation 1. We shall also assume that both installations use periodic review policies with a review period of one week. There are many similar items, but the items are handled individually.

Assume now that site 1 estimates that the final demand for all items has increased by 20 percent. Normally the site orders about one week's demand each week. The increase in customer demand means that the estimated lead-time demand is also increased by 20 percent, which corresponds to the original demand during one week. The reorder points are raised correspondingly. The orders from installation 1 will increase due both to the higher reorder points and due to the increase in the demand. The resulting demand at installation 2 in the first week will increase by more than 100 percent. If installation 2 adjusts its reorder points with respect to that change, the effects for its supplier will be devastating.

Note that the demand at installation 2 in the long run will only increase by 20 percent. The very high demand just after the change is due to the adjustments of the reorder points at installation 1. Furthermore, if installation 2 has information about the final customer demand, or if the dynamic effects that we have described are understood, the adjustments of the reorder points at installation 2 can be expected to be less dramatic.

The high demand variations at upstream stages have been analyzed and discussed by several authors. The first well-known contribution was by Forrester (1961). The phenomenon is today often denoted the *bullwhip effect*. For further details see Baganha and Cohen (1998), Lee et al. (1997a,b), and Towill and Del Vecchio (1994). It is common to illustrate the bullwhip effect by a game concerning a distribution system for beer, the Beer Game. Different variations of this game exist; see Sterman (1989), and Chen and Samroengraja (2000).

It is important to note that both the information delays and the problems of large demand variations at upstream facilities are largely due to long lead-times and large batch quantities. With small batch quantities, the times between orders will be short, and the information will be transmitted relatively rapidly between the echelons even when using an installation stock reorder point system. If the lead-times are short, adjustments of the reorder points due to demand changes will also become small, and the demand variations at the next upstream stage will be limited. Consequently, for systems with small batch quantities and short lead-times, installation stock reorder point policies and KANBAN policies will perform very well in most cases.

References

Alfredsson, P., and J. H. C. M. Verrijdt. 1999. Modeling Emergency Supply Flexibility in a Two-Echelon Inventory System, *Management Science*, 45, 1416-1431.

Archibald, A. W., S. A. Sassen, and L. C. Thomas. 1997. An Optimal Policy for a Two Depot Inventory Problem with Stock Transfer, *Management Science*, 43, 173-183.

Axsäter, S. 1997. On Deficiencies of Common Ordering Policies for Multi-Level Inventory Control, *OR Spektrum*, 19, 109-110.

Axsäter, S. 1990. Modelling Emergency Lateral Transshipments in Inventory Systems, *Management Science*, 36, 1329-1338.

Axsäter, S. 2003. A New Decision Rule for Lateral Transshipments in Inventory Systems, *Management Science*, 49, 1168-1179.

Axsäter, S. 2005. A Heuristic for Triggering Emergency Orders in an Inventory System, *European Journal of Operational Research* (to appear).

Axsäter, S., and L. Juntti. 1996. Comparison of Echelon Stock and Installation Stock Policies for Two-Level Inventory Systems, *International Journal of Production Economics*, 45, 303-310.

Axsäter, S., and K. Rosling. 1993. Installation vs. Echelon Stock Policies for Multi-level Inventory Control, *Management Science*, 39, 1274-1280.

Axsäter, S., and K. Rosling. 1994. Multi-Level Production-Inventory Control: Material Requirements Planning or Reorder Point Policies?, *European Journal of Operational Research*, 75, 405-412.

Axsäter, S., and K. Rosling. 1999. Ranking of Generalised Multi-Stage KANBAN Policies, *European Journal of Operational Research*, 113, 560-567.

Baganha, M. P., and M. A. Cohen. 1998. The Stabilizing Effect of Inventory in Supply Chains, *Operations Research*, 46, S572-S583.

Baker, K. R. 1993. Requirements Planning, in S. C. Graves et al. Eds. *Handbooks in OR & MS Vol. 4*, North Holland, Amsterdam, 571-627.

Bassok, Y., R. Anupindi, and R. Akella. 1999. Single-Period Multiproduct Inventory Models with Substitution, *Operations Research*, 47, 632-642.

Chen. F. 2000. Optimal Policies for Multi-Echelon Inventory Problems with Batch Ordering, *Operations Research*, 48, 376-389.

Chen, F., and R. Samroengraja. 2000. The Stationary Beer Game, *Production and Inventory Management*, 9, 19-30.

Clark, A. J., and H. Scarf. 1960. Optimal Policies for a Multi-Echelon Inventory Problem, *Management Science*, 5, 475-490.

Dada, M. 1992. A Two-Echelon Inventory System with Priority Shipments, *Management Science*, 38, 1140-1153.

Das, C. 1975. Supply and Redistribution Rules for Two-Location Inventory Systems: One Period Analysis, *Management Science*, 21, 765-776.

Dekker, R., M. Fleischmann, K. Inderfurth, L. N. van Wassenhove. Eds. 2004. *Reverse Logistics, A Quantitative Approach*, Springer, Berlin.

Forrester, J. W. 1961. *Industrial Dynamics*, MIT Press, Cambridge.

Grahovac, J., and A. Chakravarty 2001. Sharing and Lateral Transshipment of Inventory in a Supply Chain with Expensive, Low-Demand Items, *Management Science*, 47, 579-594.

Guide, Jr., V. D. R., R. H. Teunter, L. N. van Wassenhove. 2003. Matching Demand and Supply to Maximize Profits from Remanufacturing, *Manufacturing & Service Operations Management*, 5, 303-316.

Johansen, S. G. and A. Thorstenson. 1998. An Inventory Model with Poisson demands and Emergency Orders, *International Journal of Production Economics*, 56-57, 275-289.

Kukreja, A., C. P. Schmidt, and D. M. Miller. 2001. Stocking Decisions for Low-Usage Items in a Multilocation Inventory System, *Management Science*, 47, 1371-1383.

Lee, H. L. 1987. A Multi-Echelon Inventory Model for Repairable Items with Emergency Lateral Transshipments, *Management Science*, 33, 1302-1316.

Lee, H. L., P. Padmanabhan, and S. Whang. 1997a. The Bullwhip Effect in Supply Chains, *Sloan Management Review*, 38, 93-102.

Lee, H. L., P. Padmanabhan, and S. Whang. 1997b. Information Distortion in a Supply Chain: The Bullwhip Effect, *Management Science*, 43, 546-558.

Minner, S., E. A. Silver, and D. J. Robb. 2003. An Improved Heuristic for Deciding on Emergency Transshipments, *European Journal of Operational Research*, 148, 384-400.

Minner, S. 2003. Multiple-Supplier Inventory Models in Supply Chain Management: A Review, *International Journal of Production Economics*, 81-82, 265-279.

Moinzadeh, K. and S. Nahmias. 1988. A Continuous Review Model for an Inventory System with Two Supply Modes, *Management Science*, 34, 761-773.

Muckstadt, J. A. 2005. *Analysis and Algorithms for Service Parts Supply Chains*, Springer, New York.

Muharremoglu, A., and J. N. Tsitsiklis. 2003. A Single-Unit Decomposition Approach to Multi-Echelon Inventory Systems, Operations Research Center, MIT.

Nahmias, S. 1981. Managing Repairable Item Inventory Systems: A Review, in L. B. Schwarz Ed. *Multi-Level Production/Inventory Control Systems: Theory and Practice*, North Holland Amsterdam, 253-277.

Olsson, F. 2005. Optimal Asymmetric Policies for Inventory Systems with Lateral Transshipments, Lund University.

Orlicky, J. 1975. *Material Requirements Planning*, McGraw-Hill, New York.

Robinson, L. W. 1990. Optimal and Approximate Policies in Multiperiod, Multilocation Inventory Models with Transshipments, *Operations Research*, 38, 278-295.

Rudi, N., S. Kapur, and D. F. Pyke. 2001. A Two-Location Inventory Model with Transshipment and Local Decision Making, *Management Science*, 47, 1668-1680.

Sherbrooke, C. C. 1968. METRIC: A Multi-Echelon Technique for Recoverable Item Control, *Operations Research*, 16, 122-141.

Sherbrooke, C. C. 1992. Multiechelon Inventory Systems with Lateral Supply, *Naval Research Logistics*, 39, 29-40.

Silver, E. A., D. F. Pyke, and R. Peterson. 1998. *Inventory Management and Production Planning and Scheduling*, 3rd edition, Wiley, New York.

Spearman, M. L., D. L. Woodruff, and D. L. Hopp. 1990. CONWIP: A Pull Alternative to KANBAN, *International Journal of Production Research*, 28, 879-894.

Sterman, J. 1989. Modeling Managerial Behavior: Misperceptions of Feedback in a Dynamic Decision Making Experiment, *Management Science*, 35, 321-339.

Tagaras, G., and M. A. Cohen. 1992. Pooling in Two-Location Inventory Systems with Non-Negligible Replenishment Lead Times, *Management Science*, 38, 1067-1083.

Towill, D., and A. Del Vecchio. 1994. The Application of Filter Theory to the Study of Supply Chain Dynamics, *Production Planning & Control*, 5, 82-96.

Van der Laan, E. A., M. Fleischmann, R. Dekker, and L. N. van Wassenhove. 1998. Inventory Control for Joint Manufacturing and Remanufacturing, in S. Tayur et al. Eds. *Quantitative Models for Supply Chain Management*, Kluwer Academic Publishers, Boston, 807-835.

Van der Laan, E. A., M. Salomon, R. Dekker, and L. N. van Wassenhove. 1999. Inventory Control in Hybrid Systems with Remanufacturing, *Management Science*, 45, 733-747.

Veatch, M. H., and L. M. Wein. 1994. Optimal Control of a Two-Station Tandem Production/Inventory System, *Operations Research*, 42, 337-350.

Vidgren, S. 2005. Evaluation of a Class of Lateral Transshipment Policies, Lund University.

Vollman, T. E., W. L. Berry, and D. C. Whybark. 1997. *Manufacturing Planning and Control Systems*, 4th edition, Irwin, Boston.

Problems

8.1 Consider Example 8.1. Assume that the installation stock inventory positions for items 1-4 are 2, -2, 5, and 7, respectively. Determine the echelon stock inventory positions.

8.2 Show that (8.4) and (8.6) are equivalent.

8.3* Consider the serial system

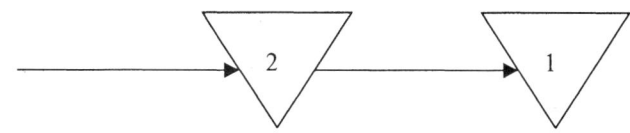

---

* Answer and/or hint in Appendix 1.

The initial installation stock inventory positions are $IP_1^{i0} = 15$ and $IP_2^{i0} = 12$. The system is controlled by installation stock reorder point policies with $R_1^i = 8$, $Q_1 = 10$, $R_2^i = 9$, and $Q_2 = 20$.

a) Determine an equivalent echelon stock reorder point policy.
b) Consider constant continuous demand, one unit per time unit, and verify that the two policies order at the same times.
c) Increase R_2^e by one unit. At what times will installation 2 order? Does an equivalent installation stock policy exist?

8.4 Consider again the serial inventory system in Problem 8.3. The production control manager has been visiting Japan. Influenced by what he has seen there he wants to use a KANBAN system. For installation 1 he has one card and a container for 50 units. For installation 2 he has also one card but a container for 100 units. Both lead-times are zero so no shortages can occur. Initially there are 50 units in stock at each installation. The demand at installation 1 is one unit per time unit.

a) The assistant production control manager thinks that the new ideas are unnecessary and suggests an equivalent installation stock (R, Q) system. How should R and Q be chosen for the two installations?
b) Is it possible to get the same control by an echelon stock policy? How should in that case R and Q be chosen for the two installations?

8.5 Consider again the serial inventory system in Problem 8.3. Let the batch quantities be $Q_1 = 10$ and $Q_2 = 20$. The installation stock reorder points are $R_1^i = 22$ and $R_2^i = 35$. The initial installation stock inventory positions are $IP_1^{i0} = 27$ and $IP_2^{i0} = 50$. Determine an equivalent echelon stock policy. Use an example to verify that your result is correct.

8.6 Consider the distribution system in Figure 8.2. Prove Proposition 8.3 for this system.

8.7* Consider the product structure below. Other data are:

Item	A	B	C
Lead-time, periods	1	1	1
Initial inventory	22	34	12
Order quantity	30	30	90

MULTI-ECHELON – STRUCTURES AND ORDERING

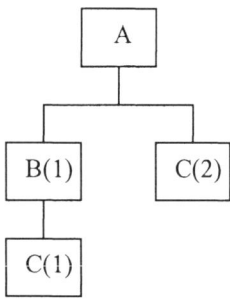

The gross requirements of item A over 8 periods are: 7, 9, 28, 30, 18, 16, 24, 13. There are no safety stocks. Carry out the complete material requirements planning.

8.8* Consider the same product structure as in Problem 8.7 and the following data:

Item	A	B	C
Lead-time, periods	1	1	1
Initial inventory	27	34	75
Order quantity	10	10	20
Safety stock	10	10	20

The gross requirements of item A over 6 periods are: 5, 26, 0, 13, 12, 0.

a) Carry out the material requirements planning. Are any orders delayed?
b) Change the safety stocks to one period safety times. Carry out the material requirements planning. Are any orders delayed?

9 MULTI-ECHELON SYSTEMS: LOT SIZING

In Chapter 8 we have discussed why and how multi-echelon inventory systems appear in practice, and how the most common ordering systems work. We shall now turn to the problem of choosing batch quantities for a multi-echelon inventory system. Note that this important problem must be solved independently of which type of ordering system we use.

First of all, when determining order quantities for a multi-echelon inventory system, it is not optimal to handle each site individually. The choice of batch quantity at a certain site will namely affect the demand structure at the next upstream site. This dependency makes the determination of batch quantities much more complex. Although it is not optimal to determine batch quantities individually for each site, this is very common in practice because it is so easy. Generally this will lead to order quantities that are too small.

When dealing with order quantities for multi-echelon systems we shall assume that the customer demand is known and deterministic, i.e., we disregard random variations. In case of stochastic demand, it is normally reasonable to replace the stochastic demand by its mean and use a deterministic model when determining batch quantities. The stochastic demand variations, are however, taken into account, in a second step, when we determine safety stocks and reorder points. This approximation, which is common also for single-echelon systems, is usually more or less necessary for multi-echelon systems.

Furthermore, we make the standard assumption that all lead-times are zero. Recall that for a single-echelon system with a constant lead-time and deterministic demand, we can make this assumption without any lack of generality. In case of a positive lead-time L, we just need to order L time units earlier. When dealing with multi-echelon systems it is generally possible to handle serial and assembly systems similarly, i.e., we can determine batch quantities for zero lead-times and still use the results also for positive lead-times. For distribution systems (Figure 8.2) and general systems (Figure 8.4), however, positive lead-times may be more difficult to deal with.

9.1 Identical order quantities

9.1.1 Infinite production rates

Consider first the simple assembly system and the corresponding product structure in Figure 9.1.

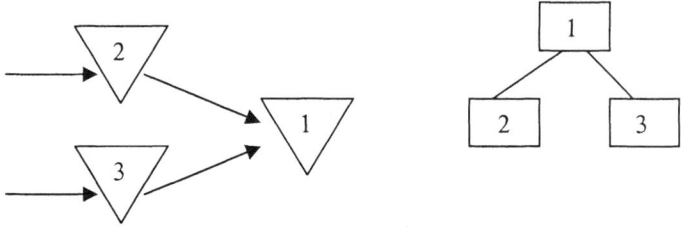

Figure 9.1 A simple assembly system.

Let us think of a production system. Assume that we have chosen the units so that when producing one unit of item 1, exactly one unit of item 2 and one unit of item 3 are needed as components. (Recall that this is always possible.)

Item 1 is a final product with constant continuous demand d. Let the setup costs be A_1, A_2, and A_3, and the holding costs h_1, h_2, and h_3.

Furthermore, we assume infinite production rates, and that for some reason, the same batch quantity is used for all three items. Infinite production rate means that the whole batch is delivered at the same time. (As discussed above we also assume that the lead-times are zero.)

Given that the batch quantities are the same, it is obviously optimal to produce batches of items 2 and 3 just-in-time for the production of item 1. This means that there are never any stocks of items 2 and 3, and all three items are produced at the same time. But if the items are produced at the same time, we can replace the considered three items by a single item with setup cost $A = A_1 + A_2 + A_3$, and holding cost $h = h_1$. In case of constant de-

MULTI-ECHELON – LOT SIZING

mand, we can then simply use the classical economic order quantity model (4.3) for determining the optimal batch quantity.

Our conclusions for the simple assembly system in Figure 9.1 can easily be generalized to a general assembly system and to time-varying demand.

9.1.2 Finite production rates

If we have finite production rates but still identical lot sizes, it is more difficult to determine the optimal lot size for a multi-echelon system. However, for serial and assembly systems and constant demand, it is still usually possible to determine the optimal batch size from a model, which can be seen as a variation of the classical economic order quantity model. We shall illustrate this by an example.

Example 9.1 Consider a product that is produced from raw material in two machines, 1 and 2. The machines have production rates $p_1 > p_2 > d$, where d is the constant customer demand. At the beginning of a production cycle, a batch of raw material is delivered to machine 1. The output from machine 1 is immediately (continuously) transferred to machine 2. When the whole batch has completed production in machine 2, it is transported to a warehouse facing the constant customer demand. The machines have setup costs A_1 and A_2. There is also a fixed cost A_3 associated with the transportation of a batch of goods from machine 2 to the warehouse. The holding cost before machine 1 is h_1 per unit and time unit. After being processed in machine 1 the holding cost is h_2, and after machine 2 it is h_3 (valid also at the warehouse). We wish to determine the optimal batch size.

First of all, we note that we can disregard the holding costs during transportation since each unit is in transport the same time independent of the batch size. Still we have four echelons where the holding costs are affected by the batch size, i.e., before machine 1 (1), between machines 1 and 2 (2), after machine 2 (3) and at the warehouse (4). Figure 9.2 illustrates the development of the stock levels over time. From the figure it is easy to determine the average stock over time at the different stock points. Consider, for example, the stock level before machine 1. The time between two deliveries is Q/d. During this time we have stock in front of the machine while the machine is producing, i.e., during the time Q/p_1. This means that the fraction of time with a positive stock level is d/p_1. The average stock level while the stock is positive is $Q/2$. The average stock level over all times is therefore $(Q/2)(d/p_1)$. Other stocks can be handled similarly. Each batch will give ordering costs in both machines and a fixed cost for transportation. Consequently we can express the total relevant costs as:

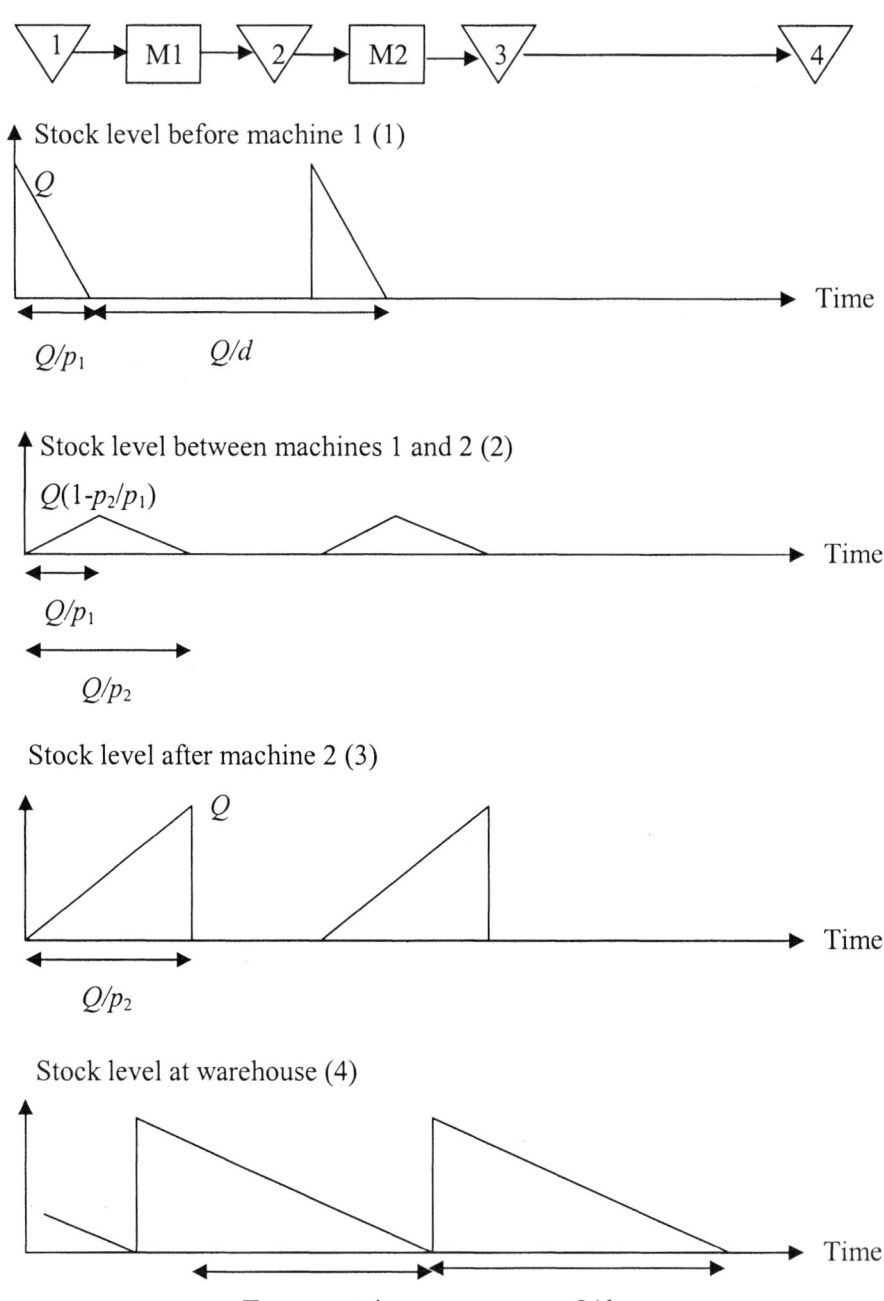

Figure 9.2 Material flow and development of stock levels over time in Example 9.1.

$$C = \frac{Q}{2}\left(\frac{d}{p_1}h_1 + (\frac{d}{p_2} - \frac{d}{p_1})h_2 + (\frac{d}{p_2} + 1)h_3\right) + \frac{d}{Q}(A_1 + A_2 + A_3). \quad (9.1)$$

By setting the derivative equal to zero we obtain the optimal batch quantity as

$$Q^* = \sqrt{\frac{2(A_1 + A_2 + A_3)d}{\frac{d}{p_1}h_1 + (\frac{d}{p_2} - \frac{d}{p_1})h_2 + (\frac{d}{p_2} + 1)h_3}}. \quad (9.2)$$

Note that if we let the production rates approach infinity, we are back in the situation considered in Section 9.1.1 and we only need to consider a single echelon with setup cost $A = A_1 + A_2 + A_3$, and holding cost $h = h_3$. The only stock that we need to consider is then the stock at the warehouse.

9.2 Constant demand

We shall now consider the more general case when the batch quantities at different installations do not need to be identical. We assume constant continuous demand for the end product and infinite production rates.

9.2.1 A simple serial system with constant demand

Consider the simple serial system in Figure 9.3.

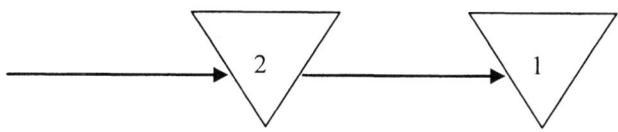

Figure 9.3 A simple serial inventory system.

Let us assume that item 1 is a final product which is produced from one unit of the component 2, which in turn, is obtained from an external supplier. As we have discussed above, we assume that the lead-times are zero. The demand, d, for item 1 is constant and continuous. Both items have ordering costs, A_1 and A_2, as well as holding costs, h_1 and h_2. When delivering a batch, the whole quantity is delivered at the same time. No backorders are allowed. We want to determine optimal batch quantities, Q_1 and Q_2. Note that except for the two echelons, the assumptions are exactly the same as in the classical economic order quantity model, see Section 4.1.

Example 9.2 Let the problem parameters be $d = 8$, $A_1 = 20$, $A_2 = 80$, $h_1 = 5$, and $h_2 = 4$.

Item 1 faces constant continuous demand and the costs are consequently obtained as

$$C_1 = h_1 \frac{Q_1}{2} + A_1 \frac{d}{Q_1}. \tag{9.3}$$

One approach to solving the problem could be to apply the classical economic order quantity model to item 1, i.e., to minimize (9.3). We obtain

$$Q_1 = \sqrt{\frac{2A_1 d}{h_1}} = 8,$$

and the corresponding costs

$$C_1 = \sqrt{2A_1 d h_1} = 40.$$

The demand for item 2 is then discrete and not continuous, 8 units each time unit. Given a constant Q_1, we should always choose Q_2 as a multiple of Q_1,

$$Q_2 = k Q_1, \tag{9.4}$$

where k is a positive integer. This is easy to show in the same way as we showed *Property 1* in Section 4.5.

Figure 9.4 illustrates the behavior of the inventory levels for $k = 3$. Since the lead-times are zero there is no difference between inventory levels and inventory positions. The figure shows both the installation stock and the echelon stock. The echelon stock of item 2 is simply the sum of the stock at both installations. For item 1 the echelon stock is equal to the installation stock since there is no downstream stock.

MULTI-ECHELON – LOT SIZING

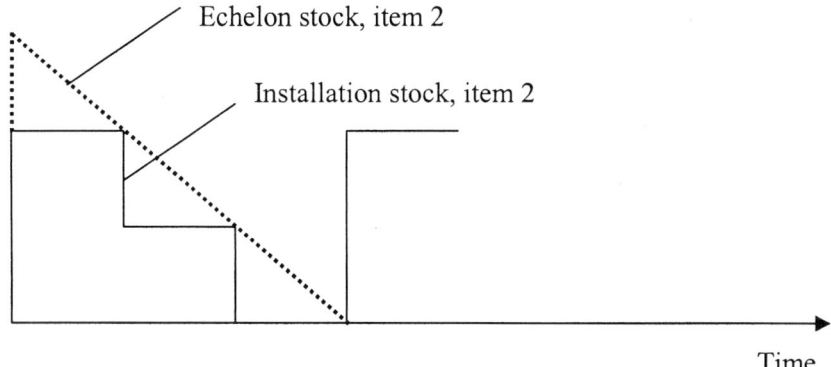

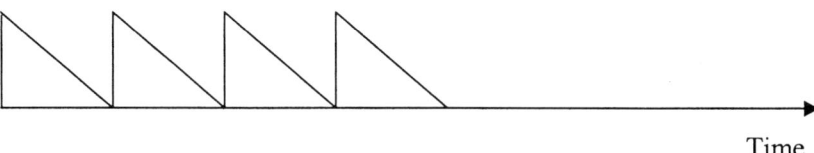

Figure 9.4 Inventory levels for $k = 3$.

The costs for item 2 are obtained as:

$$C_2 = h_2 \frac{(k-1)Q_1}{2} + A_2 \frac{d}{kQ_1}, \quad (9.5)$$

since the average installation stock of item 2 is $(k - 1)Q_1/2$. C_2 is evidently convex in k. If we disregard that k_2 has to be an integer when minimizing C_2 we obtain:

$$k^* = \frac{1}{Q_1} \sqrt{\frac{2A_2 d}{h_2}}.$$

If $k^* < 1$ it is optimal to choose $k = 1$. Assume that $k^* > 1$. Let k' be the largest integer less or equal to k^*, i.e., $k' \le k^* < k' + 1$. It is easy to show that it is optimal to choose $k = k'$ if $k^*/k' \le (k' + 1)/k^*$. Otherwise $k = k'+ 1$ is optimal. (See Section 4.1.2.)

We obtain $k^* \approx 2.24$ and it is optimal to have $k = 2$. The corresponding costs are obtained from (9.5) as $C_2 = 56$, and we get the total costs per time unit as $C = C_1 + C_2 = 40 + 56 = 96$. As we shall show below, this is not the optimal solution. It is not feasible to optimize item 1 separately without considering the consequences for item 2, since the batch size for item 1 will affect the demand for item 2.

Let us now derive the optimal solution. It can be shown that (9.4) must be satisfied in the optimal solution for a two-level system. (As illustrated by Example 9.4 in Section 9.2.2, this is not necessarily true if there are more than two levels.) The total costs are obtained by adding (9.3) and (9.5),

$$C = C_1 + C_2 = (h_1 + (k-1)h_2)\frac{Q_1}{2} + (A_1 + \frac{A_2}{k})\frac{d}{Q_1}. \qquad (9.6)$$

Alternatively, we can represent the costs in terms of the echelon stock. Since the echelon stock of item 2 includes the stock of item 1, the holding cost for item 1 should only represent the value added when producing item 1 from item 2. This means that we shall use the *echelon holding costs* $e_1 = h_1 - h_2$, and $e_2 = h_2$. We obtain:

$$C_1^e = e_1 \frac{Q_1}{2} + A_1 \frac{d}{Q_1}, \qquad (9.7)$$

and

$$C_2^e = e_2 \frac{kQ_1}{2} + A_2 \frac{d}{kQ_1}. \qquad (9.8)$$

The costs (9.7) and (9.8) are not the same as the costs (9.3) and (9.5) but it is easy to verify that the total costs,

$$C = C_1^e + C_2^e = (e_1 + k e_2)\frac{Q_1}{2} + (A_1 + \frac{A_2}{k})\frac{d}{Q_1}, \qquad (9.9)$$

are the same, i.e., the cost expressions (9.6) and (9.9) are equivalent. The representation in (9.7)-(9.9) is, in general, easier to handle since the structures of (9.7) and (9.8) are exactly the same. This is also illustrated by Figure 9.4.

For a given k the optimal Q_1 is obtained from (9.9) by setting the derivative equal to zero.

$$Q_1 = \sqrt{\frac{2(A_1 + \frac{A_2}{k})d}{e_1 + k e_2}} \, . \tag{9.10}$$

Inserting (9.10) in (9.9) we obtain the optimal costs for a given k as

$$C(k) = \sqrt{2(A_1 + \frac{A_2}{k})d(e_1 + k e_2)} \, . \tag{9.11}$$

It is easy to see that

$$C^2(k) = 2d(A_1 e_1 + A_2 e_2 + A_1 e_2 k + \frac{A_2 e_1}{k}) \tag{9.12}$$

is a convex function of k. The optimization of (9.12) with respect to k is similar to the optimization of (9.5) above. First we disregard that k has to be an integer. We get

$$k^* = \sqrt{\frac{A_2 e_1}{A_1 e_2}} \, . \tag{9.13}$$

If $k^* < 1$ it is optimal to choose $k = 1$. Assume that $k^* > 1$. Let k' be the largest integer less or equal to k^*, i.e., $k' \leq k^* < k' + 1$. It is optimal to choose $k = k'$ if $k^*/k' \leq (k' + 1)/k^*$. Otherwise $k = k' + 1$ is optimal.

If $A_1/e_1 \geq A_2/e_2$ we get $k = 1$. This means that we should use the same batch size for the two items, and consequently, each time we produce a batch of item 2, the batch should immediately be used for production of item 1. This implies that we do not need any stock of item 2 and that our two-stage system can be replaced by a single-stage system. It can be shown that the condition $A_1/e_1 \geq A_2/e_2$ is also sufficient in case of time-varying demand, see Axsäter and Nuttle (1987) and Section 9.3.3.

Example 9.3 Consider again the problem in Example 9.2, i.e., $d = 8$, $A_1 = 20$, $A_2 = 80$, $h_1 = 5$, and $h_2 = 4$.

We obtain $e_1 = h_1 - h_2 = 1$, and $e_2 = h_2 = 4$. From (9.13) we get $k^* = 1$ which is, consequently, the optimal value of k. Applying (9.10) and (9.11) we get $Q_1^* \approx 17.89$, $Q_2^* = kQ_1^* = Q_1^* \approx 17.89$, and $C^* \approx 89.44$, i.e., about 7 percent lower than the costs obtained in Example 9.2. The batch quantity of item 1 has increased by more than 100 percent. This is rather typical. If we take the multi-echelon structure into account, this will very often lead to a larger order quantity for the end product.

Note that (9.10) and (9.11) are essentially equivalent to the corresponding expressions (4.3) and (4.4) for the classical economic order quantity model. The only difference is that the ordering cost A is replaced by:

$$A'_1 = A_1 + A_2/k, \qquad (9.14)$$

and the holding cost h by:

$$h'_1 = e_1 + ke_2. \qquad (9.15)$$

Note also that it is not clear how the "real" holding costs should be determined in different situations. Assume that there is plenty of stock of item 2. If we produce one unit of item 1, the holding costs per time unit for item 1 will increase by h_1, but since the stock of item 2 is reduced by one unit, the holding costs per time unit for item 2 are reduced by h_2. The increase in holding costs therefore corresponds to the value added instead of the total value. The additional holding cost is the echelon holding cost $e_1 = h_1 - h_2$. If, on the other hand, we need to replenish item 2 in order to produce item 1, the holding costs will increase by the normal holding cost rate h_1. The average additional holding costs when producing one unit of item 1 will consequently be somewhere between e_1 and h_1.

9.2.2 Roundy's 98 percent approximation

Even if customer demand is constant, multi-echelon lot sizing problems may be very difficult to solve. Optimal solutions are sometimes surprisingly complex. For example, the optimal solution for a serial system with constant demand can mean order quantities that vary over time, see Williams (1982) and Example 9.4 below. Roundy (1985, 1986), however, has shown that it is often possible to use rather simple approximations with a guaranteed cost performance of at most 2 percent above the optimal costs. We shall describe how Roundy's technique can be applied to a serial system, but it is also possible to use the same approach for assembly and distribution systems. An overview is given in Muckstadt and Roundy (1993). Recall that we have already used Roundy's approach in Section 7.3.1.2 in connection with joint replenishments.

Consider a serial system with N stages, see Figure 9.5.

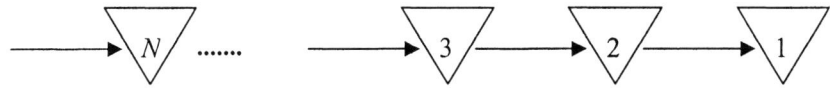

Figure 9.5 Considered serial system with N installations.

The final demand at installation 1 is constant and continuous. There are nonnegative ordering and echelon holding costs at all stages. No shortages are allowed. When producing one unit of item i - 1, one unit of item i is needed as a component (i = 2, 3, ... , N). We disregard the lead-times. It is easy to see that an optimal solution is *nested*, i.e., a batch at stage i should exactly cover an integer number of batches at stage i - 1. (See Section 4.5 and the explanation of *Property 1*.) As we will see in Example 9.4, however, the optimal batch sizes are not necessarily constant over time.

Roundy's approximation means first of all that the batch quantities for different items are restricted to be constant over time. Let us introduce the following notation:

d = final demand per time unit for item 1,
A_i = ordering cost for item i,
e_i = echelon holding cost for item i,
Q_i = batch quantity for item i.

Furthermore, the batch quantity for item i is restricted to be a power of two times the batch quantity for item i - 1, i.e.,

$$Q_i = 2^{k_i} Q_{i-1}, \qquad i = 2, 3, ..., N, \tag{9.16}$$

where k_i is a nonnegative integer. This implies that the solution must be nested, i.e., if a batch of item i is produced at some time, we must also have production of items $i - 1, i - 2, ... , 1$.

Our objective is to minimize the total holding and ordering costs. In analogy with (9.7)-(9.9) we have:

$$\min_{Q_1, Q_2, ..., Q_N} \sum_{i=1}^{N} (e_i \frac{Q_i}{2} + A_i \frac{d}{Q_i}), \tag{9.17}$$

subject to the constraints that the batch quantities are nonnegative and satisfy (9.16).

What is particularly interesting about Roundy's solution is that the costs are at most 2 percent higher than the optimal costs, i.e., without the constraints (9.16).

Roundy's approach is to first relax the constraints (9.16) and instead use the milder constraints:

$$Q_i \geq Q_{i-1}, \qquad i = 2, 3, ..., N. \tag{9.18}$$

Note that (9.16) implies (9.18) while the opposite is not true. Therefore, by using (9.18) we will get a lower bound for the costs when using the constraints (9.16). The objective function (9.17) is convex and the constraints (9.18) are linear. The unique solution of the relaxed problem can therefore be obtained through the Lagrangean relaxation:

$$\max_{\lambda_2, \lambda_3, ..., \lambda_N} \min_{Q_1, Q_2, ..., Q_N} \sum_{i=1}^{N} (e_i \frac{Q_i}{2} + A_i \frac{d}{Q_i}) + \sum_{i=2}^{N} \lambda_i (Q_{i-1} - Q_i), \tag{9.19}$$

where the multipliers λ_i are also restricted to be nonnegative.

Assume that we have determined the optimal multipliers. Define

$$e'_1 = e_1 + 2\lambda_2,$$

$$e'_2 = e_2 - 2\lambda_2 + 2\lambda_3,$$

$$\vdots \tag{9.20}$$

$$e'_{N-1} = e_{N-1} - 2\lambda_{N-1} + 2\lambda_N,$$

$$e'_N = e_N - 2\lambda_N.$$

It is possible to show that the optimal solution of the relaxed problem must give all $e'_i \geq 0$. To see this note that the optimal costs for the relaxed problem can be obtained as the solution of:

$$\min_{Q_1, Q_2, ..., Q_N} \sum_{i=1}^{N} (e'_i \frac{Q_i}{2} + A_i \frac{d}{Q_i}), \tag{9.21}$$

MULTI-ECHELON – LOT SIZING

without any constraints on the batch quantities, i.e., a problem with independent items, ordering costs A_i, and echelon holding costs e'_i. If $e'_i < 0$ we can let $Q_i \to \infty$ which means that the costs in (9.21) approach $-\infty$. This is clearly impossible.

Let us now compare the original holding costs e_i to the holding costs e'_i, see (9.20). Recall that the multipliers are nonnegative. Consider first the effect of λ_2. The echelon holding cost rate at stage 2 is reduced by $2\lambda_2$ and the echelon holding cost rate at stage 1 is increased by $2\lambda_2$. More generally, the effect of λ_i is that the echelon holding cost rate at stage i is reduced by $2\lambda_i$ and the echelon holding cost rate at stage i - 1 is increased by $2\lambda_i$.

What is very interesting is that this implies that (9.21) provides a lower bound for the optimal costs for our original problem where the batch quantities can be chosen in any way. They can, for example, vary over time. Yet the echelon stock at stage i - 1 is always less than or equal to the echelon stock at stage i. A change from the echelon holding costs e_i to the echelon holding costs e'_i can therefore not increase the costs for the optimal solution. Furthermore, given the echelon holding costs e'_i, (9.21) provides a lower bound for the costs since there are no constraints at all on the batch quantities.

Assume that we have solved the relaxed problem. This means that we have a lower bound for the optimal costs and batch quantities satisfying (9.18). We can now adjust this solution by rounding the batch quantities so that they can be expressed as

$$Q_i = 2^{m_i} q. \qquad (9.22)$$

We know from Section 7.1 that if q is given, the maximum cost increase is at most 6 percent, and if we can also adjust q to get a better approximation the cost increase is at most 2 percent. Due to (9.18), we know that $m_i \geq m_{i-1}$ and the batch quantities obtained must consequently satisfy (9.16). We have now obtained Roundy's solution of the problem. We have feasible constant batch quantities and the cost increase compared to the optimal solution is at most 2 percent, since it is at most 2 percent compared to the lower bound (9.21).

It is possible to use the considered Lagrangean relaxation for numerical determination of Roundy's solution, but generally it is much simpler to use the following technique. Consider the relaxed problem. Without the constraints (9.18) it would be optimal to use $Q_i^* = (2A_i d/e_i)^{1/2}$ for all i. Consequently, if A_i/e_i increases with i, we have found the optimal solution since the resulting batch quantities will satisfy (9.18). Assume that for some i, A_i/e_i < A_{i-1}/e_{i-1}, or equivalently that $Q_i^* < Q_{i-1}^*$. Also assume that in the optimal solution of the relaxed problem $Q_i > Q_{i-1}$. This implies that $Q_i \leq Q_i^*$ since we would otherwise reduce Q_i, and similarly that $Q_{i-1} \geq Q_{i-1}^*$ since we would otherwise increase Q_{i-1}. But this means that $Q_i < Q_{i-1}$, which is a contradiction. Consequently, $Q_i = Q_{i-1}$ in the optimal solution of the relaxed problem. This means that we can aggregate items $i - 1$ and i into a single item with ordering cost $A_{i-1} + A_i$ and echelon holding cost $e_{i-1} + e_i$. Next we consider the resulting reduced problem with $N - 1$ items. If $A_i/e_i < A_{i-1}/e_{i-1}$ for some i, we can aggregate the two items; otherwise we obtain the optimal solution of the relaxed problem from the classical economic order quantity model, etc.

Example 9.4 Consider a serial system with $N = 4$ installations (items). The constant continuous demand for item 1 is $d = 1$. Ordering and echelon holding costs are given in Table 9.1.

Table 9.1. Ordering and echelon holding costs.

Item	Ordering cost, A_i	Echelon holding cost, e_i
1	50	40
2	20	100
3	3	1
4	125	10

First we want to solve the relaxed problem, i.e., we want to minimize the costs in (9.17) with respect to (9.18). First consider items 1 and 2. Note that $A_1/e_1 = 5/4 > A_2/e_2 = 2/10$. This means that $Q_1 = Q_2$ in the optimal solution, and we can aggregate items 1 and 2 into a single item with ordering cost $A_1 + A_2 = 70$ and echelon holding cost $e_1 + e_2 = 140$. We can now see that $A_4/e_4 = 12.5 \geq A_3/e_3 = 3 \geq (A_1 + A_2)/(e_1 + e_2) = 0.5$. The constraints (9.18) will therefore not be binding for the remaining three items: items 1 and 2 together, item 3, and item 4. By optimizing each batch quantity separately, we obtain the optimal solution of the relaxed problem: $Q_1 = Q_2 = (2d(A_1 + A_2)/(e_1 + e_2))^{1/2} = 1^{1/2} = 1$, $Q_3 = 6^{1/2}$, and $Q_4 = 25^{1/2} = 5$. The corresponding costs per time unit are: $C_{1+2} = (2d(A_1 + A_2)(e_1 + e_2))^{1/2} = 140$, $C_3 = 6^{1/2}$, and $C_4 = 50$. The total relaxed costs are then $C_{rel} = 190 + 6^{1/2} = 192.45$. (The corresponding Lagrangean multipliers in (9.20) are $\lambda_2 = 30$, $\lambda_3 = 0$, and $\lambda_4 = 0$. This means that $e_1' = 40 + 60 = 100$ and $e_2' = 100 - 60 = 40$. The ratios A_1/e_1' and A_2/e_2' are then both equal to 0.5, which implies identical batch quantities when optimizing (9.21).)

Next we round the order quantities to obtain Roundy's feasible solution. Let us first, for simplicity, set $q = 1$ in (9.22). We then obtain the rounded batch quantities $Q_1 = Q_2 = 1$, $Q_3 = 2$, and $Q_4 = 4$. This gives the costs $C = 193.75$, i.e., 0.68 percent above the relaxed costs and well below the guaranteed 6 percent. It turns out that an optimal q in (9.22) is $q = 1.0626$ (we omit the details). When using this q, it is optimal to use the same multiples, i.e., $Q_1 = Q_2 = q$, $Q_3 = 2q$, and $Q_4 = 4q$. The corresponding costs are 193.39, i.e., 0.49 percent above the relaxed costs to be compared to the guaranteed 2 percent.

Note, however, that this is not the optimal solution. A feasible solution is to use $Q_1 = Q_2 = 1$, $Q_4 = 5$, and to have $Q_3 = 2$ and $Q_3 = 3$ every second time. This nonstationary solution gives the costs $C = 192.5$, i.e., only 0.026 percent above the lower bound from the relaxed problem. Why is this solution so good? Note first that the costs for item 3 are very low. Furthermore, $Q_3 = 2$ and $Q_3 = 3$ work equally well. The best individual solutions for items 1, 2 and 4 is $Q_1 = Q_2 = 1$, and $Q_4 = 5$, and these batch quantities are combined with having $Q_3 = 2$ and $Q_3 = 3$ every second time.

The corresponding solution for an assembly system is similar. The solution for a distribution system is different, however. For a distribution system it is generally not possible to keep the batch quantities constant. Instead, the cycle times are kept constant and it is required that they can be expressed as powers-of-two times some basic period.

To see that constant cycle times can give time-varying batch quantities in a distribution system with constant demand, consider the system in Figure 9.6.

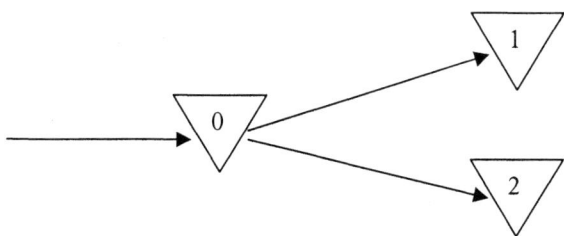

Figure 9.6 A simple distribution system.

Assume that customer demands at installations 1 and 2 are 2 units per time unit, and that the cycle times are $T_0 = 1$, $T_1 = 1/2$ and $T_2 = 2$. The batches at installations 1 and 2 are, of course, constant and equal to 1 and 4, respectively. But the batches at installation 0 are 2 and 6 every second time. During a cycle time $T_0 = 1$, there is always a demand for two batches of size 1 from installation 1. However, the demand from installation 2 is 4 units every second cycle.

9.3 Time-varying demand

Let us now instead consider time-varying discrete demand. As before, we assume that the lead-times are zero. No shortages are allowed and it is assumed that the customer demands take place in the beginning of a period immediately after a possible delivery. Except that we are dealing with more than one echelon, the assumptions are consequently exactly the same as in Sections 4.5-4.9.

9.3.1 Sequential lot sizing

Consider again the simple serial system in Figure 9.7 and the following example.

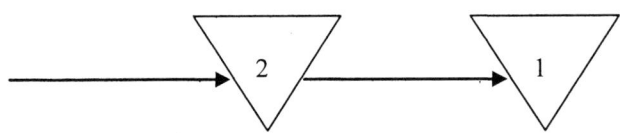

Figure 9.7 A simple serial inventory system.

Example 9.5 Assume that ordering and holding costs are the same as in Examples 9.2 and 9.3, i.e., the ordering costs are $A_1 = 20$ and $A_2 = 80$, and the holding costs per unit and time unit are $h_1 = 5$ and $h_2 = 4$. We consider a time horizon of four periods with customer demands for item 1 equal to $d_1 = 3$, $d_2 = 6$, $d_3 = 5$, and $d_4 = 2$.

As in Example 9.2 we shall first solve the problem sequentially and disregard its multi-stage structure. This is a common approach in practice. We start with the end item, item 1, for which we know the demands. After determining batch quantities for item 1 we know the demands for item 2, i.e., the orders from installation 1, and are also able to determine batch quantities for this item. It is easy to determine the optimal solution for item 1 when disregarding item 2 by applying the Wagner-Whitin algorithm with the ordering cost $A_1 = 20$ and the holding cost $h_1 = 5$. Since there are only four periods, it is also easy to derive the optimal solution by considering all feasible alternatives. Note first that due to the high holding cost, the demands in periods 1-3 should be covered by separate deliveries. Given this we can see that the batch in period 3 should also cover the demand in period 4. The optimal solution for item 1 is consequently to have three batches. The first batch has size 3 and covers the demand in period 1, the second batch has size 6 and covers the demand in period 2, and the third batch has size 7 and covers the demands in periods 3 and 4. It is also easy to see that the costs C_1 for item 1 are

$$C_1 = 3 \cdot 20 + 2 \cdot 5 = 70,$$

i.e., three ordering costs plus the holding costs during period 3 for the demand in period 4.

The solution for item 1 implies that the demands for item 2 in periods 1-4 are 3, 6, 7, and 0, respectively, i.e., the demands occur when we produce item 1. Given these demands, the ordering cost $A_2 = 80$, and the holding cost $h_2 = 4$, it is easy to conclude that the optimal solution for item 2 is to have a single batch of size 16 in period 1. The total holding costs for item 2 are then 80, and there are still some holding costs left if we add another order, which would incur the additional ordering cost 80. The costs for item 2 are obtained as

$$C_2 = 80 + 6 \cdot 4 + 2 \cdot 7 \cdot 4 = 160,$$

and the total costs as

$$C = C_1 + C_2 = 70 + 160 = 230.$$

The solution obtained is not optimal. We minimized the costs for item 1 without considering the implications for item 2. It is easy to find a better solution. Choose instead to have a single batch of size 16 for both item 1 and item 2. We then obtain the costs as

$$C_1 = 20 + 6 \cdot 5 + 2 \cdot 5 \cdot 5 + 3 \cdot 2 \cdot 5 = 130,$$

$$C_2 = 80,$$

$$C = C_1 + C_2 = 210.$$

As expected, this solution means additional costs for item 1. But the cost reduction for item 2 is larger. It is possible to show that this is the optimal solution.

9.3.2 Sequential lot sizing with modified parameters

Let us now consider an alternative relatively simple approximate technique, which was suggested by Blackburn and Millen (1982). When using this technique we still determine the lot sizes sequentially. However, we use modified cost parameters. When dealing with the two-echelon serial system in Figure 9.7, we still start with item 1. Instead of the original cost parameters, however, we use the modified parameters (9.14) and (9.15), which give the optimal solution in case of constant demand.

Example 9.6 We shall solve the problem in Example 9.5 by applying the suggested technique. Recall that $A_1 = 20$, $A_2 = 80$, $h_1 = 5$, and $h_2 = 4$. The customer demands for item 1 are $d_1 = 3$, $d_2 = 6$, $d_3 = 5$, and $d_4 = 2$.

In Example 9.3 we determined the optimal $k = 1$. Note that the optimal k is not affected by the demand, see (9.13). From (9.14) and (9.15) we get $A_1' = A_1 + A_2 = 100$, and $h_1' = e_1 + e_2 = 5$. If we use these modified cost parameters and solve the problem for item 1 by the Wagner-Whitin algorithm, we obtain the optimal solution, i.e., a single batch of size 16 in period 1. Item 2 has no upstream installation so we use the original parameters $A_2 = 80$ and $h_2 = 4$. Since there is a single demand in period 1 it is obvious that we will have a single batch in period 1.

Let us now leave the simple serial system and consider a general assembly system. See Figure 9.8.

Recall that in an assembly system each installation has at most one downstream successor. We can also, without any lack of generality, assume that one unit of each component is needed when producing one unit of a parent item. Lot sizing for an assembly system is more difficult than for a serial system. Yet it is much easier to deal with an assembly system than with a general system.

It is possible to generalize the approximate technique of Blackburn and Millen (1982) to assembly systems. Let

$s(j)$ = the installation which is an immediate successor of installation j,

$p(j)$ = the set of installations which are immediate predecessors of installation j.

In Figure 9.8 we have, for example, $s(4) = 3$, and $p(3) = \{4, 5\}$.

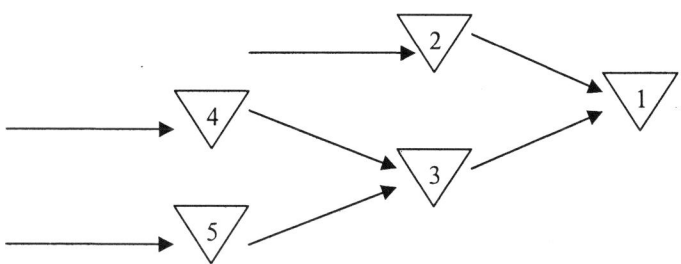

Figure 9.8 Assembly system.

We obtain the echelon holding costs as

$$e_j = h_j - \sum_{i \in p(j)} h_i. \tag{9.23}$$

The modified ordering and holding costs are obtained as

$$A'_j = A_j + \sum_{i \in p(j)} A'_i / k_i, \tag{9.24}$$

and

$$h'_j = e_j + \sum_{i \in p(j)} h'_i k_i. \tag{9.25}$$

In (9.24) and (9.25), k_i is the estimated ratio between a batch quantity for item i and a batch quantity for its successor, item $s(i)$. Blackburn and Millen (1982) obtain the ratio k_j by rounding

$$k_j^* = \sqrt{\frac{A'_j e_{s(j)}}{A_{s(j)} h'_j}} \qquad (9.26)$$

to be a positive integer. Note that (9.24) - (9.26) can be seen as generalizations of (9.13)-(9.15).

Example 9.7 Consider the assembly system in Figure 9.8. Table 9.2 illustrates how the modified ordering and holding costs are obtained. The original costs A_j and h_j are given. We get the echelon holding costs from (9.23) starting with items 2, 4, and 5, which have no predecessors. Consider again items 2, 4 and 5. For these items $A_j' = A_j$ and $h_j' = e_j$. We can then also get the ratios k_j for these items from (9.26) (rounded to integers). Next we go to item 3. We get A_3' and h_3' from (9.24) and (9.25). Thereafter we can determine k_3. Finally we obtain A_1' and h_1'.

Table 9.2 Determination of modified ordering and holding costs.

Item	A_j	h_j	e_j	A_j'	h_j'	k_j
1	1800	5	1	11600	8	-
2	1800	1	1	1800	1	1
3	3200	3	1	16000	3	2
4	6400	1	1	6400	1	1
5	6400	1	1	6400	1	1

9.3.3 Other approaches

It is also possible to solve quite large lot sizing problems exactly for assembly systems with deterministic time-varying demand by using mathematical programming techniques. See Afentakis et al. (1984) and Rosling (1986). Such an approach would, however, be difficult to use in practice if there are many items and complex product structures.

There are also various other approximate techniques. One example is Graves (1981), who suggests an iterative procedure.

Furthermore, it may sometimes be possible to simplify the lot-sizing problem by initially showing that some items should always be produced in identical batch sizes. A general assembly system can be seen as a number of subsystems with two levels. See Figure 9.9.

MULTI-ECHELON – LOT SIZING

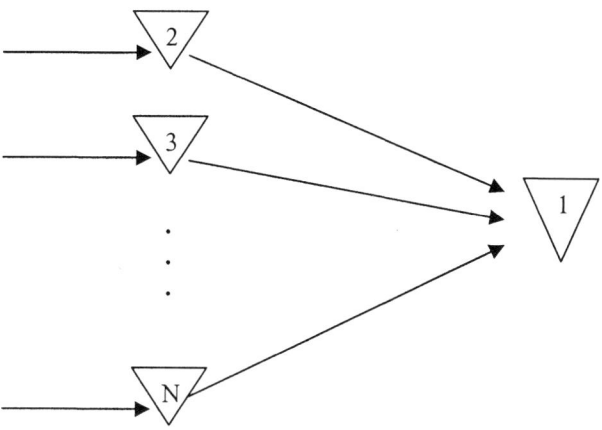

Figure 9.9 Two-level assembly subsystem.

Assume that the installations in Figure 9.9 have ordering costs A_i and echelon holding costs e_i. It is possible to show that all installations should use the same batch sizes provided that the following condition is satisfied (Axsäter and Nuttle, 1987).

$$\frac{\sum_{j \neq i} A_j}{\sum_{j \neq i} e_j} \geq \frac{A_i}{e_i}, \quad i = 2, 3, ..., N. \tag{9.27}$$

Note that the condition is sufficient but not necessary. Recall that we have already discussed a special case of this condition in connection with (9.13). If we always use the same batch quantities in the subsystem in Figure 9.9, we only need stock of item 1. This means, as we have discussed in Section 9.1.1, that the subsystem can be replaced by a single item with ordering cost $A_1 + A_2 + ... + A_N$ and echelon holding cost $e_1 + e_2 + ... + e_N$.

9.3.4 Concluding remarks

We have seen that it is difficult to determine optimal lot sizes for multi-echelon inventory systems, although we have assumed given deterministic demand. In most practical situations it is necessary to use simple approximations. However, even when using approximations it may be possible to let theoretical results have an impact on the heuristic rules.

For assembly systems and systems that resemble assembly systems, it is very often practical to let upstream batch quantities be integer multiples of downstream batch quantities, although this does not need to be optimal.

For serial and assembly systems, an optimization of the batch quantities will, in general, lead to larger lot sizes as compared to a simple sequential technique where different items are considered separately. Furthermore, we recall that stochastic demand variations will also make it advantageous to use larger batch quantities (Section 6.1). Note however, that these conclusions are only based on ordering costs, holding costs and possibly shortage costs. The conclusions do not take capacity constraints into account. In Section 7.2 we have discussed different possibilities to smooth the production load. A cyclical planning system can be used to smooth the load also in a multi-stage environment. But if it is not feasible for some reason to use a cyclical planning system, the remaining possibility is quite often to reduce the batch quantities in order to smooth the flow of orders. This means that we have conflicting goals. We wish to have large batches due to the multi-echelon structure, but on the other hand we want to have small batches to get a smooth production load. It is difficult to know how to handle such a situation, but in general, it is more important in practice to get smooth capacity utilization than to reduce ordering and holding costs. See also Axsäter (1986).

References

Afentakis, P., B. Gavish, and U. S. Karmarkar. 1984. Computationally Efficient Optimal Solutions to the Lot-Sizing Problem in Multi-Stage Assembly Systems, *Management Science*, 30, 222-239.

Axsäter, S. 1986. Evaluation of Lot-Sizing Techniques, *International Journal of Production Research*, 24, 51-57.

Axsäter, S., and H. L. W. Nuttle. 1987. Combining Items for Lot Sizing in Multi-Level Assembly Systems, *International Journal of Production Research*, 25, 795-807.

Blackburn, J. D., and R. A. Millen. 1982. Improved Heuristics for Multi-Echelon Requirements Planning Systems, *Management Science*, 28, 44-56.

Graves, S. C. 1981. Multi-Stage Lot Sizing: An Iterative Procedure, in L. B. Schwarz Ed. *Multi-Level Production/Inventory Control Systems: Theory and Practice*, North Holland Amsterdam, 95-110.

Muckstadt, J. A., and R. Roundy. 1993. Analysis of Multistage Production Systems, in S. C. Graves et al. Eds. *Handbooks in OR & MS Vol. 4*, North Holland Amsterdam, 59-131.

Rosling, K. 1986. Optimal Lot-Sizing for Dynamic Assembly Systems, in S. Axsäter et al. Eds. *Multi-Stage Production Planning and Inventory Control*, Springer Berlin Heidelberg, 119-131.

Roundy, R. 1985. 98%-Effective Integer-Ratio Lot-Sizing for One-Warehouse Multi-Retailer Systems, *Management Science*, 31, 1416-1430.

Roundy, R. 1986. 98%-Effective Lot-Sizing Rule for a Multi-Product Multi-Stage Production/Inventory System, *Mathematics of Operations Research*, 11, 699-729.

Williams, J. F. 1982. On the Optimality of Integer Lot Size Ratios in Economic Lot Size Determination in Multi-Stage Assembly Systems, *Management Science*, 28, 1341-1349.

Problems

9.1* Consider the following production process.

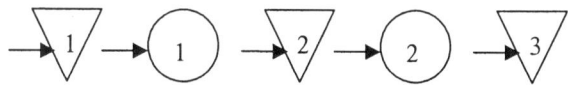

The final demand at site 3 is 100 units/day. No backorders are allowed. Deliveries at site 1 come from a supplier, which delivers a full batch immediately on request. The ordering cost is 1500. For the two production stages we have:

Stage 1: Production rate (p_1) = 300 units/day,
Stage 2: Production rate (p_2) = 200 units/day.

The setup costs for each production stage is 750. The holding costs are 1 in stock 1, 2 in stock 2, and 3 in stock 3. Determine the optimal batch quantity. Production at production stage 2 starts as soon as there are units available in stock 2. When a unit has passed production stage 2, it is immediately delivered to stock 3.

9.2 A batch Q is processed in two machines and is then transported to a warehouse. The demand at the warehouse is constant and continuous. No backorders are allowed. During a cycle Q units are placed in front of the first machine. Units which are processed in the first machine, are successively processed in the second machine. When all Q units have passed the second machine, the batch is transported to the warehouse. Use the following assumptions and notation and determine the optimal batch quantity.

---

* Answer and/or hint in Appendix 1.

d = demand per unit of time,
p_1 = $8d$ = production rate in the first machine,
p_2 = $4d$ = production rate in the second machine,
A_1 = setup cost in the first machine,
A_2 = setup cost in the second machine,
h_1 = holding cost before the first machine,
h_2 = holding cost between the two machines,
h_3 = holding cost after the second machine,
T = transportation time.

9.3 Verify that the cost expressions (9.6) and (9.9) are equivalent.

9.4 Consider Example 8.1. Assume that the lead-times are zero so that inventory positions and inventory levels are the same. Let the normal holding costs be h_1, h_2, h_3, and h_4. Determine the corresponding echelon holding costs e_1, e_2, e_3, and e_4. Use the installation stocks and the echelon stocks in Table 8.1 and verify that you get the same total holding costs when using the echelon stocks together with the echelon holding costs.

9.5 Consider Example 9.3. Let A_2 increase while other data are unchanged. What is the smallest A_2 such that it is optimal to have $Q_2 > Q_1$? Determine this solution.

9.6* Consider the serial system

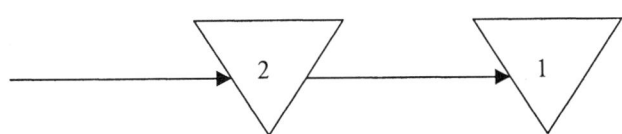

The customer demand at installation 1 is constant and continuous 10 units per time unit. The ordering costs are $A_1 = 100$ and $A_2 = 500$, and the installation holding costs $h_1 = 2$ and $h_2 = 1$.

a) Determine batch quantities by using Roundy's 98 percent approximation.
b) Determine the optimal batch quantities given that Q_2 should be an integer multiple of Q_1.

9.7 Given the solution of Problem 9.6 a), determine the corresponding Lagrangean multipliers in (9.19)-(9.20).

9.8* a) Derive the generalization of (9.14) and (9.15) for a three-level serial system.
b) Verify the results from (9.24) and (9.25).

9.9 Prove that the condition (9.27) is sufficient for $N = 2$, i.e., for a two-echelon serial system.

10 MULTI-ECHELON SYSTEMS: REORDER POINTS

This chapter deals with various techniques for determining safety stocks and reorder points in multi-echelon inventory systems. Throughout the chapter we assume that the batch quantities are given.

The methods for determining safety stocks and reorder points described in Chapter 5 are not directly applicable in a multi-echelon inventory system. The reason is, again, that different installations cannot be handled separately. A fundamental problem in connection with multi-level inventory systems is to find the best balance between upstream and downstream stock. As we have already discussed in Section 8.1, it is clear that inventories at different levels will support each other. For example, if there is a large safety stock of a final product, it is possible to accept some production delays and it may consequently be feasible to have relatively low safety stocks of different components of the product. For inventories in a distribution system there are similar dependencies. Consider a two-level distribution system with a central warehouse and a number of retailers. With a large central inventory, orders from the retailers will very seldom be delayed and it may be possible to reduce the local safety stocks at the retailers.

In general, it is very difficult to allocate safety stocks optimally in a multi-echelon system. Quite often we have to be satisfied with a reasonable "second best" solution. Still, in most situations it is important to adapt the solutions in some way to the multi-echelon environment. See Muckstadt and Thomas (1980), and Hausman and Erkip (1994). In those cases where no efficient methods are available, it is still usually possible to evaluate at least a few alternative solutions by simulation.

The allocation of safety stocks in a multi-stage inventory system concerns two related questions. The first question is how much total system safety stock is needed. The second question is how to allocate the stock among the different installations. For example, how much of the total safety stock should be placed early, respectively late, in the material flow. This distribution is affected by various factors. One such factor is the structure of the system. In a distribution system (Figure 8.2) there are few installations early in the material flow as compared to an assembly system (Figure 8.3). In general, it is advantageous to allocate safety stocks where there are few items with high and relatively stable demand. This indicates that it could be more reasonable to keep safety stock early in the material flow for a distribution system than for an assembly system.

There are, however, other factors which may point in the other direction. In production, the holding costs are often quite low early in the material flow. This means that it can be a good policy to have relatively large safety stocks of various components in an assembly system.

The allocation of safety stocks should also be affected by the lead-times. Consider again the distribution system in Figure 8.2. If the warehouse lead-time is very long as compared to the lead-times of the retailers, relatively more stock should be kept at the warehouse.

Note also that we can improve the service in different ways. Sometimes a safety time is a better alternative than a safety stock. This is the case for a component which is used only a few times per year when producing a higher level item. If we use a safety time in an MRP system, we will only stock the item during a short time just before it is needed in production.

Safety stocks and safety times in production are largely due to uncertain lead-times. One question that should always be raised is whether it is possible to reduce the uncertainty and hence avoid large safety stocks. To do that it is often necessary to have more, or possibly more flexible, production capacity, so that production delays due to various disturbances will not be too long. *Safety capacity* can be an alternative to safety stocks. There will, however, always remain some uncertainty both in the lead-times and in the demand for final items.

10.1 The Clark-Scarf model

The best-known technique for determining safety stocks in a multi-echelon inventory system was presented by Clark and Scarf (1960). This was also the origin of the echelon-inventory measure. Their method, which is exact for serial systems, can be seen as a decomposition technique. In the first step we consider the most downstream installation facing customer demand. Shortages at the next upstream installation lead to stochastic delays, which will imply certain additional costs. These additional costs are evaluated and added as shortage costs when determining an optimal policy for the next upstream installation.

10.1.1 Serial system

We shall first describe the Clark-Scarf technique for the simple two-level serial system in Figure 10.1. Our assumptions are somewhat different from those in Clark and Scarf (1960), who considered a finite time horizon. We assume an infinite horizon as in Federgruen and Zipkin (1984a,b). For more details and further aspects we refer to Van Houtum (2001) and Van Houtum et al. (2001).

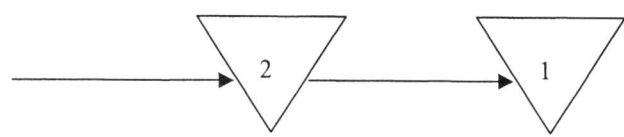

Figure 10.1 The considered two-echelon serial inventory system.

We assume periodic review. The continuous period demand at installation 1 is assumed to be normally distributed and independent across periods. (It is straightforward to generalize to other distributions.) Demand that cannot be met directly from stock is backordered. Installation 1 replenishes from installation 2, and installation 2 replenishes from an outside supplier with infinite supply. We assume that the lead-times (transportation times) are integral numbers of periods. There are standard holding and backorder costs but no ordering or setup costs.

It is assumed that all events take place in the beginning of a period in the following order:

1. Installation 2 orders.
2. The period delivery from the outside supplier arrives at installation 2.
3. Installation 1 orders from installation 2.
4. The period delivery from installation 2 arrives at installation 1.
5. The stochastic period demand takes place at installation 1.
6. Evaluation of holding and shortage costs.

We shall use the following notation:

L_j = lead-time (transportation time) for replenishments at installation j,
μ = average period demand,
σ = standard deviation of period demand,
e_j = nonnegative echelon holding cost per unit and period at installation j,
h_j = holding cost per unit and period at installation j, $h_1 = e_1 + e_2$, $h_2 = e_2$,
b_1 = shortage cost per unit and period,
IL_j^i = installation stock inventory level at installation j just before the period demand,
IL_j^e = echelon stock inventory level at installation j just before the period demand,
$D(n)$ = stochastic demand during n periods,
y_2 = echelon stock inventory position at installation 2 in period t just before the period demand,
y_1 = realized (echelon stock) inventory position at installation 1 in period $t + L_2$ just before the period demand.

The *realized* inventory position at installation 1 does not include outstanding orders, which are backordered at installation 2.

Our purpose is to minimize the total expected holding and backorder costs per period. We consider the holding costs for stock on hand at the installations, i.e., we do not include the expected holding costs for units in transportation from installation 2 to installation 1, $h_2 \mu L_1$. These costs are not affected by the control policy.

Recall that for installation 1 there is no difference between installation and echelon stock, and that $e_1 = h_1 - h_2$ can be interpreted as the holding cost on the value added when going from installation 2 to installation 1.

We use the notation $x^+ = \max(x, 0)$, $x^- = \max(-x, 0)$, and $x^+ - x^- = x$.

Let t be an arbitrary period. After ordering in period t, installation 2 has a certain echelon stock inventory position y_2. At this stage we do not assume anything about the ordering policy. Because the outside supplier has infinite supply, there is no difference between the inventory position and the realized inventory position. Consider then the echelon stock inventory level at installation 2 in period $t + L_2$ just before the period demand, IL_2^e. We can express IL_2^e as $IL_2^e = y_2 - D(L_2)$, i.e., y_2 minus the stochastic demand during the L periods $t, t + 1, \ldots, t + L_2 - 1$. We are then using the standard argument that was introduced in Section 5.3.2. All that was included in the inventory position after the order in period t has arrived in stock in period $t + L_2$. Orders after period t are still outstanding. Note that the customer demand in period $t + L_2$ has not taken place at this stage. $D(L_2)$ has mean $\mu'_2 = L_2\mu$ and standard deviation $\sigma'_2 = (L_2)^{1/2}\sigma$.

Next, we consider the realized echelon stock inventory position y_1 at installation 1 after ordering in period $t + L_2$. Recall that the difference between inventory position and realized inventory position is that the realized inventory position does not include outstanding orders that are backordered at installation 2. Again we do not assume anything about the ordering policy. Note, however, that whatever policy is followed we must have

$$y_1 \leq IL_2^e = y_2 - D(L_2). \tag{10.1}$$

In case of strict inequality, the difference is positive installation stock at installation 2.

The (installation stock) inventory level at installation 1 *after* the demand in period $t + L_1 + L_2$ is, by using our standard argument, obtained as y_1 minus the demand in periods $t + L_2, t + L_2 +1, \ldots, t + L_2 + L_1$, i.e., the demand during $L_1 + 1$ periods, $D(L_1 + 1)$. Note that whatever policy is followed, y_1 is independent of this demand, which has mean $\mu''_1 = (L_1 + 1)\mu$ and standard deviation $\sigma''_1 = (L_1 + 1)^{1/2}\sigma$.

For installation 2, we shall consider the ordering in period t and the costs in period $t + L_2$. For installation 1, we consider the ordering in period $t + L_2$ and the costs in period $t + L_2 + L_1$. Note that all these periods can be seen as arbitrary periods.

The installation stock inventory on hand at installation 2 just after the order from installation 1 is $(IL_2^i)^+ = IL_2^e - y_1$. This is then the installation inventory status at installation 2 during the whole period.

Using this, we can now express the average holding costs at installation 2 in period $t + L_2$ as

$$C_2 = h_2 E(IL_2^e - y_1) = h_2 E(y_2 - D(L_2) - y_1)$$
$$= h_2(y_2 - \mu'_2) - h_2 y_1. \tag{10.2}$$

Similarly, the average period costs at installation 1 (evaluated after the period demand has taken place) in period $t + L_2 + L_1$ are obtained as

$$C_1 = h_1 E((y_1 - D(L_1+1))^+) + b_1 E((y_1 - D(L_1+1))^-) \qquad (10.3)$$

$$= h_1(y_1 - \mu_1'') + (h_1 + b_1)E((y_1 - D(L_1+1))^-).$$

Recall that so far, we have not assumed anything about the ordering policies used at the two considered installations. Whatever ordering policy is used it must result in some y_2 in period t and some y_1 in period $t + L_2$. We note that the total expected period costs are completely determined by y_2 and y_1, so when looking for the optimal policy we do not need to consider any other parameters.

It is now convenient to reallocate the costs slightly. The last term in (10.2), $-h_2 y_1$ is transferred to stage 1. Using that $h_1 - h_2 = e_1$, we obtain:

$$\tilde{C}_2 = h_2(y_2 - \mu_2'), \qquad (10.4)$$

and

$$\tilde{C}_1 = e_1 y_1 - h_1 \mu_1'' + (h_1 + b_1)E((y_1 - D(L_1+1))^-). \qquad (10.5)$$

Of course this reallocation does not affect the total costs. Since the transferred term is normally negative, $\tilde{C}_1$ may very well also become negative.

Observe now that (10.4) is independent of y_1. The costs (10.5) at stage 1 may depend on y_2 due to (10.1). However, it turns out that the optimal policy at installation 1 is independent of y_2. To see this, let us for a moment forget that the realized inventory position y_1 may depend on y_2. We simply assume that we can choose any value of y_1. We denote this value $\hat{y}_1$. This means that we replace (10.5) by

$$\hat{C}_1(\hat{y}_1) = e_1 \hat{y}_1 - h_1 \mu_1'' + (h_1 + b_1)E((\hat{y}_1 - D(L_1+1))^-)$$

$$= e_1 \hat{y}_1 - h_1 \mu_1'' + (h_1 + b_1) \int_{\hat{y}_1}^{\infty} (u - \hat{y}_1) \frac{1}{\sigma_1''} \varphi\left(\frac{u - \mu_1''}{\sigma_1''}\right) du$$

$$= e_1 \hat{y}_1 - h_1 \mu_1'' + (h_1 + b_1) \sigma_1'' \int_{\frac{\hat{y}_1 - \mu_1''}{\sigma_1''}}^{\infty} (v - \frac{\hat{y}_1 - \mu_1''}{\sigma_1''}) \varphi(v) dv$$

$$= e_1 \hat{y}_1 - h_1 \mu_1'' + (h_1 + b_1) \sigma_1'' G\left(\frac{\hat{y}_1 - \mu_1''}{\sigma_1''}\right). \tag{10.6}$$

Recall that the loss function $G(v)$, which is tabulated in Appendix 2, is convex and that $G'(v) = \Phi(v) - 1$.

This implies that it is easy to get the optimal $\hat{y}_1$ from the first order condition

$$\frac{d\hat{C}_1(\hat{y}_1)}{d\hat{y}_1} = e_1 + (h_1 + b_1)(\Phi\left(\frac{\hat{y}_1 - \mu_1''}{\sigma_1''}\right) - 1) = 0, \tag{10.7}$$

or equivalently,

$$\Phi\left(\frac{\hat{y}_1 - \mu_1''}{\sigma_1''}\right) = \frac{e_2 + b_1}{h_1 + b_1}. \tag{10.8}$$

Note that the optimization of (10.6) is essentially a newsboy problem. See Section 5.13. Let $\hat{y}_1^*$ be the optimal solution.

Consider now (10.1). If $y_2 - D(L_2) \geq \hat{y}_1^*$ we obtain the optimal solution if we have $y_1 = \hat{y}_1^*$. But if $y_2 - D(L_2) < \hat{y}_1^*$, the best possible value of y_1 is $y_1 = y_2 - D(L_2)$ due to the convexity of (10.6). But this optimal policy can be realized if we apply an (echelon stock) order-up-to-S policy with $S_1^e = \hat{y}_1^*$. Note that this optimal policy at installation 1 is completely independent of y_2. Furthermore, the optimal policy is easy to obtain from (10.8). (Recall that there is no difference between echelon stock and installation stock at installation 1, so we can also see the obtained policy as an installation stock policy with $S_1^i = \hat{y}_1^*$.)

Note that if $e_1 = 0$, or equivalently if $h_1 = e_2$, (10.8) implies that $S_1^e = \hat{y}_1^* \to \infty$. This means that installation 2 will never carry any stock. To understand this, note that in a serial system the whole consumption takes place at installation 1. If there is no difference in holding costs, we can just as well move all stock to installation 1. In a distribution system this is no longer the case. It can be advantageous to retain stock at an upstream installation because it is then still possible to allocate the stock to alternative lower level sites. See Section 10.1.2.

It remains to determine the optimal policy at installation 2. The only remaining decision parameter is y_2. We consider the *total costs* assuming that we use the optimal policy at installation 1. We obtain the costs from (10.4) and (10.6). Since $y_1 = S_1^e$ if $D(L_2) \leq y_2 - S_1^e$, and $y_1 = y_2 - D(L_2)$ for $D(L_2) > y_2 - S_1^e$, we have

$$\hat{C}_2(y_2) = h_2(y_2 - \mu_2') + \hat{C}_1(S_1^e)$$

$$+ \int_{y_2 - S_1^e}^{\infty} \left[\hat{C}_1(y_2 - u) - \hat{C}_1(S_1^e)\right] \frac{1}{\sigma_2'} \varphi\left(\frac{u - \mu_2'}{\sigma_2'}\right) du.$$

(10.9)

Note that the costs in (10.9) include the costs at installation 1 under the assumption that installation 1 uses the optimal policy. The last term in (10.9) can be seen as the shortage costs at installation 2 induced by its inability to deliver on time to installation 1. It is easy to verify that $\hat{C}_2(y_2)$ is convex in y_2, which means that we only have to look for a local minimum. Denote the optimal solution by y_2^*. Because the outside supplier has infinite supply it is obvious that it is optimal to apply an echelon stock order-up-to-S policy with $S_2^e = y_2^*$.

Example 10.1 Consider a system with the following data: $L_1 = L_2 = 5$, $\mu = 10$, $\sigma = 5$, $e_1 = 0.5$, $e_2 = 1$, and $b_1 = 10$. In the first step we determine $\hat{C}_1(\hat{y}_1)$ according to (10.6). A graph is given in Figure 10.2.

The optimal $S_1^e = \hat{y}_1^* = 81.0$ is obtained from (10.8). The corresponding costs are $\hat{C}_1 = -47.0$.

In the next step we determine the total costs $\hat{C}_2(y_2)$ from (10.9), see Figure 10.3. The optimal echelon stock order-up-to level is $S_2^e = y_2^* = 129.7$, giving the total costs $\hat{C}_2 = 39.4$.

Note that the total costs are relatively flat, and particularly that the costs do not increase especially rapidly for low values of y_2. This is typical for multi-echelon inventory systems. Furthermore, optimal upstream stocks are, in general, surprisingly low. The average optimal inventory level at installation 2 is $S_2^e - \mu_2' - S_1^e = 129.7 - 50 - 81.0 = -1.3$.

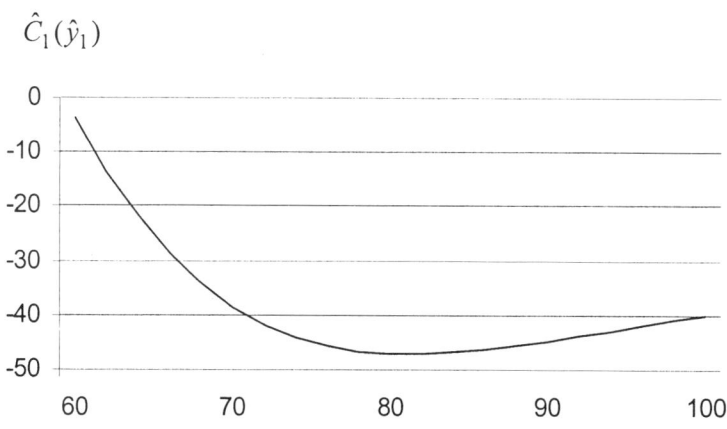

Figure 10.2 $\hat{C}_1$ as a function of $\hat{y}_1$.

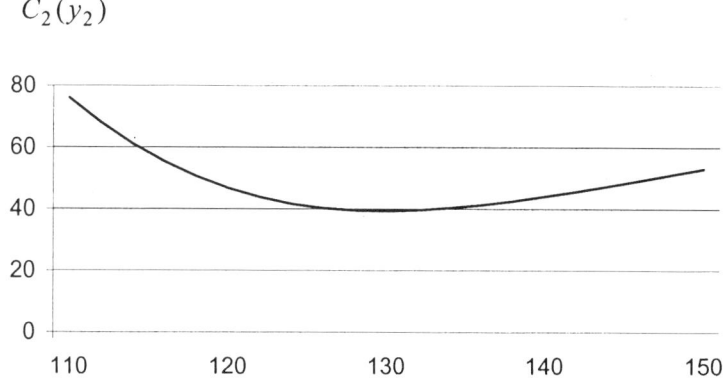

Figure 10.3 Total costs as a function of y_2.

The described technique is easy to generalize to more than two echelons. The additional costs at installation 2 due to insufficient supply are then used as the shortage costs at installation 3, etc. It is also easy to generalize to a batch-ordering policy at the most upstream installation (installation 2 in our case).

We have assumed echelon stock policies since this simplifies the derivation. Otherwise this is not important, though. As shown in Section 8.2.3, we can, under general conditions, replace echelon stock S policies by equivalent installation stock S policies. The corresponding installation stock inventory positions are $S_2^i = S_2^e - S_1^e$, and $S_1^i = S_1^e$. However, if we start with initial inventory positions that are above the optimal order-up-to positions, the echelon stock policy is still optimal while the installation stock policy may fail.

For two levels and normally distributed demand, i.e., the case that we have considered, the exact solution is relatively easy to obtain, see Federgruen and Zipkin (1984a,b). However, in a general case the computations can be very time consuming. Van Houtum and Zijm (1991) have developed approximate techniques. One approximation is exact for demand distributions of mixed Erlang type. A different type of approximation based on separate single-stage problems has been suggested by Shang and Song (2003).

It can be shown that the Clark-Scarf approach can also be used for assembly systems. Rosling (1989) has demonstrated that an assembly system can be replaced by an equivalent serial system when carrying out the computations. Chen and Song (2001) have generalized the Clark-Scarf model to demand processes where demands in different time periods may be correlated.

10.1.2 The Clark-Scarf approach for a distribution system

Let us now instead consider the distribution system in Figure 10.4. It is common to use the Clark-Scarf approach also for such systems, but the approach is then no longer exact. The approximation is based on a so-called "*balance*" assumption, which essentially means that an upstream installation is allowed to make negative allocations to its downstream neighbors. This implies that the total stock at the downstream neighbors can be optimally distributed between the sites in any period. The technique was sketched in the original paper by Clark and Scarf (1960). Eppen and Schrage (1981) used the approach in a model with identical retailers and normally distributed demand where the central warehouse was not allowed to carry any stock. Their model was extended in a number of ways, including non-identical retailers and stock at the warehouse, by Federgruen and Zipkin (1984a,c). See also Federgruen (1993), and Diks and de Kok (1998).

As in Section 10.1.1, let us assume periodic review and normally distributed period demands at the retailers. The demands at the retailers are independent across periods and retailers. The retailers replenish their stocks from the warehouse, and the warehouse replenishes its stock from an outside supplier with infinite stock. We assume that all installations apply echelon stock order-up-to-S policies.

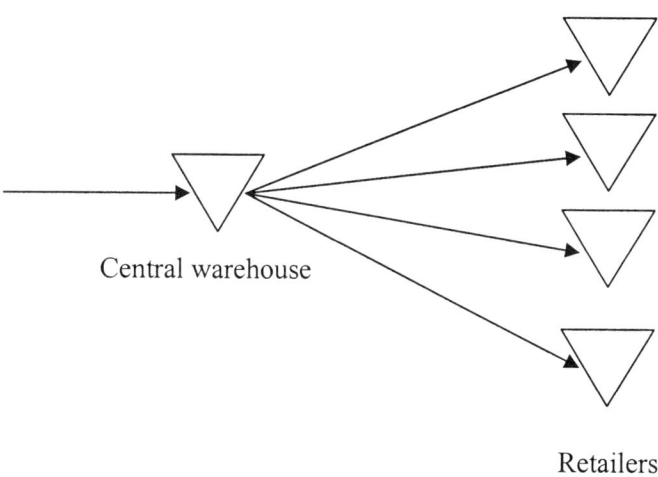

Figure 10.4 Distribution inventory system.

The warehouse is installation 0 and the retailers installations 1, 2, ... , N. We shall use the following notation:

L_0 = lead-time (integral number of periods) for an order generated by the warehouse,

L_j = transportation time (integral number of periods) for a delivery from the warehouse to retailer j,

$D_j(n)$ = stochastic demand at installation j over n periods,

μ_j = average demand per period at retailer j,

σ_j = standard deviation of the demand per period at retailer j,

e_j = echelon holding cost per unit and period at installation j, $e_j \geq 0$,

h_j = holding cost per unit and period at installation j, i.e., $h_0 = e_0$, and $h_j = e_0 + e_j$ for $j > 0$,

b_j = backorder cost per unit and period at retailer j,

S_j^e = echelon stock order-up-to position at installation j,

y_j = realized (echelon stock) inventory position at retailer j in period $t + L_0$ just before the period demand.

A major difference compared to Section 10.1.1 is that we now have a number of parallel installations at the downstream level. Otherwise we can say that the warehouse corresponds to installation 2 and the retailers to installation 1 in Section 10.1.1. We assume that all events take place in the same order as in Section 10.1.1.

Again we consider an arbitrary period t. The echelon stock level at the warehouse at time $t + L_0$, just before the period demand, is obtained as S_0^e minus the total system demand during L_0 periods, $D_0(L_0)$.

We consider the costs at retailer j in period $t + L_0 + L_j$. Using the same cost allocation technique as in Section 10.1.1, we obtain the costs at the warehouse and at retailer j in complete analogy with (10.4) and (10.5). (Recall that y_j is the realized inventory position.)

$$\tilde{C}_0(S_0^e) = h_0(S_0^e - \mu_0'), \qquad (10.10)$$

and

$$\tilde{C}_j(y_j) = e_j y_j - h_j \mu_j'' + (h_j + b_j) E(y_j - D_j(L_j + 1))^-. \qquad (10.11)$$

In (10.10), $\mu_0' = L_0 \sum_{j=1}^{N} \mu_j$, and in (10.11), $\mu_j'' = (L_j + 1)\mu_j$. We can determine the value of y_j that minimizes (10.11) in analogy with (10.8)

$$\Phi\left(\frac{\hat{y}_j - \mu_j''}{\sigma_j''}\right) = \frac{e_0 + b_j}{h_j + b_j}, \qquad (10.12)$$

where $\sigma_j'' = (L_j + 1)^{1/2} \sigma_j$. Let $\hat{y}_j^*$ be the optimal solution.

In analogy with the optimal solution in the serial case, it may now seem reasonable to allocate $S_j^e = \hat{y}_j^*$ to each retailer provided that the sum of all S_j^e does not exceed the echelon stock inventory level at the warehouse. This is also what we will do, but it is important to understand that this is just an approximation.

Assume, for example, that we have only two identical retailers and that in a certain period we are able to allocate $S_1^e = S_2^e$ to each of them. After this allocation we get a large period demand at retailer 2 and no demand at all at retailer 1. This means that we would like to allocate up to S_2^e at retailer 2 in the next period. But this may not be possible due to insufficient supply at the warehouse. In that case we will get unequal realized inventory positions at the two retailers. However, it is rather obvious that it would be better to distribute the inventory positions equally. This might have been possible if we had saved some more stock at the warehouse in the preceding period, i.e., if we had not allocated $S_1^e = S_2^e$ to both retailers. Due to this "balance" problem the decision rule that was optimal in the serial case is now only approximate.

The considered approximation is often denoted the "balance" approximation since it means that we disregard the possibility of unbalanced retailer inventory positions. Another interpretation is that we also allow negative allocations from the warehouse to the retailers. An implication is that in a two-echelon system, given this approximation and if the holding costs at all installations are the same, then there is no reason to keep stock at the warehouse.

Let us now accept the "balance" approximation and also take into account that we cannot always allocate S_j^e to each retailer. Assume that the available amount is $v \leq S_r^e = \sum_{j=1}^{N} S_j^e$. Given that negative allocations are possible it is "optimal" to solve the following myopic allocation problem:

$$\hat{C}_r(v) = \min_{\sum_{j=1}^{N} \hat{y}_j \leq v} \sum_{j=1}^{N} \tilde{C}_j(\hat{y}_j), \qquad (10.13)$$

and $\hat{C}_r(v)$ provides the corresponding retailer costs. If we relax the constraint $\sum_{j=1}^{N} \hat{y}_j \leq v$ by a Lagrange multiplier $\lambda \geq 0$, the solution of (10.13) can be obtained from

$$\Phi\left(\frac{\hat{y}_j - \mu_j''}{\sigma_j''}\right) = \frac{e_0 + b_j - \lambda}{h_j + b_j}, \qquad (10.14)$$

which is a slight variation of (10.12).

Finally, in analogy with (10.9), we obtain S_0^e by minimizing

$$\hat{C}_0(S_0^e) = h_0(S_0^e - \mu_0') + \hat{C}_r(S_r^e)$$

$$+ \int_{S_0^e - S_r^e}^{\infty} \left[\hat{C}_r(S_0^e - u) - \hat{C}_r(S_r^e) \right] \frac{1}{\sigma_0'} \varphi\left(\frac{u - \mu_0'}{\sigma_0'}\right) du, \tag{10.15}$$

where $\sigma_0' = (L_0 \sum_{j=1}^{N} \sigma_j^2)^{1/2}$.

The "balance" assumption means that we assume that the warehouse stock can be used more efficiently than what is possible in reality. Therefore, $\hat{C}_0(S_0^e)$ is a lower bound for the real costs, and the optimal S_0^e is lower than what would be optimal.

When implementing the solution we cannot allocate negative quantities to the retailers. Therefore the actual allocations are determined by solving a modified version of the myopic allocation problem (10.13). Let x_j denote the realized inventory position at retailer j just before the allocation.

$$\hat{C}_r'(v) = \min_{\substack{\sum_{j=1}^{N} \hat{y}_j \leq v \\ \hat{y}_j \geq x_j}} \sum_{j=1}^{N} \tilde{C}_j(\hat{y}_j), \tag{10.16}$$

Note that $\hat{C}_r'(v) \geq \hat{C}_r(v)$, since (10.16) is more constrained than (10.13).

The "balance" assumption has been used extensively in the inventory literature and has been shown to produce solutions of very good quality in many different situations, see for example Eppen and Schrage (1981), Federgruen and Zipkin (1984a,b,c), Federgruen (1993), Van Houtum et al (1996), Verrijdt and de Kok (1996), Van der Heijden et al. (1997), Diks and de Kok (1998, 1999), Dogru et al. (2005) and references therein. Still the "balance" assumption may be less appropriate in situations with large differences between the retailers in terms of service requirements and demand characteristics. Other allocation approaches that are not based on the "balance" assumption have been suggested by e.g., Erkip (1984), Jackson (1988), Jackson and Muckstadt (1989), McGavin et al. (1993), Graves (1996), and Axsäter et al. (2002).

10.2 The METRIC approach for distribution systems

In multi-echelon inventory theory there are two classical approaches. One is the Clark-Scarf technique that we considered in Section 10.1. The other is the so-called METRIC approximation of Sherbrooke (1968, 2004), which has also been of great importance for the development of the field. In this section we will describe how this simple technique can be applied to a two-level distribution system with independent Poisson demand processes at the retailers, complete backordering, and one-for-one replenishments. Warehouse backorders are filled on a first come-first served basis. The structure of the inventory system is illustrated in Figure 10.4.

A difference compared to Section 10.1 is that we consider continuous review policies instead of periodic review policies. We shall choose to deal with installation stock $(S-1, S)$ policies, or equivalently order-up-to-S policies. The considered policy means that the installation stock inventory position at an installation is kept equal to some S at all times. Because of the continuous review, each demand for a unit at a retailer will immediately trigger an order, i.e., a corresponding demand for a unit at the warehouse. The Poisson demand processes at the retailers are assumed to be independent. The warehouse demand is consequently a superposition of these independent Poisson processes and is therefore also a Poisson process.

Our assumptions of Poisson demand, continuous review, and one-for-one replenishments are often reasonable for items with relatively low demand and high holding costs. Such items are, for example, common among spare parts.

We introduce the following notation:

N = number of retailers,
L_i = transportation time for an item to arrive at retailer i from the warehouse, i.e., the lead-time if the warehouse has stock on hand,
L_0 = replenishment lead-time for the warehouse,
λ_i = demand intensity at retailer i,
λ_0 = $\sum_{i=1}^{N} \lambda_i$ = demand intensity at the warehouse,
S_i = order-up-to installation inventory position at installation i,
IL_i = stochastic installation stock inventory level at installation i in steady state,
W_0 = stochastic delay at the warehouse in steady state due to stock-outs.

For simplicity, we suppress superscript i for installation stock, e.g., we write S_0 instead of S_0^i.

As explained in Section 8.2.3 installation stock and echelon stock policies are equivalent for the considered inventory system. As a consequence, we can just as well use the corresponding echelon stock order-up-to positions: $S_0^e = \sum_{i=0}^{N} S_i$, and $S_i^e = S_i$ for $i > 0$.

Our purpose is to evaluate the distributions of the inventory levels for given ordering policies. The optimization of ordering policies is considered in Section 10.4. For standard cost structures with holding and backorder costs, it can be shown that the optimal S_i-levels ($i = 0, 1, \ldots, N$) are nonnegative. We shall therefore assume that this is the case.

When applying the METRIC approximation we first consider the warehouse. It is easy to determine the distribution of the warehouse inventory level exactly. The inventory position is always S_0. We use our standard approach and consider an arbitrary time, t. At time $t + L_0$, all that is included in the inventory position at time t has reached the warehouse, and we can express the inventory level as S_0 minus the demand in the interval $(t, t + L_0]$, $D_0(L_0)$,

$$IL_0(t + L_0) = S_0 - D_0(L_0). \qquad (10.17)$$

The demand in $(t, t + L_0]$ has a Poisson distribution with mean $\lambda_0 L_0$, and we obtain the exact steady state distribution as

$$P(IL_0 = j) = P(D_0(L_0) = S_0 - j) = \frac{(\lambda_0 L_0)^{S_0 - j}}{(S_0 - j)!} e^{-\lambda_0 L_0}, \quad j \leq S_0. \qquad (10.18)$$

Note that this is also true if the warehouse lead-times are independent stochastic variables with mean L_0 due to Palm's theorem, see Section 5.15.3.

Given the distribution of the inventory level, it is easy to determine the average inventory on hand as

$$E(IL_0^+) = \sum_{j=1}^{S_0} j \cdot P(IL_0 = j), \qquad (10.19)$$

and the average number of backorders as

$$E(IL_0^-) = \sum_{j=-\infty}^{-1}(-j) \cdot P(IL_0 = j), \quad (10.20)$$

or

$$E(IL_0^-) = E(IL_0^+) - E(IL_0) = E(IL_0^+) - (S_0 - \lambda_0 L_0). \quad (10.21)$$

It is normally more efficient to use (10.21) instead of (10.20), because we avoid an infinite sum.

The warehouse can be interpreted as an (M/D/∞) queuing system, where the backorders are the waiting customers. This means that we can get the average delay at the warehouse according to Little's formula:

$$E(W_0) = E(IL_0^-) / \lambda_0. \quad (10.22)$$

Because of the Poisson demand at the warehouse and the first come-first served assumption, the average delay is the same for all retailers.

Consider now retailer i. The retailer lead-time is stochastic due to the delays at the warehouse. However, we already know the average lead-time $\overline{L}_i$, i.e., the transportation time plus the average delay at the warehouse:

$$\overline{L}_i = L_i + E(W_0). \quad (10.23)$$

The METRIC approximation simply means that the real stochastic lead-time is replaced by its mean according to (10.23). (It may appear that Palm's theorem implies that this is an exact approach. This is not true, though, because successive stochastic delays at the warehouse depend on the inventory status at the warehouse and are not independent.) Given this approximation, we have a constant lead-time and can determine the distribution of the inventory level precisely as we did for the warehouse in (10.18)

$$P(IL_i = j) = P(D_i(\overline{L}_i) = S_i - j) = \frac{(\lambda_i \overline{L}_i)^{S_i - j}}{(S_i - j)!} e^{-\lambda_i \overline{L}_i}, \quad j \leq S_i. \quad (10.24)$$

The determination of the average stock on hand and the average number of backorders parallels completely (10.19) - (10.21). Thus the approximation enables us to decompose the problem so that all installations can be handled as single-echelon systems. Note that the approach starts with the upstream installation. When using the decomposition approach in Section 10.1 we start with the downstream installations.

Example 10.2 Consider a system with the following data: $N = 2$, $L_0 = L_1 = L_2 = 2$, $\lambda_1 = 0.1$, $\lambda_2 = 0.9$, $S_0 = 2$, $S_1 = 1$, and $S_2 = 4$. Note first that $\lambda_0 = \lambda_1 + \lambda_2 = 1$. Applying (10.18), (10.19), and (10.21) we obtain

$$E(IL_0^+) = \sum_{j=1}^{S_0} j \cdot P(IL_0 = j) = 1 \cdot \frac{2^1}{1!} e^{-2} + 2 \cdot \frac{2^0}{0!} e^{-2} = 0.541,$$

and

$$E(IL_0^-) = E(IL_0^+) - (S_0 - \lambda_0 L_0) = 0.541 - 0 = 0.541.$$

From (10.22) we have $E(W_0) = E(IL_0^-)/\lambda_0 = 0.541$, and applying (10.23) we get

$$\overline{L}_i = L_i + E(W_0) = 2.541.$$

For retailer 1 we obtain

$$E(IL_1^+) = \sum_{j=1}^{S_1} j \cdot P(IL_1 = j) = 1 \cdot \frac{0.2541^0}{0!} e^{-0.2541} = 0.776,$$

and

$$E(IL_1^-) = E(IL_1^+) - (S_1 - \lambda_1 \overline{L}_1) = 0.776 - (1 - 0.254) = 0.030.$$

For retailer 2 we similarly obtain $E(IL_2^+) = 1.837$ and $E(IL_2^-) = 0.124$. Recall that the results for the retailers are approximate.

The corresponding exact results are $E(IL_1^+) = 0.777$, $E(IL_1^-) = 0.031$, $E(IL_2^+) = 1.866$, and $E(IL_2^-) = 0.153$.

It is straightforward to extend the METRIC approach to more than two levels. The analysis is then exact at the most upstream level, while all downstream installations are handled approximately. It is also easy to handle compound Poisson demand.

Furthermore, it is easy to generalize to a batch-ordering policy at the warehouse. Assume that the warehouse applies an installation stock (R_0, Q_0) policy. The inventory position at the warehouse is then in steady state uniformly distributed on the integers $R_0 + 1, R_0 + 2, \ldots, R_0 + Q_0$. This means that (10.18) is replaced by

$$P(IL_0 = j) = \frac{1}{Q_0} \sum_{k=\max(j,R_0+1)}^{R_0+Q_0} \frac{(\lambda_0 L_0)^{k-j}}{(k-j)!} e^{-\lambda_0 L_0}, \quad j \leq R_0 + Q_0,$$

(10.25)

where we average over the uniformly distributed inventory position. Thereafter we can use exactly the same procedure.

The METRIC approximation has many advantages. It is simple and computationally efficient. At this stage it seems to be the multi-echelon technique that has been applied most widely in practice. There are especially many military applications involving high-value repairable items.

It is more difficult, though, to handle *nonidentical* batch-ordering retailers. It is still relatively easy to determine the average delay at the warehouse. The problem is that the retailers normally face different average delays, and it is more difficult to determine the average delay for a specific retailer. One possibility is, of course, to disregard that the average delays are different for the retailers. This means an additional approximation.

Quite a few approximate techniques for evaluation of general batch-ordering policies in distribution inventory systems are similar in spirit to the METRIC approach. In a two-level system a retailer is handled as a single-echelon system but the parameters describing this system depend on the warehouse. See further Section 10.5.2.2.

Muckstadt (1973) has shown how the METRIC framework can be extended to include a hierarchical or indented-parts structure (MOD-METRIC). To illustrate the basic idea in Muckstadt's approach, consider a group of n items ($j = 1, 2, \ldots, n$) that are controlled by $(S-1, S)$ policies in a two-level inventory system. We can use the METRIC approach to estimate the average waiting time for item j at retailer i, $E(W_{i,j})$. Let us furthermore assume that the considered items are exclusively used as modules when repairing a certain assembly. Each retailer also has a stock of such assemblies. It is assumed that the assembly is repaired by exchanging exactly one of the items. The demand for item j at retailer i is $\lambda_{i,j}$, and the demand for repairs at retailer i is consequently $\lambda_i^a = \sum_{j=1}^n \lambda_{i,j}$. Assume that the repair time at retailer i is T_i. We can now estimate the average repair time, including waiting time, for modules as

$$\overline{T_i} = T_i + \frac{1}{\lambda_i^a} \sum_{j=1}^n \lambda_{i,j} E(W_{i,j}). \quad (10.26)$$

When determining a suitable inventory policy for the assembly, we can then replace the stochastic total repair time by its mean $\overline{T_i}$. See also Rustenburg et al. (2003).

10.3 Two exact techniques

10.3.1 Disaggregation of warehouse backorders

We shall now show how we can analyze the problem in Section 10.2 exactly. We refer to that section for the necessary notation. Our derivation essentially follows Graves (1985). The first derivation of exact steady-state inventory level distributions was carried out by Simon (1971).

From (10.18) we have for $k > 0$

$$P(IL_0^- = k) = P(IL_0 = -k) = P(D_0(L_0) = S_0 + k) = \frac{(\lambda_0 L_0)^{S_0+k}}{(S_0+k)!} e^{-\lambda_0 L_0}.$$

(10.27)

Now let

B_i = number of backorders from retailer i at the warehouse in steady state, stochastic variable.

Evidently $\sum_{i=1}^N B_i = IL_0^-$. When a new backorder occurs, the probability that it emanates from retailer i is λ_i/λ_0 due to the Poisson demand. Consequently, the conditional distribution of B_i for a given IL_0^- is binomial and we have

$$P(B_i = j) = \sum_{k=j}^{\infty} P(IL_0^- = k) \binom{k}{j} \left(\frac{\lambda_i}{\lambda_0}\right)^j \left(\frac{\lambda_0 - \lambda_i}{\lambda_0}\right)^{k-j}, \quad j > 0,$$

(10.28)

$$P(B_i = 0) = 1 - \sum_{j=1}^{\infty} P(B_i = j).$$

Now consider retailer i at some arbitrary time t when the system is in steady state. $S_i - B_i$ units are on their way to, or already at, retailer i. At time $t + L_i$ all these units have reached retailer i, while ordered units backordered at the warehouse at time t and all orders that have been triggered after time t have still not reached the retailer. Consequently, using our standard approach we have

$$IL_i(t + L_i) = S_i - B_i - D_i(L_i),$$

(10.29)

or in other words, to get the inventory level at time $t + L_i$ we subtract the demand at retailer i in the interval $(t, t + L_i]$ from the amount on route to, or already at, retailer i at time t. Note that the demand $D_i(L_i)$ is independent of B_i. The exact distribution of the inventory level at retailer i can therefore be obtained as

$$P(IL_i = j) = P(B_i + D_i(L_i) = S_i - j)$$

(10.30)

$$= \sum_{k=0}^{S_i - j} P(B_i = k) \frac{(\lambda_i L_i)^{S_i - j - k}}{(S_i - j - k)!} e^{-\lambda_i L_i}, \quad j \leq S_i.$$

The considered derivation works very well for Poisson demand and one-for-one replenishments. It is difficult, though, to use the approach in more general situations with batch-ordering retailers. The problem is that we can no longer use the binomial disaggregation of the warehouse backorders in (10.28). See Section 10.5.2.3.

10.3.2 A recursive procedure

We shall now again solve the problem in Section 10.2 exactly, but in a different way. The described procedure is a special case of a more general technique in Axsäter (2000) that provides an exact solution for installation stock (R, Q) policies and independent compound Poisson processes at the retailers.

Consider retailer i and the lead-time for an ordered unit. The lead-time is evidently stochastic due to possible delays at the warehouse. The stochastic lead-time for an ordered unit is independent of the lead-time demand. Furthermore, orders cannot cross in time due to the first come-first served assumption. We have dealt with such lead-times in Section 5.15.2. Let

J = stochastic lead-time demand at retailer i when averaging over the lead-time.

It was shown in Section 5.15.2 that for this type of stochastic lead-time, the distribution of the inventory level can be determined by using the relationship

$$P(IL_i = j) = P(J = S_i - j), \qquad (10.31)$$

i.e., essentially in the same way as for a constant lead-time.

It remains to determine the distribution of J. Since we are dealing with (S - 1, S) policies, each system demand triggers a retailer order for a unit that in turn, triggers a warehouse order for a unit. Consider a warehouse order for a unit at some time t. Because of the first come-first served assumption, the ordered unit will fill the S_0-th retailer order for a unit at the warehouse after the order. (If $S_0 = 0$ the ordered unit will fill the retailer order that triggered the warehouse order.) This means that the considered ordered unit will be assigned to the retailer order triggered by the S_0-th system demand after the warehouse order. We are interested in the lead-time demand for an order from retailer i. Therefore we consider a situation where the S_0-th system demand occurs at retailer i.

The S_0-th system demand after time t can occur either before or after the ordered unit has reached the warehouse. If the ordered unit has reached the warehouse the retailer lead-time is L_i, otherwise it is longer. Let

$u(S_0, j)$ = $P(S_0$-th system demand occurs *before* the order has reached the warehouse and $J \leq j$),

$v(S_0, j)$ = $P(S_0$-th system demand occurs *after* the order has reached the warehouse and $J \leq j$).

We can express the distribution function for J as

$$P(J \le j) = u(S_0, j) + v(S_0, j). \quad (10.32)$$

Given $u(S_0, j)$ and $v(S_0, j)$, we can determine the distribution of the inventory level from (10.31) and (10.32).

$$P(IL_i = j) = P(J = S_i - j)$$
$$= u(S_0, S_i - j) - u(S_0, S_i - j - 1) + v(S_0, S_i - j) - v(S_0, S_i - j - 1). \quad (10.33)$$

It remains to determine the probabilities $u(S_0, j)$ and $v(S_0, j)$.

It is easy to determine the probability $v(S_0, j)$. The corresponding event occurs if there are first at most S_0 - 1 system demands during the time L_0, and then at most j demands at retailer i during the lead-time L_i. Because there are at most S_0 - 1 system demands during the time L_0, the item is in stock at the warehouse when ordered by retailer i. The lead-time is therefore L_i. The two considered probabilities are independent. Consequently,

$$v(S_0, j) = \left(\sum_{k=0}^{S_0-1} \frac{(\lambda_0 L_0)^k}{k!} e^{-\lambda_0 L_0} \right) \left(\sum_{k=0}^{j} \frac{(\lambda_i L_i)^k}{k!} e^{-\lambda_i L_i} \right). \quad (10.34)$$

We shall now show how to derive $u(S_0, j)$ recursively. If the S_0-th demand arrives before the time $t + L_0$, there are three mutually exclusive possibilities. In all three cases the unit ordered by retailer i will arrive at the retailer at time $t + L_0 + L_i$, i.e., it will not stop at the warehouse. In other words, it is ordered by retailer i before it has reached the warehouse.

1. One possibility is that there are exactly S_0 system demands during L_0, and at most j demands at retailer i during the time L_i. After the item has been ordered by retailer i there is no more system demand until the item has reached the warehouse. This probability is given by the first term in (10.35) below.

2. The second possibility is that there are at least $S_0 + 1$ system demands during L_0, the $(S_0 + 1)$-th system demand is also at retailer i, and there are at most j - 1 additional demands at retailer i before the time $t + L_0 + L_i$. The probability for this can be expressed as $(\lambda_i/\lambda_0)u(S_0 + 1, j - 1)$. This is the second term in (10.35).

3. The third possibility is that there are at least $S_0 + 1$ system demands during L_0, the $(S_0 + 1)$-th system demand is not at retailer i, and there

are at most j additional demands at retailer i before the time $t + L_0 + L_i$. The probability for this can be expressed as $((\lambda_0 - \lambda_i)/\lambda_0)u(S_0 + 1, j)$. This is the third term in (10.35).

Consequently, we have for $j > 0$

$$u(S_0, j) = \left(\frac{(\lambda_0 L_0)^{S_0}}{S_0!} e^{-\lambda_0 L_0} \right) \left(\sum_{k=0}^{j} \frac{(\lambda_i L_i)^k}{k!} e^{-\lambda_i L_i} \right)$$

$$+ \frac{\lambda_i}{\lambda_0} u(S_0 + 1, j - 1) + \frac{\lambda_0 - \lambda_i}{\lambda_0} u(S_0 + 1, j).$$

(10.35)

For $j = 0$, Case 2. above does not exist and (10.35) degenerates to

$$u(S_0, 0) = \left(\frac{(\lambda_0 L_0)^{S_0}}{S_0!} e^{-\lambda_0 L_0} \right) e^{-\lambda_i L_i} + \frac{\lambda_0 - \lambda_i}{\lambda_0} u(S_0 + 1, 0). \quad (10.36)$$

Note that $u(S_0, j) \to 0$ as $S_0 \to \infty$ for any value of j. This follows from the definition. We can therefore first set $u(S_0' + 1, j) = 0$ for some large S_0' and all values of j. Thereafter we can first apply (10.36) for $j = 0$, and then (10.35) for $j > 0$, recursively for $S_0 = S_0', S_0'-1, \ldots, 0$. Since the coefficients λ_i/λ_0 and $(\lambda_0 - \lambda_i)/\lambda_0$ in (10.35) and (10.36) are both strictly smaller than 1, the suggested numerical procedure is always stable.

A related iterative procedure for the costs in case of Poisson demand and $(S - 1, S)$ policies is given in Axsäter (1990). Forsberg (1995) and Axsäter and Zhang (1996) generalize this procedure to compound Poisson demand. See also Axsäter (1993c), which deals with periodic review.

Normally, it is considerably more difficult to deal with models where all installations apply (R, Q) policies as compared to the simpler $(S -1, S)$ policies, which we have considered in Sections 10.2 - 10.3. However, the analysis of (R, Q) policies is simplified significantly by the important fact that the inventory positions of all installations are uniformly distributed and independent in steady state, as in the single-echelon case. See Sections 10.5.2.1 and 10.5.2.4.

10.4 Optimization of ordering policies

When applying the Clark-Scarf approach in Section 10.1 we obtained the optimal policy as part of the approach. In Sections 10.2 and 10.3, however, we have focused on deriving the distributions of the inventory levels. We shall now add a standard cost structure to the considered model and show how the policy can be optimized. We assume that we have either obtained the exact probability distributions as in Section 10.3, or approximate distributions as in Section 10.2. Let

S_i = installation stock order-up-to inventory position at installation i, ($i = 0, 1, \ldots, N$),
h_i = holding cost per unit and time unit at installation i, ($i = 0, 1, \ldots, N$),
b_i = backorder cost per unit and time unit at retailer i, ($i = 1, 2, \ldots, N$),
C = average system costs per time unit.

The holding costs at the warehouse are not affected by the retailer inventory positions because the structure of the demand at the warehouse is not affected if we change S_i. Furthermore, the costs at retailer i are not affected by the order-up-to-inventory positions at the other retailers. Recall that these inventory positions do not affect the demand at the warehouse. We therefore have

$C_0(S_0)$ = average holding costs per time unit at the warehouse,

$C_i(S_0, S_i)$ = average holding and backorder costs per time unit at retailer i, ($i = 1, 2, \ldots, N$).

Note that we have no explicit backorder costs at the warehouse. Backorders at the warehouse will, however, cause delays for retailer orders and thus indirectly influence the costs at the retailers.

The warehouse costs are obtained as

$$C_0(S_0) = h_0 E(IL_0^+), \qquad (10.37)$$

and the costs at retailer i as

$$C_i(S_0, S_i) = h_i E(IL_i^+) + b_i E(IL_i^-). \qquad (10.38)$$

The total costs can then be expressed as

$$C = C_0(S_0) + \sum_{i=1}^{N} C_i(S_0, S_i) . \qquad (10.39)$$

If we have determined the distribution of the inventory levels approximately as in Section 10.2, or exactly as in Section 10.3.1 or 10.3.2, it is easy to evaluate these costs. See (10.19) - (10.21). Recall that is more efficient to apply (10.21) than (10.20). It can be shown that the optimal order-up-to-levels must be nonnegative.

Consider now the costs at retailer i for a given value of S_0. It follows from the results in Section 5.9 that if the retailer lead-time were constant, the retailer costs would be convex in S_i. This implies directly that $C_i(S_0, S_i)$ is convex in S_i when using the approximate technique in Section 10.2, since we are then replacing the stochastic retailer lead-time by its mean. But this convexity is also true in the exact case. To see this, recall that the lead-time for an ordered unit is independent of the lead-time demand, and that orders cannot cross in time. This means, as explained in Section 5.15.2, that the costs $C_i(S_0, S_i)$ can be seen as an average of costs for different constant lead-times, which implies convexity in S_i.

Furthermore, it is easy to find lower and upper bounds for the optimal S_i. A lower bound, S_i^ℓ, can be found by optimizing S_i for the shortest possible deterministic lead-time, L_i, and an upper bound, S_i^u, by optimizing S_i for the longest possible lead-time, $L_0 + L_i$.

For a given S_0, we can find the corresponding optimal $S_i^*(S_0)$ by a simple local search where, starting with $S_i = S_i^\ell$, we successively increase S_i by one unit at a time until we find a local minimum of $C_i(S_0, S_i)$. Note that this optimization can be done separately for each retailer.

The optimization of C with respect to S_0 is not that simple, though, since

$$C(S_0) = C_0(S_0) + \sum_{i=1}^{N} C_i(S_0, S_i^*(S_0)), \qquad (10.40)$$

is not necessarily convex in S_0. We can, however, find a lower bound, S_0^ℓ, by optimizing (10.39) with respect to S_0 for $S_i = S_i^u$, and similarly, an upper bound, S_0^u, by optimizing S_0 for $S_i = S_i^\ell$. To find the optimal S_0 we then need to evaluate the costs (10.40) for all values of S_0 within these bounds. See Axsäter (1990) for details.

An optimization of the reorder points when the installations apply installation stock (R, Q) policies can normally be carried out in essentially the same way. An optimization of echelon stock (R, Q) policies requires some modifications. See Axsäter (1997) or Chen and Zheng (1997).

A joint optimization of reorder points and batch quantities is more complex, though. It is therefore most common to determine the batch quantities in advance from a deterministic model. Recall that this is also the standard procedure for single-echelon systems and that the resulting errors are usually very limited.

10.5 Batch-ordering policies

10.5.1 Serial system

We shall now generalize the model in Section 10.1.1 by considering more general echelon stock (R, Q) policies. As in Section 10.1.1, we shall deal with normally distributed demand, periodic review, and the two-level serial system in Figure 10.1. The events in a period also take place in the same order, i.e.,

1. Installation 2 orders.
2. The period delivery from the outside supplier arrives at installation 2.
3. Installation 1 orders from installation 2.
4. The period delivery from installation 2 arrives at installation 1.
5. The stochastic period demand takes place at installation 1.
6. Evaluation of holding and shortage costs.

Let

L_i = lead-time (transportation time) for replenishments at installation i,
μ = average period demand,
σ = standard deviation of period demand,
R_i = echelon stock reorder point at installation i, (Note that we write R_i instead of R_i^e.)
Q_i = order quantity at installation i,
$F_i(x)$ = distribution function of the echelon stock inventory level at installation i,
$f_i(x)$ = density of the echelon stock inventory level at installation i.

We make the assumption that Q_2 as well as the initial installation stock at installation 2 are integer multiples of Q_1. This means that the installation stock inventory level at installation 2 is a multiple of Q_1 at all times. Recall from Section 8.2.3 that, given these assumptions, the class of echelon stock reorder point policies contains the class of installation stock reorder point policies as a subset. Furthermore, Chen (2000) has shown that echelon stock (R, Q) policies are optimal for serial systems under quite general conditions.

We shall derive the probability distributions of the echelon stock inventory levels. The derivation essentially follows Chen and Zheng (1994).

Note first that the echelon stock inventory status of installation 2 completely parallels the status of a single-echelon inventory, see Section 5.3. Consider an arbitrary period, t. The echelon stock inventory position just after a possible order in period t (before the demand), $IP_2^{e-}(t)$, is in steady state uniformly distributed on the interval $(R_2, R_2 + Q_2)$. The echelon stock inventory level in period $t + L_2$ after the period demand, $IL_2^e(t + L_2)$, is obtained by subtracting the demands in periods $t, t + 1, \ldots, t + L_2$, i.e., the demands during $L_2 + 1$ periods, $D(L_2 + 1)$,

$$IL_2^e(t + L_2) = IP_2^{e-}(t) - D(L_2 + 1). \tag{10.41}$$

The demand during $L_2 + 1$ periods has mean $\mu_2'' = (L_2 + 1)\mu$ and standard deviation $\sigma_2'' = (L_2 + 1)^{1/2} \sigma$. In complete analogy with (5.42) we now obtain

$$F_2(x) = P(IL_2^e \leq x) = \frac{\sigma_2''}{Q} \left[G\left(\frac{R_2 - x - \mu_2''}{\sigma_2''} \right) - G\left(\frac{R_2 + Q_2 - x - \mu_2''}{\sigma_2''} \right) \right]. \tag{10.42}$$

Given $F_2(x)$, we can, for example, evaluate the echelon holding costs at installation 2.

Installation 1 orders from installation 2 before the demand in period $t + L_2$. We are therefore also interested in the distribution of the inventory level at this point in time. Let IL_2^{e-} be the inventory level and $F_2^-(x)$ the corresponding distribution function. The only difference compared to (10.42) is that μ_2'' and σ_2'' are replaced by $\mu_2' = L_2 \mu$ and $\sigma_2' = L_2^{1/2} \sigma$, and we obtain

$$F_2^-(x) = \frac{\sigma_2'}{Q}\left[G\left(\frac{R_2 - x - \mu_2'}{\sigma_2'}\right) - G\left(\frac{R_2 + Q_2 - x - \mu_2'}{\sigma_2'}\right)\right]. \quad (10.43)$$

The corresponding density is (recall that $G'(x) = \Phi(x) - 1$)

$$f_2^-(x) = \frac{1}{Q}\left[\Phi\left(\frac{R_2 + Q_2 - x - \mu_2'}{\sigma_2'}\right) - \Phi\left(\frac{R_2 - x - \mu_2'}{\sigma_2'}\right)\right]. \quad (10.44)$$

The inventory position at installation 1, IP_1, after the review with the possibility to order is always in the interval $(R_1, R_1 + Q_1]$. Consequently, if $IL_2^{e-} \leq R_1$, we know that there is no installation stock at installation 2 after the review of the inventory position at installation 1. All stock included in IL_2^{e-} must be on its way to, or already at, installation 1. The other possibility is that $IL_2^{e-} > R_1$. Recall now that the installation inventory level at installation 2 after the review at installation 1 is always a multiple of Q_1, say jQ_1. We have $jQ_1 = IL_2^{e-} - IP_1 > R_1 - (R_1 + Q_1) = -Q_1$ since $IP_1 \leq R_1 + Q_1$. But this means that $j \geq 0$. Consequently, there are no backorders at installation 2 and all stock included in IP_1 must be on its way to, or already at, installation 1. Note that given IL_2^{e-}, we can obtain IP_1 uniquely from the conditions $IP_1 = IL_2^{e-} - jQ_1$ and $R_1 < IP_1 \leq R_1 + Q_1$ since the second condition determines j uniquely. Let us denote the resulting j by $\hat{j}$.

We can conclude that the realized inventory position, i.e., the stock on its way to, or already at, installation 1 can be expressed as

$$O(IL_2^{e-}) = \begin{cases} IL_2^{e-}, & IL_2^{e-} \leq R_1, \\ IL_2^{e-} - \hat{j}Q_1, & \text{otherwise.} \end{cases} \quad (10.45)$$

Consider next the inventory level at installation 1, IL_1, after the demand in period $t + L_1 + L_2$. We obtain IL_1 as $O(IL_2^{e-})$ minus the demands in periods $t + L_2, t + L_2 + 1, \ldots, t + L_1 + L_2$, i.e., the demand during $L_1 + 1$ periods, $D(L_1 + 1)$ with mean $\mu_1'' = (L_1 + 1)\mu$ and standard deviation $\sigma_1'' = (L_1 + 1)^{1/2}\sigma$. We know the density of IL_2^{e-} from (10.44). Consequently we obtain the distribution function for the inventory level at installation 1, $F_1(x)$, as

$$F_1(x) = \int_{-\infty}^{R_2+Q_2} f_2^-(u)\left[1 - \Phi\left(\frac{O(u) - x - \mu_1''}{\sigma_1''}\right)\right] du \ . \tag{10.46}$$

Given $F_1(x)$, we can evaluate holding and backorder costs at installation 1.

It is straightforward to extend the procedure to more than two levels. The numerical computations are easier in case of a discrete demand distribution.

10.5.2 Distribution system

Let us now again consider the two-echelon distribution system in Figure 10.5.

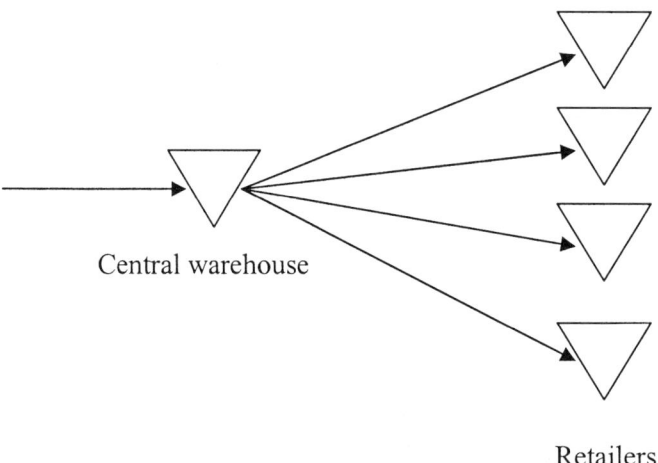

Figure 10.5 Distribution inventory system.

We have considered the evaluation of order-up-to-S policies in such inventory systems in Sections 10.1.2, 10.2, and 10.3. We shall now, as in Section 10.5.1, consider more general batch-ordering policies. But in contrast to Section 10.5.1, we shall mainly deal with continuous review policies.

We start by providing some basic results in Section 10.5.2.1. Thereafter we discuss, in Sections 10.5.2.2 - 10.5.2.5, different approaches for evaluation of batch-ordering policies.

10.5.2.1 Some basic results

Let us first consider a simple special case. Assume that all retailers apply S-policies. This means that their inventory positions are kept constant, and there is no difference between an echelon stock (R_0^e, Q_0) policy and an installation stock (R_0^i, Q_0) policy at the warehouse. The difference between the echelon stock inventory position and the installation stock inventory position at the warehouse is simply the sum of the constant retailer inventory positions. So although we shall consider installation stock policies, the following is also true for echelon stock policies.

Consider first the case when also the warehouse applies an installation stock S policy. Let

$C(S_0^i, S_1^i, ..., S_N^i)$ = total system costs per time unit for a given installation stock S policy.

Assume that these costs can be evaluated for different policies using the techniques in Sections 10.1 - 10.3. Consider then instead an (R_0^i, Q_0) policy at the warehouse. For discrete integral demand this means that the warehouse inventory position is uniform on [$R_0^i + 1, R_0^i + 2, ..., R_0^i + Q_0$] and we obtain the total costs for the (R_0^i, Q_0) policy at the warehouse by simply averaging over these inventory positions.

$$C(R_0^i, Q_0, S_1^i, ..., S_N^i) = \frac{1}{Q_0} \sum_{k=R_0^i+1}^{R_0^i+Q_0} C(k, S_1^i, ..., S_N^i). \qquad (10.47)$$

This means that it is easy to evaluate and optimize the ordering policy. Note also that we can, in a sense, force the retailers to order in "batches" while still using a simple S policy, if we use periodic review with a suitably long review period.

Although multi-echelon inventory systems with general batch-ordering policies are difficult to analyze, there are some important properties of single-echelon inventory systems that carry over to the multi-echelon case. One such property concerns the uniform distribution of the inventory position when using (R, Q) policies.

Consider first a distribution system with discrete demand (for example compound Poisson demand), where all installations apply installation stock (R, Q) policies. Let q be the largest common factor of all batch quantities Q_0,

$Q_1, \ldots, Q_N$. It is evident that all replenishments and all demands at the warehouse are multiples of q. Therefore it is natural to assume that also the warehouse reorder point R_0^i, as well as the initial inventory position at the warehouse are multiples of q. Let us furthermore assume that not all customer demands are multiples of some unit larger than one. Under these assumptions we have the following simple result (see e.g., Axsäter, 1998).

Proposition 10.1 (Installation stock (R, Q) policies.) In steady state the installation stock inventory positions are independent. The warehouse installation stock inventory position is uniform on $[R_0^i + q, R_0^i + 2q, \ldots, R_0^i + Q_0]$, and the retailer installation stock inventory positions are uniform on $[R_j^i + 1, R_j^i + 2, \ldots, R_j^i + Q_j]$.

Consider then the same system with echelon stock (R, Q) policies instead of installation stock policies. We then have

Proposition 10.2 (Echelon stock (R, Q) policies.) In steady state all installations have their echelon inventory positions uniformly distributed on $[R_j^e + 1, R_j^e + 2, \ldots, R_j^e + Q_j]$, i.e., also the warehouse. The retailer inventory positions are still independent of each other, but the warehouse echelon stock inventory position is correlated to the retailer inventory positions.

The coupling between the warehouse echelon stock inventory position and the retailer inventory positions depends on the initial stock at the warehouse. See Axsäter (1997), and Chen and Zheng (1997).

Recall that when dealing with batch-ordering policies a standard approach is to determine the batch quantities from a deterministic model in an initial step. This is a major simplification. The remaining main difficulty is the policy evaluation. Given a suitable evaluation method we can normally optimize the reorder points in the same way as the order-up-to inventory positions when dealing with S policies. See Section 10.4.

In the following we give an overview of how different researchers have approached the policy evaluation problem for distribution systems with general batch-ordering policies. To avoid getting involved in too many technical details, we shall only sketch the basic ideas behind different approaches.

10.5.2.2 METRIC type approximations

As we have already discussed in Section 10.2, several authors have tried to generalize the METRIC approach to batch ordering retailers. Recall that the

METRIC approach means that we evaluate the average delay for retailer orders due to shortages at the warehouse. This average delay is added to the retailer transportation times to get exact average lead-times for the retailers. When evaluating the costs at the retailers, these averages are, as an approximation, used instead of the real stochastic lead-times.

When dealing with different batch-ordering retailers it is still easy to determine the average backorder level at the warehouse, and to obtain the average delay at the warehouse by applying Little's formula. This delay, however, is an average over all retailers. A problem in connection with this approach is that the average delays may vary substantially between the retailers. Of course the average delays are the same in the special case of identical retailers.

The first METRIC type approach for batch-ordering retailers was by Deuermeyer and Schwarz (1981). They first approximate the mean and variance of the warehouse lead-time demand and fit a normal distribution to these parameters. Next they estimate holding costs and the average delay at the warehouse. The retailer lead-times are, as in METRIC, obtained as the transportation times plus the average delay at the warehouse. Svoronos and Zipkin (1988) use a different type of second moment approximation. Except for the average warehouse delay, they also derive an approximate estimate of the variance of the delay. Using these parameters they then fit a negative binomial distribution to the retailer lead-time demand. Another related technique is suggested in Axsäter (2002).

A METRIC type approximation is quite simple. It means essentially that a multi-echelon system is decomposed into a number of single-echelon systems. Although the errors may be substantial in some cases, the approach is also often reasonable in practical applications.

10.5.2.3 Disaggregation of warehouse backorders

Another possibility is to use the approach in Section 10.3.1. This approach is similar to our standard approach for single-echelon systems. For a single-echelon system we get the inventory level at time $t + L$ as the inventory position at time t minus the lead-time demand. When dealing with retailer i in the considered two-echelon distribution inventory system, we also need to subtract the backorders at the warehouse at time t, which emanates from retailer i. These backorders cannot be delivered before time $t + L_i$. This means that we need to disaggregate the total number of backorders to obtain the distribution of the backorders that emanate from a certain retailer. In the case of Poisson demand and one-for-one replenishments, a binomial disaggregation is exact and it is relatively easy to evaluate a policy exactly as shown in Section 10.3.1. In the case of batch-ordering retailers, an exact disaggregation is much more complicated. Chen and Zheng (1997) have still been able to derive an exact solution this way for echelon stock (R, Q) policies and Poisson demand. Lee and Moinzadeh (1987a,b), and Moinzadeh and Lee (1986) approximate the demand process at the warehouse by a Poisson process. Using this approximation they determine the distribution of the total number of outstanding orders at all retailers. Next the mean and variance of the number of outstanding orders at a single retailer are obtained by an approximate disaggregation procedure. They then use different two-moment approximations of the distribution of outstanding orders at the retailer. Axsäter (1995) is another related approach.

10.5.2.4 Following supply units through the system

A third possible approach is to try to keep track of each supply unit as it moves through the system. A batch is then seen as a "package" of individual units. This approach is related to the technique for one-for-one ordering policies described in Section 10.3.2 and to Axsäter (1990). Using this technique Axsäter (2000) derives the exact probability distributions for the inventory levels for a quite general continuous review multi-echelon inventory system with batch-ordering retailers facing compound Poisson demand. Related papers, which also deal with batch-ordering policies, are Axsäter (1993a,b, 1997, 1998, 2003), Forsberg (1997a,b), and Andersson (1997). See also Muharremoglu and Tsitsiklis (2003).

Marklund (2002, 2005), Cachon (2001), Cachon and Fisher (2000), and Axsäter and Marklund (2005) use related approaches for evaluating different types of policies. The policy considered in Axsäter and Marklund (2005) will always outperform both installation stock and echelon stock (R, Q) policies.

Using this type of techniques exact policy evaluations are possible as long as we deal with relatively small problems characterized by low demand and corresponding small batch quantities. For larger problems the computational effort grows quite rapidly and it is more realistic to rely on approximate techniques.

10.5.2.5 Practical considerations

It is clear that the research concerning evaluation and optimization of distribution inventory systems with stochastic demand has made a lot of progress during the past two decades. It is possible to handle a wider range of systems exactly, and we have better approximate techniques for systems which cannot be analyzed exactly.

Numerical experimentation with various test problems in many research papers has also illustrated what a typical optimal solution looks like. See e.g., Muckstadt and Thomas (1980), Axsäter (1993b, 2000), Hausman and Erkip (1994), Andersson (1997), and Gallego and Zipkin (1999). The optimal solution of a particular problem will, of course, depend on problem parameters like holding costs, backorder costs (or service constraints), transportation times, and demand processes. Although it is difficult to generalize, it is striking that the optimal solutions very often look the same in a certain sense. What is typical is that upstream installations should have very low stocks compared to downstream installations. Consider, for example, a distribution inventory system with a central warehouse and about five retailers. Let us assume that the retailers are required to provide a fill rate of 95% towards the final customers. The optimal solution may then typically show that we should have a negative safety stock at the warehouse, and that the warehouse fill rate for orders from the retailers should be something like 50%. A first important question is how this knowledge can be used for more efficient control of practical supply chains.

In practice it is still most common to handle different stocks in a supply chain by single-echelon techniques. As discussed in Section 10.5.2.2, it may then be possible to use some simple type of METRIC approximation.

Quite often practitioners handle upstream and downstream stock points in a similar way. This means generally that the distribution of the total stock between upstream and downstream installations is far from optimal. One way to achieve better control is, of course, to implement exact or approximate techniques for multi-echelon inventory systems. However, it may be difficult to replace an existing simple control system with a relatively advanced multi-echelon technique. The computational effort will also grow considerably. In the short run it may be more realistic to implement the new knowledge in the following way: (i) Analyze a small number of representative items by multi-echelon techniques, and (ii) Adjust the service levels in the existing control system so that they roughly correspond to the optimal solution.

10.6 Other assumptions

10.6.1 Guaranteed service model approach

We shall now consider a relatively simple approximate approach for setting order-up-to inventory positions in a multi-echelon inventory system. The advantage of the approach is its simplicity, while it is difficult to see how the approximations will affect the results. The general approach was first suggested by Kimball in 1955. (This manuscript was later reprinted in Kimball, 1988.) The most well-known early paper dealing with this topic is Simpson (1958), who used the technique for determining safety stocks in a serial inventory system. Various generalizations have been provided by Graves (1988), Inderfurth (1991), and Inderfurth and Minner (1998). The presentation here follows essentially Graves and Willems (2003).

We assume continuous review. Let

μ_j = mean demand per unit of time at installation j,
σ_j = standard deviation of demand per unit of time at installation j,
S_j = order-up-to inventory position at installation j,
s_j^{out} = service time provided by installation j to its downstream installations,
s_j^{in} = service time for installation j from upstream installations,
L_j = processing time at installation j,
k_j = safety factor for installation j,
$p(j)$ = the set of installations which are immediate predecessors of installation j,
I_j = on-hand inventory at installation j.

Consider installation j. It is assumed that the service times provided by installations that are immediate predecessors, s_i^{out} for $i \in p(j)$, have been determined. The service time, or delay, for installation j from upstream installations is then

$$s_j^{in} = \max_{i \in p(j)} \{s_i^{out}\}. \qquad (10.48)$$

The corresponding lead-time for replenishments is obtained as $L_j + s_j^{in}$. The average demand during this time is $\mu_j(L_j + s_j^{in})$, and the standard deviation $\sigma_j(L_j + s_j^{in})^{1/2}$. A standard procedure for setting the order-up-to inventory position so that a delivery from stock (i.e., $s_j^{out} = 0$) is possible with a high probability is

$$S'_j = \mu_j(L_j + s_j^{in}) + k_j \sigma_j (L_j + s_j^{in})^{1/2}, \qquad (10.49)$$

where k_j is the so-called safety factor. In case of unbounded demand, we will of course, not always be able to deliver immediately from stock, but we shall still interpret (10.49) as the order-up-to inventory position that will provide $s_j^{out} = 0$.

Let us now instead assume that we can accept a positive service time towards downstream installations, i.e., some $s_j^{out} > 0$. This is equivalent to reducing the lead-time by s_j^{out} in (10.49). This simple result, which was first pointed out by Hariharan and Zipkin (1995), is easy to derive. We obtain the corresponding S_j as

$$S_j = \mu_j(L_j + s_j^{in} - s_j^{out}) + k_j\sigma_j(L_j + s_j^{in} - s_j^{out})^{1/2}. \quad (10.50)$$

In (10.50) we must require that $s_j^{out} \leq L_j + s_j^{in}$.

Given (10.50) we can approximate the expected inventory on hand at installation j as

$$E(I_j) = k_j\sigma_j(L_j + s_j^{in} - s_j^{out})^{1/2}. \quad (10.51)$$

Using the considered approach in a decentralized setting, each site can successively set its service time towards downstream sites and determine the corresponding order-up-to inventory position S_j.

We can also include (10.51) in a safety stock optimization. Assume that the service times provided by external suppliers are given. Furthermore, for sites facing customer demand there are required service times towards the customers. Using (10.48) and (10.51) we can then, for example, minimize the total holding costs with respect to the internal service times, s_j^{out}, provided by sites not facing demand from external customers.

10.6.2 Coordination and contracts

In practice it is common with supply chains, which consist of several independent firms. So far in this book we have assumed that all information concerning the supply chain is available and that our purpose is to optimize the total supply chain. So we have adopted the perspective of a central planner.

However, it is obvious that this perspective is not always relevant. A company that is part of a supply chain quite often wishes to limit the information that is available for other companies. An interesting question is then to which extent this limits the possibilities for improved control. It is clear that extensive information sharing becomes more and more common in supply chains. An example is the implementation of so-called vendor managed inventory (VMI) programs. In such implementations the supplier may have access to real-time sales data at its downstream installation. This puts, in a sense, the supplier in the position of a central planner.

Furthermore, a company in a supply chain may be more interested in its own objectives than in the objectives of the total supply chain. This may lead to additional costs due to sub-optimization. A question here is whether such incentive conflicts can be resolved by agreements or contracts that guarantee good coordination, while still allowing decentralized planning and control.

There is a rich and rapidly growing literature on information sharing, coordination, and contracts. We refer to two recent overviews in Cachon (2003) and Chen (2003) for more details and additional references.

We shall in this section illustrate modeling and analysis of contracts by considering a simplified version of an example that is discussed in Cachon (2003).

10.6.2.1 The newsboy problem with two firms

We return to the newsboy problem that we dealt with in Section 5.13. However, we now assume that the distribution is handled by two independent firms, a supplier and a retailer. The two firms want to maximize their own profits.

The supplier has a certain given production cost per unit. The retailer orders from the supplier and faces a newsboy problem. There is a single season and the demand is normally distributed. (Other distributions can be handled in the same way.) There is a certain given market price per unit. The retailer is not paid anything for unsold units. Let us introduce the following notation:

c = given production cost per unit for the supplier,
p = given market price per unit,
S = amount ordered by the retailer,
μ = mean of stochastic demand,
σ = standard deviation of stochastic demand,
q = expected sales.

It is evident that we must assume that $p > c$. Otherwise no business will take place.

10.6.2.2 Wholesale-price contract

Let us first consider a so-called wholesale-price contract. This means that the supplier decides on a price per unit that the retailer has to pay. Let us introduce the following additional notation.

w = price per unit charged by the supplier and paid by the retailer,
π_s = expected profit for the supplier,
π_r = expected profit for the retailer,
Π = expected total profit.

Consider first the retailer profit:

$$\pi_r = p(\int_{-\infty}^{S} x \frac{1}{\sigma}\varphi(\frac{x-\mu}{\sigma})dx + S\int_{S}^{\infty}\frac{1}{\sigma}\varphi(\frac{x-\mu}{\sigma})dx) - Sw$$

$$= p(\mu - \int_{S}^{\infty}(x-S)\frac{1}{\sigma}\varphi(\frac{x-\mu}{\sigma})dx) - Sw = p(\mu - \sigma G(\frac{S-\mu}{\sigma})) - Sw, \quad (10.52)$$

where

$$q = \mu - \sigma G(\frac{S-\mu}{\sigma}), \quad (10.53)$$

is the expected sales. The expected profit for the supplier is simply

$$\pi_s = S(w-c), \quad (10.54)$$

and the expected total profit is

$$\Pi = \pi_r + \pi_s = p(\mu - \sigma G(\frac{S-\mu}{\sigma})) - Sc. \quad (10.55)$$

Note that we can assume $c < w < p$. If $c \geq w$ there is no supplier profit, and if $w \geq p$ it is optimal for the retailer to choose $S = 0$.

Let us now first consider the total profit. The problem to maximize the total profit is equivalent to the original newsboy problem in Section 5.13. Note that $\Pi = (p-c)\mu - C$, where C is the cost in (5.89) with $c_o = c$ and $c_u =$

$p - c$. So, profit maximization with respect to S is equivalent to cost minimization.

The optimal S is obtained from

$$\Phi(\frac{S-\mu}{\sigma}) = \frac{p-c}{p}. \qquad (10.56)$$

This condition is easy to derive from (10.55) and equivalent to (5.91). The optimal S is denoted $S^*(c)$.

Consider then the expected retailer profit. The structure of (10.52) is completely equivalent to (10.55) so we obtain the minimizing $S^*(w)$ from

$$\Phi(\frac{S-\mu}{\sigma}) = \frac{p-w}{p}. \qquad (10.57)$$

It is easy to see that $S^*(w) < S^*(c)$ because $w > c$. The considered contract will therefore fail in the sense that it will not lead to a maximization of the total profit.

Assuming that the retailer uses $S^*(w)$, the supplier can maximize her profit by maximizing

$$\pi_s = (w-c)S^*(w). \qquad (10.58)$$

10.6.2.3 Buyback contract

Let us now consider a different type of contract. The supplier still charges the retailer a certain price per unit purchased. However, he also pays the retailer a certain amount per unit of leftover inventory. Let

w_b = price per unit charged by the supplier and paid by the retailer,
b = compensation per unit for leftover inventory,
$\tilde{\pi}_s$ = expected profit for the supplier,
$\tilde{\pi}_r$ = expected profit for the retailer,
Π = expected total profit.

It is obvious that we should require $b \leq w_b$. Otherwise, the retailer earns profit from leftover inventory. We shall choose the parameters b and w_b as

$$b = (1-\lambda)p,$$
$$w_b = (1-\lambda)p + \lambda c,$$
(10.59)

where $0 < \lambda < 1$. We get

$$\tilde{\pi}_r = pq - Sw_b + b(S-q) = \lambda pq - \lambda Sc = \lambda \Pi,$$
(10.60)

and

$$\tilde{\pi}_s = S(w_b - c) - b(S-q) = (1-\lambda)pq - (1-\lambda)Sc = (1-\lambda)\Pi.$$

(10.61)

We can conclude that the considered contract coordinates the supply chain in the sense that both the supplier and the retailer wish to maximize the total profit. For both firms it is rational to use $S^*(c)$ obtained from (10.56). We can interpret λ as the retailer's share of the total supply chain's profit. It is reasonable for both firms to accept the cost structure (10.59) and only negotiate about the value of λ.

References

Andersson, J. 1997. Exact Evaluation of General Performance Measures in Multi-Echelon Inventory Systems, Lund University.

Axsäter, S. 1990. Simple Solution Procedures for a Class of Two-Echelon Inventory Problems, *Operations Research*, 38, 64-69.

Axsäter, S. 1993a. Exact and Approximate Evaluation of Batch-Ordering Policies for Two-Level Inventory Systems, *Operations Research*, 41, 777-785.

Axsäter, S. 1993b. Continuous Review Policies for Multi-Level Inventory Systems with Stochastic Demand, in S. C. Graves et al. Eds. *Handbooks in OR & MS Vol. 4*, North Holland Amsterdam, 175-197.

Axsäter, S. 1993c. Optimization of Order-up-to-*S* Policies in 2-Echelon Inventory Systems with Periodic Review, *Naval Research Logistics*, 40, 245-253.

Axsäter, S. 1995. Approximate Evaluation of Batch-Ordering Policies for a One-Warehouse, *N* Non-Identical Retailer System Under Compound Poisson Demand, *Naval Research Logistics*, 42, 807-819.

Axsäter, S. 1997. Simple Evaluation of Echelon Stock (*R, Q*) Policies for Two-Level Inventory Systems, *IIE Transactions*, 29, 661-669.

Axsäter, S. 1998. Evaluation of Installation Stock Based (*R, Q*)-Policies for Two-Level Inventory Systems with Poisson Demand, *Operations Research*, 46, S135-S145.

Axsäter, S. 2000. Exact Analysis of Continuous Review (R, Q) Policies in Two-Echelon Inventory Systems with Compound Poisson Demand, *Operations Research* 48, 686-696.

Axsäter, S. 2002. Approximate Optimization of a Two-Level Distribution Inventory System, *International Journal of Production Economics*, 81-2, 545-553.

Axsäter, S. 2003. Supply Chain Operations: Serial and Distribution Inventory Systems, in A. G. de Kok, and S. C. Graves Eds. *Handbooks in OR & MS Vol. 11*, North Holland Amsterdam, 525-559.

Axsäter, S., and J. Marklund. 2005. Optimal "Position-Based" Warehouse Ordering in Divergent Two-Echelon Inventory Systems, Lund University.

Axsäter, S., J. Marklund, and E. A. Silver. 2002. Heuristic Methods for Centralized Control of One-Warehouse N-Retailer Inventory Systems, *Manufacturing & Service Operations Management*, 4, 75-97.

Axsäter, S., and W. F. Zhang. 1996. Recursive Evaluation of Order-up-to-S Policies for Two-Echelon Inventory Systems with Compound Poisson Demand, *Naval Research Logistics*, 43, 151-157.

Cachon, G. P. 2001. Exact Evaluation of Batch-Ordering Inventory Policies in Two-Echelon Supply Chains with Periodic Review, *Operations Research*, 49, 79-98.

Cachon, G. P. 2003. Supply Chain Coordination with Contracts, in A. G. de Kok, and S. C. Graves Eds. *Handbooks in OR & MS Vol. 11*, North Holland Amsterdam, 229-339.

Cachon, G. P., and M. Fisher. 2000. Supply Chain Inventory Management and the Value of Shared Information, *Management Science*, 46, 1032-1048.

Chen. F. 2000. Optimal Policies for Multi-Echelon Inventory Problems with Batch Ordering, *Operations Research*, 48, 376-389.

Chen, F. 2003. Information Sharing and Supply Chain Coordination, in A. G. de Kok, and S. C. Graves Eds. *Handbooks in OR & MS Vol. 11*, North Holland Amsterdam, 341-421.

Chen, F., and J. S. Song. 2001. Optimal Policies for Multi-Echelon Inventory Problems with Markov-Modulated Demand, *Operations Research*, 49, 226-234.

Chen, F., and Y. S. Zheng. 1994. Evaluating Echelon Stock (R, nQ) Policies in Serial Production/Inventory Systems with Stochastic Demand, *Management Science*, 40, 1262-1275.

Chen, F., and Y. S. Zheng. 1997. One-Warehouse Multi-Retailer Systems with Centralized Stock Information, *Operations Research*, 45, 275-287.

Clark, A. J., and H. Scarf. 1960. Optimal Policies for a Multi-Echelon Inventory Problem, *Management Science*, 5, 475-490.

Deuermeyer, B., and L. B. Schwarz. 1981. A Model for the Analysis of System Service Level in Warehouse/Retailer Distribution Systems: The Identical Retailer Case, in L. B. Schwarz Ed. *Multi-Level Production/Inventory Control Systems: Theory and Practice*, North Holland Amsterdam, 163-193.

Diks, E. B., and A. G. de Kok. 1998. Optimal Control of a Divergent Multi-Echelon Inventory System, *European Journal of Operational Research*, 111, 75-97.

Diks, E. B., and A. G. de Kok. 1999. Computational Results for the Control of a Divergent N-Echelon Inventory System, *International Journal of Production Economics*, 59, 327-336.

Dogru, M. K., A. G. de Kok, and G. J. van Houtum. 2005. A Numerical Study on the Effect of the Balance Assumption in One-Warehouse Multi-Retailer Inventory Systems, Eindhoven University of Technology.

Eppen, G. D., and L. Schrage. 1981. Centralized Ordering Policies in a Multi-Warehouse System with Leadtimes and Random Demand, in L. B. Schwarz Ed. *Multi-Level Production/Inventory Control Systems: Theory and Practice*, North Holland Amsterdam, 51-69.

Erkip N. K. 1984. Approximate Policies in Multi-Echelon Inventory Systems, unpublished Ph.D. dissertation, Department of Industrial Engineering and Engineering Management, Stanford University.

Federgruen, A. 1993. Centralized Planning Models for Multi-Echelon Inventory Systems under Uncertainty, in S. C. Graves et al. Eds. *Handbooks in OR & MS Vol. 4*, North Holland Amsterdam, 133-173.

Federgruen, A., and P. H. Zipkin. 1984a. Approximations of Dynamic Multilocation Production and Inventory Problems, *Management Science*, 30, 69-84.

Federgruen, A., and P. H. Zipkin. 1984b. Computational Issues in an Infinite-Horizon Multiechelon Inventory Model, *Operations Research*, 32, 818-836.

Federgruen A., and P. H. Zipkin. 1984c. Allocation Policies and Cost Approximations for Multilocation Inventory Systems, *Naval Research Logistics Quarterly*, 31, 97-129.

Forsberg, R. 1995. Optimization of Order-up-to-S Policies for Two-Level Inventory Systems with Compound Poisson Demand, *European Journal of Operational Research*, 81, 143-153.

Forsberg, R. 1997a. Exact Evaluation of (R, Q)-Policies for Two-Level Inventory Systems with Poisson Demand, *European Journal of Operational Research*, 96, 130-138.

Forsberg, R. 1997b. Evaluation of (R, Q)-Policies for Two-Level Inventory Systems with Generally Distributed Customer Inter-Arrival Times, *European Journal of Operational Research*, 99, 401-411.

Gallego, G., and P. H. Zipkin. 1999. Stock Positioning and Performance Estimation in Serial Production-Transportation Systems, *Manufacturing & Service Operations Management*, 1, 77-88.

Graves, S. C. 1985. A Multi-Echelon Inventory Model for a Repairable Item with One-for-One Replenishment, *Management Science*, 31, 1247-1256.

Graves, S. C. 1988. Safety Stocks in Manufacturing Systems, *Journal of Manufacturing and Operations Management*, 1, 67-101.

Graves, S. C. 1996. A Multiechelon Inventory Model with Fixed Replenishment Intervals, *Management Science*, 42, 1-18.

Graves, S. C., and S. P. Willems. 2003. Supply Chain Design: Safety Stock Placement and Supply Chain Configuration, in A. G. de Kok, and S. C. Graves Eds. *Handbooks in OR & MS Vol. 11*, North Holland Amsterdam, 95-132.

Hariharan, R., and P. H. Zipkin. 1995. Customer-Order Information Leadtimes, and Inventories, *Management Science*, 41, 1599-1607.

Hausman, W. H., and N. K. Erkip. 1994. Multi-echelon vs. Single-echelon Inventory Control Policies for Low-demand Items, *Management Science*, 40, 597-602.

Inderfurth, K. 1991. Safety Stock Optimization in Multi-Stage Inventory Systems, *International Journal of Production Economics*, 24, 103-113.

Inderfurth, K., and S. Minner. 1998. Safety Stocks in Multi-Stage Inventory Systems under Different Service Levels, *European Journal of Operational Research*, 106, 57-73.

Jackson P.L. 1988. Stock Allocation in a Two-Echelon Distribution System or What to do Until Your Ship Comes In, *Management Science*, 34, 880-895.

Jackson P.L., and J.A. Muckstadt. 1989. Risk Pooling in a Two-Period, Two-Echelon Inventory Stocking and Allocation Problem, *Naval Research Logistics*, 36, 1-26.

Kimball, G. E. 1988. General Principles of Inventory Control, *Journal of Manufacturing and Operations Management*, 1, 119-130.

Lee, H. L., and K. Moinzadeh. 1987a. Two-Parameter Approximations for Multi-Echelon Repairable Inventory Models with Batch Ordering Policy, *IIE Transactions*, 19, 140-149.

Lee, H. L., and K. Moinzadeh. 1987b. Operating Characteristics of a Two-Echelon Inventory System for Repairable and Consumable Items under Batch Ordering and Shipment Policy, *Naval Research Logistics Quarterly*, 34, 365-380.

Marklund, J. 2002. Centralized Inventory Control in a Two-Level Distribution System with Poisson Demand, *Naval Research Logistics*, 49, 798-822.

Marklund, J. 2005. Controlling Inventories in Divergent Supply Chains with Advance-Order Information, *Operations Research* (to appear).

McGavin, E. J., L. B. Schwarz, and J. E. Ward. 1993. Two-Interval Inventory-Allocation Policies in a One-Warehouse *N*-Identical-Retailer Distribution System, *Management Science*, 39, 1092-1107.

Moinzadeh, K. and H. L. Lee. 1986. Batch Size and Stocking Levels in Multi-Echelon Repairable Systems, *Management Science*, 32, 1567-1581.

Muckstadt, J. A. 1973. A Model for a Multi-Item, Multi-Echelon, Multi-Indenture Inventory System, *Management Science*, 20, 472-481.

Muckstadt, J. A., and L. J. Thomas. 1980. Are Multi-Echelon Inventory Models Worth Implementing in Systems with Low-Demand-Rate Items?, *Management Science*, 26, 483-494.

Muharremoglu, A., and J. N. Tsitsiklis. 2003. A Single-Unit Decomposition Approach to Multi-Echelon Inventory Systems, Operations Research Center, MIT.

Rosling, K. 1989. Optimal Inventory Policies for Assembly Systems under Random Demands, *Operations Research*, 37, 565-579.

Rustenburg, W. D., G. J. Van Houtum, and W. H. M. Zijm. 2003. Exact and Approximate Analysis of Multi-Echelon, Multi-Indenture Spare Parts System with Commonality, in J. G. Shantikumar, D. D. Yao, and W. H. M. Zijm Eds. *Stochastic Modeling and Optimization of Manufacturing Systems and Supply Chains*, Kluwer Boston, 143-176.

Shang, K., and J. S. Song. 2003. Newsvendor Bounds and Heuristic for Optimal Policies in Serial Supply Chains, *Management Science*, 49, 618-638.

Sherbrooke, C. C. 1968. METRIC: A Multi-Echelon Technique for Recoverable Item Control, *Operations Research*, 16, 122-141.

Sherbrooke, C. C. 2004. *Optimal Inventory Modeling of Systems, Multi-Echelon Techniques*. 2nd edition, Kluwer Academic Publishers, Boston.
Simon, R. M. 1971. Stationary Properties of a Two-Echelon Inventory Model for Low Demand Items, *Operations Research*, 19, 761-777.
Simpson, K. F. 1958. In-Process Inventories, *Operations Research*, 6, 863-873.
Svoronos, A., and P. H. Zipkin. 1988. Estimating the Performance of Multi-Level Inventory Systems, *Operations Research*, 36, 57-72.
Van der Heijden M. C., E. B. Diks, and A. G. de Kok. 1997. Stock Allocation in General Multi-Echelon Distribution Systems with (R, S) Order-up-to-policies, *International Journal of Production Economics*, 49, 157-174.
Van Houtum, G. J. 2001. Analysis of the Clark-Scarf Model, Eindhoven University of Technology.
Van Houtum, G. J., K. Inderfurth, and W. H. M. Zijm. 1996. Materials Coordination in Stochastic Multi-Echelon Systems, *European Journal of Operational Research*, 95, 1-23.
Van Houtum, G. J., A. Scheller-Wolf, and J. Yi. 2001. Optimal Control of Serial Multi-Echelon Inventory/Production Systems with Periodic Batching, Eindhoven University of Technology.
Van Houtum, G. J., and W. H. M. Zijm. 1991. Computational Procedures for Stochastic Multi-Echelon Production Systems, *International Journal of Production Economics*, 23, 223-237.
Verrijdt, J. H. C. M., and A. G. De Kok. 1996. Distribution Planning for Divergent Depotless Two-Echelon Network under Service Constraints, *European Journal of Operational Research*, 89, 341-354.

Problems

10.1   Show that $\hat{C}_2(y_2)$ in (10.9) is convex in y_2.

10.2* Consider Example 10.1. Assume that $L_2 = 0$ instead of 5. Derive the corresponding expression for $\hat{C}_2(y_2)$. What is the optimal solution?

10.3   Consider Example 10.1. Set $e_1 = 0$ and determine the corresponding optimal policy and costs.

10.4* Consider a two-level distribution system with one central warehouse, two retailers, and Poisson demand at the retailers. All installations apply installation stock $(S - 1, S)$ policies. We use the notation in Section 10.2. The transportation times are $L_0 = L_1 = L_2 = 1$, and the demand intensities at the retailers are $\lambda_1 = 1$ and $\lambda_2 = 2$. The holding costs per unit and time unit are the same at all sites, $h_0 = h_1 = h_2 = 1$. The backorder costs per unit and time unit are the same at both retailers $b_1 = b_2 = 10$. The optimal policy is to use S_0

---

* Answer and/or hint in Appendix 1.

= 3, S_1 = 3, and S_2 = 5. The exact holding costs per time unit are then 0.672 at the warehouse, 1.830 at retailer 1, and 2.627 at retailer 2. The backorder costs are 0.539 at retailer 1, and 0.752 at retailer 2. The total costs are 6.420 per time unit. The exact fill rates are 42.3 percent at the warehouse, 87.0 percent at retailer 1, and 88.8 percent at retailer 2. Note the typical low warehouse fill rate.

a) Verify the exact holding costs and the exact fill rate at the warehouse.
b) Evaluate holding costs, backorder costs and fill rates at the retailers using the METRIC approximation.

10.5 Consider Problem 10.4. Introduce the constraint $S_0 = 0$.
a) What does the constraint mean?
b) What is the optimal solution under this constraint?

10.6 We consider a two-level distribution system with one central warehouse and two retailers. The warehouse lead-time is 3 and the retailer lead-times are both equal to 1. The retailers face Poisson demand with intensities 0.1 (retailer 1) and 0.2 (retailer 2). All installations apply $(S - 1, S)$ policies. The warehouse has $S_0 = 1$, and the retailers have $S_1 = 0$, and $S_2 = 1$. Use the METRIC approximation to estimate the average number of backorders at the retailers in steady state.

10.7 Consider the serial inventory system in Figure 10.1. The demand at installation 1 is Poisson with intensity 1. Both lead-times (transportation times) are equal to 2. The installations apply continuous review $(S - 1, S)$ policies with $S_1 = 2$, and $S_2 = 1$. Use the METRIC approximation for estimating the average waiting time for a demand at installation 1.

10.8 Consider again the serial system in Figure 10.1. Otherwise the assumptions are as in Section 10.2.

a) Reformulate the technique in Section 10.3.1 for this system.
b) Reformulate the technique in Section 10.3.2 for this system.
c) Set $S_2 = 0$. Use the results in a) and b) for determining $P(IL_1 = 0)$. Verify that you get the same result. How can the solution be interpreted?

10.9 Make a computer program to determine the exact results in Problem 10.4 using the technique in Section 10.3.1.

10.10 Make a computer program to determine the exact results in Problem 10.4 using the technique in Section 10.3.2.

10.11 Consider the cost structure in Section 10.4. Explain why it can never be optimal to use $S_i < 0$ for retailer i.

10.12 Consider the data in Example 5.8 in Section 5.13. Change the assumptions as in Section 10.6.2.3. Determine a buyback contract, which will maximize the total profit and give each firm 50% of the total profit.

11 IMPLEMENTATION

In Chapters 2 - 10 we have described different methods for forecasting and inventory control. An inventory control system is usually based on a suitable selection of these methods. However, efficient algorithms cannot guarantee successful control. It is also necessary to create a good environment for practical application of the methods. This means, for example, correct inventory records and sound objectives for the control. Section 11.1 deals with various prerequisites for implementation of inventory control. In order to succeed, it is also necessary to understand how different control parameters like holding costs and service levels can be used for adjusting the control system. The implementation process, as well as adjustments of the control system, should be planned carefully and it is very important to monitor the results closely. We discuss issues concerning implementation and adjustments in Section 11.2.

11.1 Preconditions for inventory control

Due to the development of modern information technology, an inventory control system is today usually computerized. There still exist, though, environments where it is more suitable to use simple manual systems. One example can be an inventory control system for items of low value such as office supplies or bolts. It can also be reasonable to use a manual control system when dealing with very few items. When discussing various practical aspects of inventory control, we shall, however, primarily think of a computerized system with many items.

11.1.1 Inventory records

To be able to use an inventory control model it is first of all necessary to have the required data available. This includes stock on hand, stock on order, backorders, various costs, lead-times, etc. If these data contain errors, the inventory control system will not be able to trigger orders as intended. When implementing an inventory control system it is therefore vital to have good procedures for keeping the inventory records accurate.

In general, it is especially difficult to keep track of the inventory position due to frequent receipts and deliveries of material. All such transactions are possible sources of errors. When an error has occurred, it can often affect the inventory control for a long time. There are two ways to overcome such problems. The first way is to improve the procedures for updating the records in connection with various transactions. In that way we can avoid having errors occur. The second way is to develop a good system for auditing and correcting inventory records.

11.1.1.1 Updating inventory records

To keep the records accurate it is, of course, necessary to have correct procedures for updating the records in connection with different transactions. However, this is usually not the main problem. Instead the difficulty often lies in making sure that the procedures are followed. The most common source of error is, without any doubt, that transactions take place without being recorded properly.

The behavior of the personnel involved is obviously very important. First of all, the stockroom supervisors must take full responsibility for maintaining record accuracy. All personnel must be instructed and trained in stockroom operating procedures. The training should also provide motivation for keeping the records accurate. It is important to set accuracy goals and to measure the fulfillment of these goals. Errors that occur should be traced and analyzed. It is usually efficient to limit and control access to the storeroom. This can prevent most undocumented transactions.

Various types of automatic identification systems may reduce the errors considerably. The less you rely on human intervention to input information the more accurate your records will be. Automatic identification will also reduce the time needed for updating records. A major tool is *bar coding*, which is an optical method that relies on light being reflected off of a printed pattern. In connection with inventory control it is most common with one dimensional, linear bar coding. A system includes the bar code itself, a reading device (scanner), and a printer. Bar coding is not the only automated technique of identifying inventory. There is also, for example, optical character reading, machine vision, magnetic stripe, and radio frequency tags.

All parts should be identified by number and location in the storeroom. A well-ordered storage room will reduce problems involving lost and misplaced items.

It is easier to keep the records accurate if the items have *fixed locations*. This means that each item has a home and that nothing else can live there. On the other hand, such a system can lead to inefficient space utilization. This is because you need to allocate space for each item corresponding to the largest quantity of the item that may be in the facility at one time. With *random location* the same item can be stored in more than one location. However, it is then necessary to accurately note the location in a computer database. Although space can be used very efficiently when using a random locator system, it is obvious that such a system will further increase the need for good operating procedures. In practice it is very common to use combination systems, where some items are given specific locations, while most items are randomly located.

For items where the stock always consists of an integral number of units, it should, in principle, be possible to eliminate nearly all sources of errors. This can be more difficult for items which can have a continuous stock level that is obtained, for example, by weighing. Each transaction may then cause a small error in the stock level due to measurement errors.

Another type of problem occurs, when we have, for example, a stock of tubes of a certain dimension. The tubes in stock may have different lengths. Furthermore, it may be the case that short tubes cannot cover certain types of demands. Therefore, in situations with many short tubes in stock, the inventory on hand may appear to be much larger than it really is.

For certain types of items, thefts may be a major problem. Apart from the loss in value, thefts will also lead to inaccurate inventory records. Thefts can be another strong reason for limiting access to the storeroom.

11.1.1.2 Auditing and correcting inventory records

All inventory balances must also be checked by counting. The counting frequency should, in general, be higher for items with high demand. For such items the transactions are more frequent, and it is usually also more important to have accurate balances.

Traditionally it has been most common to use *periodic counting*, for example to take an annual physical inventory. This usually means a complete count of more or less all items over a short time period. It is common that the stock of each item is counted twice by different persons.

It has become more and more common, though, to replace the periodic counting by *cycle counting*. Each day a limited number of items are checked. This method means a continuous counting of stock throughout the year. As a consequence, personnel can be assigned to counting on a full-time basis which permits the use of specialists. It is practical to let the inventory control system trigger counts at suitable times. It is, of course, easier to check the stock level when it is low. Therefore it may be suitable to count the items just before a replenishment is received. The stock level is then, on average, equal to the safety stock. It is also common to perform a count when the balance on hand becomes zero or negative, or after a certain number of transactions.

The inventory records are not always updated exactly when the physical movements take place. The resulting differences in records must be taken into account. Therefore it is often practical to count when there is no movement of paper or product, for example, at end of business day, prior to start of day, or over the weekend.

Initially, or when having problems with inventory records, it may be a good idea to count a control group of items very often. Let, for example, the control group consist of 100 items that constitute a good cross-section of the total population of items, and let the cycle for each item be 10 days. By following the control group in detail you will usually be able to identify major errors that may occur. This will, in turn, enable adjustments of the procedures that will correct these errors.

11.1.2 Performance evaluation

The purpose of an inventory control system is, in general, to reduce holding and ordering costs while still maintaining satisfactory customer service. To achieve this it is necessary to be able to continuously evaluate the performance of the system. A performance evaluation is needed both for creating motivation for efficient application of the system and for being able to adjust the control when various changes in the operating conditions occur.

A good system for performance evaluation should satisfy several important requirements. First of all it is necessary to measure costs and service in a well-defined way so that measurements at different times can be compared. Some performance measures are not always easy to define suitably. Customer service, for example, can be evaluated in many ways. It is, of course, a clear advantage if the measured service is defined in the same way as the service level used when determining the reorder levels in the inventory control system. It is then much easier to check that the control system works as intended.

Except for the customer service, it is in general necessary to monitor the capital tied up in inventories and the ordering costs. It is also common to evaluate other performance measures like inventory turnover. Usually it is not very difficult to determine suitable performance measures. It is more difficult to determine the level of aggregation for the measurements. There are often many thousands of items. It is therefore not practical to follow individual items separately. Neither is it appropriate to aggregate all items. It should be possible to use the evaluation system as a tool for identifying various problems that may exist. For example, we want to know whether the control is inadequate for certain types of items, and we would also like to identify major savings that can be achieved by improved control.

To obtain a reasonable overview it is necessary to aggregate and consider suitable groups of items instead of individual items. It is normally advantageous if the items in such a group are related and if the inventory control is similar. See also Section 11.2.2. The performance evaluation should initially be carried out for relatively large groups of items which are then, if required, divided into smaller groups. Even with a well-designed system for performance evaluation, it may sometimes be quite difficult to evaluate the efficiency of the inventory control system. The measurements are often affected by various changes in the operating environment. Such changes can be due to new suppliers, major changes in the demands for different items, business cycles, etc. It is important to be aware of possible measurement errors and take them into account when interpreting the results of the evaluation. Even more difficult is to compare various performance measures across different companies. Two companies may, at a first glance, seem to have very similar conditions for inventory control. However, when looking at the conditions in more detail, it often turns out that it is very difficult to compare performance measures like inventory turnover.

11.2 Development and adjustments

An industrial computerized inventory system generally contains much more than the inventory control, e.g., customer billing and supplier payments. The total system is often more an information system than a control system, and the inventory control module is only a minor part of the entire system. An example is an inventory control module in a modern *ERP* system. (See Section 8.2.4.) Furthermore, it is not uncommon that major parts of the inventory system already exist when starting to develop or install a module for inventory control. In such situations the development and installation costs can be surprisingly low.

11.2.1 Determine the needs

Developing and implementing inventory control is a customizing operation. The requirements can vary substantially between different companies and between different types of items within a company. Sometimes it can, at least initially, be reasonable to leave items with very special requirements outside the system.

At this stage it is recommended to question more or less everything. The very first question is what items that should be kept in stock at all. Maybe it is advantageous to deliver some items only after receiving customer orders. Nearly everything can be changed. For example, investments in production can reduce the setup times. This means that smaller batch quantities are feasible and it may also be possible to use a simpler replenishment system. The evaluation of such investments has been considered by e.g., Porteus (1985, 1986a, 1986b), Spence and Porteus (1987), and Silver (1992).

Furthermore, better information can sometimes reduce the service needs. If the inventory system can provide customers with precise information concerning the lengths of various delays, this can enable the customers to revise their plans in such a way that the additional costs in connection with delays are relatively low. This will consequently reduce the need for a high service level and thereby make it possible to obtain lower carrying costs.

When designing an inventory control system it is also necessary to determine which control methods should be used for different groups of items. A first important question is whether we are dealing with a single installation and items that can be controlled independently of each other. If this is the case, at least approximately, it is possible to use the methods which we have considered in Chapters 4 - 6, and it is, in general, relatively easy to obtain efficient inventory control.

If the operating conditions are more complex so that we need to use methods from Chapters 7 - 10, the implementation of inventory control will become substantially more difficult. First of all the information system must be able to coordinate data for different items or installations. Normally this means additional costs. Furthermore, the control system will be considerably more complex. Obviously, it is always necessary to compare the additional costs for a better control system to the possible advantages of improved inventory control.

When designing an inventory control system it is also important to evaluate potential savings for different types of items. If possible the implementation should be planned so that large savings are obtained as early as possible.

11.2.2 Selective inventory control

In principle each individual item can be controlled in a unique way. But since there are, in general, several thousands of items, this is not a practical solution. It is much easier to divide the items into a number of groups and use the same forecasting and inventory control techniques for all items in the same group. See also Section 6.3. We have already discussed grouping of items in connection with performance evaluation (Section 11.1.2). It is a clear advantage if the same groups can be used both when selecting control techniques and when evaluating the performance.

A grouping of the items can be done in many different ways and must depend on the needs of the organization. It should, in general, be affected both by marketing and production factors.

A type of grouping that is common in many companies is so-called *ABC inventory analysis*. This means a grouping by dollar volume. Usually a relatively small percentage of the items accounts for a large share of the total volume. Typically 20 percent of the items can account for about 80 percent of the dollar volume. This means, of course, that the remaining 80 percent of the items only account for 20 percent of the volume. Items with a high volume are often more important, and it is therefore reasonable to expect that such items should require a more precise control and performance evaluation. Consequently, many companies initially group their items in three classes, A, B, and C. The A class consists of items with very high dollar volume. Typically the A class contains 10 percent of the items. Likewise about 30 percent of the items with intermediate dollar volumes can be classified as class B items, and finally the remaining 60 percent of the items with low dollar volumes are referred to as the C class items.

Especially if the inventory control is expensive, e.g., if the control system contains manual elements, it is important to concentrate the control efforts on the A items. In a completely computerized system it does not, in general, require much additional effort to include more or less all items once the system has been implemented. It is always very important, however, to monitor the performance of the A items closely.

When dealing with a single installation and independent items, it can sometimes be reasonable to initially use an ABC classification and let the control techniques be the same for all items in the same class. In general, though, there will successively emerge needs for dividing these initial groups into smaller subgroups. When using the more complicated methods in Chapters 7 - 10, a grouping of the items must normally also depend on couplings between items and facilities.

11.2.3 Model and reality

In this book we have presented different models and corresponding methods which can be used for forecasting demand and controlling inventories. There are also many other models available in the inventory literature. Although there are many models to choose from, it is still nearly always very difficult or impossible to find a model that completely describes a certain real-life situation. Most real settings are indeed very complicated. Furthermore, in order to be a useful tool, a model should be reasonably simple. In most situations we are forced to accept that our model is a rather rough approximation of the real system. When working with inventory control it is very important to remember this when dealing, for example, with methods for lot sizing and determination of safety stocks.

The relationship between model and reality is illustrated in Figure 11.1.

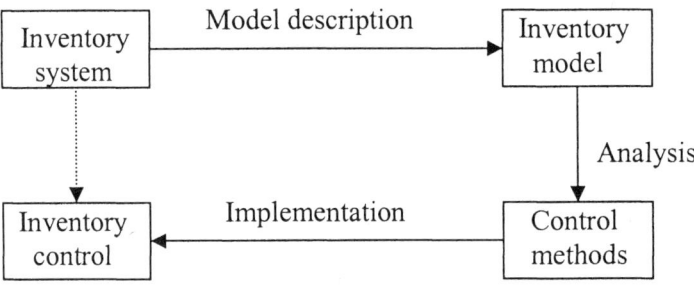

Figure 11.1 Using inventory models in practice.

The control methods used are based on an approximate description of the real system, the inventory model. This means, in other words, that the methods take some aspects of reality into account but ignore others. The resulting control will not always be as intended. Quite often it is therefore necessary to adjust cost parameters and service constraints in order to compensate for the incomplete model description. To understand how this can be done, and be able to use an inventory control system efficiently, it is very important to understand the underlying basic assumptions of different models. It is usually very fruitful to think over carefully how these basic assumptions differ from the real inventory system. For example, what can the consequences be if the real demand distribution differs substantially from that in the model?

On the other hand, it is not necessary for successful application of a certain control method to be able to follow the mathematical analysis that is used when deriving the control method from the inventory model.

Because a model is an approximate description of reality, we cannot just choose cost parameters (Section 3.1) corresponding to what we think are the real cost factors and hope to get an adequate balance between different types of costs. To obtain efficient control it is normally necessary to monitor different performance measures carefully (Section 11.1.2) and adjust the control parameters with respect to the actual outcome.

Assume, for example, that in the performance evaluation we find that the capital tied up in stock is higher that what was expected. If we want to reduce the carrying costs, there are several possibilities. We can use smaller batch quantities and/or smaller safety stocks. One way to obtain this is to increase the carrying charge when determining holding costs. This will reduce the batch quantities and also the safety stocks if the holding costs are balanced against certain backorder costs. The safety stocks, however, are not affected in case of given service levels. A lower service level will, of course, always reduce the safety stocks. Since the model is approximate, the cost parameters of an inventory model do not need to correspond to real cost factors. The cost parameters of an inventory model should instead be regarded as control parameters that can be adjusted to obtain a satisfactory overall control as measured in the performance evaluation. Parameter adjustments should, however, as we shall discuss in Section 11.2.6, be carried out slowly while continuously monitoring the outcome.

11.2.4 Step-by-step implementation

An inventory control system can be more or less automatic. At least in principle, the system can generate complete purchasing and production orders without any manual interaction. It is possible, however, that this is too dangerous because of possible errors. Especially in the implementation phase and when adjusting the control, it is rational to be extremely careful.

An implementation as well as adjustments of the control system should be carried out successively under thorough supervision. Two main rules can be recommended:

- Introduce changes step-by-step

As we have discussed earlier, the same control techniques are usually used for a group of items. The economic outcome is also, in general, monitored for each group separately. When implementing the control system it is recommendable to take one group at a time and continue the implementation with other groups first when feeling confident concerning already implemented parts of the system.

- Use the new and the old system in parallel

In an initial phase it is reasonable to compare a new or revised inventory control system to the previous decision rules. This makes it easier to discover and correct various errors and shortcomings of a new system.

An inventory control system should be continuously adjusted to changing operating conditions. In practice it is not uncommon that the personnel using the system discover that orders for some items are triggered in a less suitable way. In such a situation it is far too common that the orders are changed manually without adjusting the system itself. This means, of course, that such changes must be carried out also in the future, and the benefits of the control system are reduced substantially. Nearly always in such situations it is possible to adjust cost parameters or make minor changes in the methodology so that the control system will work adequately. It is therefore important to document all errors and other problems initially.

11.2.5 Simulation

Sometimes it can be difficult to choose suitable inventory control methods, and also to find appropriate values of the various control parameters that affect system control. One problem in this context is that it is normally not feasible to evaluate a control policy by practical application, since such an evaluation would take considerable time. For example, if the service level is adjusted for a group of items, it can take several months before the consequences can be clearly observed. If the change turns out to have mainly negative consequences the experiment can be quite expensive.

One way to evaluate different control methods in advance is to use simulation. When using simulation we are also, like in Figure 11.1, replacing the real system by a mathematical model. Simulation experiments with the model which can be carried out very quickly are then used for analyzing the system. Using simulation, it is in general, possible to evaluate the development during several years in a few seconds, and it is easy to make various test runs with different inventory control techniques.

Technically it is quite easy to simulate an inventory system. It is possible to use special simulation software, or to develop a model using a standard programming language. A simulation model is, however, in principle associated with the same limitations as all other mathematical models and can never give a complete illustration of the real system. Note especially that if a simulation model is based on the same assumptions as a certain inventory model which can be evaluated exactly, the results of the simulation will, of course, be exactly the same and will therefore not give any additional insight. Simulations can give valuable additional information, though, if the simulation model illustrates the real system better in some way. Assume, for example, that the determination of safety stocks in an inventory control system is based on the assumption of a certain demand distribution, and that it has not been possible to analyze in detail how well this type of demand corresponds to the real demand distribution. A simulation model based on real historical demand data can then be used for analyzing whether the demand distribution used is reasonable.

11.2.6 Short-run consequences of adjustments

Adjustments of the inventory control have consequences both in the short run and in the long run. The objective when making adjustments is usually to obtain good control in the long run. Furthermore, it is normally only the long-run steady state situation that is reflected in different inventory models. Quite often, though, it is important to consider also the short-run, transient behavior of an inventory system.

To illustrate the development in the short run we shall discuss the consequences when changing reorder points and batch quantities. When increasing the reorder points, due to, for example, longer lead-times or higher safety stocks, this will result in a lot of new orders since many items will reach their reorder points directly or relatively soon. If the increase of the reorder points corresponds to, for example, one week's demand, the *additional* orders in the short run will also correspond to about one week's demand. The consequences are similar but in the other direction when reducing the reorder points. When reducing the reorder points the change in the flow of orders is slower, though, since it cannot be seen until the items successively reach their original reorder points. (See also Section 8.2.5.) An increase of the batch quantities will, of course, mean larger orders when the items reach their reorder points and will consequently increase the total flow of orders in the short run. From this discussion we can formulate two important conclusions:

- Changes in the inventory control will nearly always first result in a higher total stock. When changing the control, some reorder points are increased while others are reduced. Because the effects of the increased reorder points can be observed earlier, the total stock will increase in the short run.

- In the short run, adjustments of the control can cause major changes in the flow of orders. This can, for example, result in a production load that is far too high. It is therefore important to adjust the control continuously and under careful supervision.

When adjusting the inventory control system it is important to be aware of the short-run effects and understand that it can take a relatively long time until the system reaches steady state. It is not uncommon that changes in the inventory control, due to insufficient understanding of these effects, can create a situation that is completely opposite of what was intended. An example is when increasing the system lead-times due to a *temporary* overload in production. Although it may be correct that the real lead-times have increased because of the overload, increased system lead-times will mean higher reorder points and a short-run increase in the triggered orders. This will obviously just increase the overload.

11.2.7 Education

Successful inventory control requires both knowledge and motivation. The importance of education cannot be overemphasized. First of all it is, of course, very important that the inventory control methods implemented are based on adequate inventory models. It is surprisingly common that inefficient techniques are implemented by people who do not know the inventory control area sufficiently well. However, the need for education concerns not only an initial implementation phase. Too often an inventory control system deteriorates when a key person leaves the company and the person taking over does not have adequate training. The inventory control system will then no longer be adjusted properly to fit a changing operating environment.

Poor training in connection with inventory control can sometimes emanate from the misunderstanding that the control system completely replaces manual decisions. An inventory control system should instead be seen as a tool that can be applied more or less professionally. How the tool is used is often even more important than the design of the tool. To gain the advantages of a more advanced tool, the user needs a better education.

The most important part of the education concerns inventory models. It is necessary to understand the basic assumptions in the models used and how system control is affected by different parameters. Education programs frequently focus too much on the information system aspects and do not give sufficient insight concerning the stock control. Good training is needed not only for being able to adjust system control but also for obtaining satisfactory motivation. When clearly understanding how the control system works, it is both challenging and stimulating to work for improved system control.

When implementing a new system, a broad education program for all involved personnel is recommended. The persons who have the main responsibility for the inventory control need continuous training. They should be active in professional societies, attend conferences, and read journals in the field.

References

Axsäter, S. 1986. Evaluation of Lot-Sizing Techniques, *International Journal of Production Research*, 24, 51-57.
Muller, M. 2003. *Essentials of Inventory Management*, AMACOM, New York.
Porteus, E. L. 1985. Investing in Reducing Setups in the *EOQ* Model, *Management Science*, 31, 998-1010.
Porteus, E. L. 1986a. Investing in New Parameter Values in the Discounted *EOQ* Model, *Naval Research Logistics*, 33, 39-48.
Porteus, E. L. 1986b. Optimal Lot Sizing, Process Quality Improvement and Setup Cost Reduction, *Operations Research*, 34, 137-144.
Silver, E. A. 1992. Changing the Givens in Modelling Inventory Problems: The Example of Just-in-Time Systems, *International Journal of Production Economics*, 26, 347-351.
Silver, E. A., D. F. Pyke, and R. Peterson. 1998. *Inventory Management and Production Planning and Scheduling*, 3rd edition, Wiley, New York.
Spence, A. M., and E. L. Porteus. 1987. Setup Reduction and Increased Effective Capacity, *Management Science*, 33, 1291-1301.
Tersine, R. J. 1988. *Principles of Inventory and Materials Management*, 3rd edition, North-Holland, New York.
Vollman, T. E., W. L. Berry, and D. C. Whybark. 1997. *Manufacturing Planning and Control Systems*, 4th edition, Irwin, Boston.

APPENDIX 1

ANSWERS AND HINTS TO SELECTED PROBLEMS

2.1

Period	3	4	5	6	7	8	9	10
$\hat{a}$	456.7	460.0	468.7	476.0	485.3	492.7	496.0	492.7

2.4 a) Define

$$f_t = \hat{a}_t - c - d \cdot t, \ g_t = \hat{b}_t - d.$$

We obtain

$$f_t = (1-\alpha)f_{t-1} + (1-\alpha)g_{t-1},$$

$$g_t = -\alpha\beta f_{t-1} + (1-\alpha\beta)g_{t-1}.$$

It follows that both f_t and g_t will approach zero.

b) Define $h_t = \hat{a}_t - c - d \cdot (t+1)$. We obtain

$$h_t = (1-\alpha)h_{t-1} - d = (1-\alpha)h_{t-1} + \alpha(-d/\alpha),$$

and from our results on exponential smoothing it follows that h_t will approach $-d/\alpha$ in the long run.

2.7 $(1000+1\cdot10)\hat{F}_1 = 808 \qquad \hat{F}_1 = 0.8$
$(1000+2\cdot10)\hat{F}_2 = 1020 \qquad \hat{F}_2 = 1.0$
$(1000+3\cdot10)\hat{F}_3 = 1648 \qquad \hat{F}_3 = 1.6$
$(1000+4\cdot10)\hat{F}_4 = 624 \qquad \hat{F}_4 = 0.6$

$\hat{a}_{05.1} = 0.8\cdot(1000+10) + 0.2\cdot(795/0.8) = 1006.75$
$\hat{b}_{05.1} = 0.8\cdot 10 + 0.2\cdot(1006.75 - 1000) = 9.35$
$\hat{F}'_1 = 0.8\cdot 0.8 + 0.2\cdot(795/1006.75) = 0.7979$

$\overline{T} = \sum_{t=1}^{4} \hat{F}'_t = 0.7979 + 1.0 + 1.6 + 0.6 = 3.9979 \quad \hat{F}_t = (4/\overline{T})\hat{F}'_t$

$\hat{F}_1 = 0.79835, \ \hat{F}_2 = 1.0005, \ \hat{F}_3 = 1.6008, \ \hat{F}_4 = 0.6003$

$\hat{a}_{05.2} = 0.8\cdot(1006.75+9.35) + 0.2\cdot(1023/1.0005) = 1017.38$
$\hat{b}_{05.2} = 0.8\cdot 9.35 + 0.2\cdot(1017.38 - 1006.75) = 9.606$
$\hat{F}'_2 = 0.8\cdot 1.0005 + 0.2\cdot(1023/1017.38) = 1.0015$
$\overline{T} = 4.001$
$\hat{F}_1 = 0.7981, \ \hat{F}_2 = 1.0013, \ \hat{F}_3 = 1.6004, \ \hat{F}_4 = 0.6002$
$\hat{x}_{05.2,05.3} = (1017.38+1\cdot 9.606)\cdot 1.6004 = 1643.60$
$\hat{x}_{05.2,05.4} = (1017.38+2\cdot 9.606)\cdot 0.6002 = 622.16$
$\hat{x}_{05.2,06.1} = (1017.38+3\cdot 9.606)\cdot 0.7981 = 834.97$
$\hat{x}_{05.2,06.2} = (1017.38+4\cdot 9.606)\cdot 1.0013 = 1057.18$

2.12 Use that

$$\hat{a}_t = x_t - (1-\alpha)\varepsilon_t,$$
$$\hat{b}_t = x_t - x_{t-1} - (1-\alpha\beta)\varepsilon_t + (1-\alpha)\varepsilon_{t-1}.$$

2.13 a)

Month	1	2	3	4	5	6
$\hat{a}$	727.6	731.1	738.3	736.2	746.6	755.9
MAD	18.4	18.2	21.7	19.4	25.9	30.0

ANSWERS AND HINTS

b)

Month	1	2	3	4	5	6
$\hat{a}$	727.6	730.7	738.1	737.5	748.5	760.0
$\hat{b}$	-0.48	0.24	1.68	1.21	3.18	4.83

Forecast for month 5 after month 2: $730.7 + 3 \cdot 0.24 = 731.4$.

2.14 $\hat{a}_5 = 234$, $\hat{b}_5 = 10.4$, $MAD_5 = 33.5$, $\sigma = 41.9$,
Months 6 - 8: mean = 764.4, standard deviation = 72.53.

4.1 a) $Q_A = 1000$, $Q_B = 939$.
b) The relative cost increase is given by

$$\frac{C}{C^*} = \frac{1}{2}(\sqrt{125/200} + \sqrt{200/125}) = 1.02774.$$

For product A the correct cost is $C^* = \sqrt{2 \cdot 200 \cdot 48000 \cdot 0.24 \cdot 50} = \$15179$
so the cost increase is \$421.

4.2
$$\frac{C}{C^*} = \frac{1}{2}(\sqrt{0.6} + \sqrt{1/0.6}) = 1.0328.$$

4.6
$$C = \frac{hQ}{2}(1 + d/p) + \frac{Ad}{Q}, \quad Q^* = \sqrt{\frac{2Ad}{h(1 + d/p)}}, \quad C^* = \sqrt{2Adh(1 + d/p)}.$$

4.11 a) We get $Q^* = 4$ and $x^* = 1/4$.
b) The average waiting time for a customer is $Qx^2/2d = 1/8$.
c) The considered problem is to minimize

$$C = \frac{Q(1-x)^2}{2}h + \frac{d}{Q}A,$$

under the constraint $Qx^2/2d \leq 1/8$. Let us first show that the constraint is binding. If not, we get for a given x, $C = \sqrt{2Adh}(1-x)$ and we can see that the costs will decrease with x. Consequently, $Qx^2/2d = 1/8$, or $Q = 1/(4x^2)$. Inserting in C we get

$$C = \frac{1}{8}(\frac{1}{x^2} - \frac{2}{x} + 1) + 24x^2.$$

It is easy to verify that C is minimized for $x^* = 1/4$. This implies $Q^* = 4$. (An alternative way to see that the original solution solves the new problem is to note that b_1 serves a Lagrange multiplier for the considered constraint.)

4.16 a)

Period t	1	2	3	4	5
d_t	25	40	40	40	90
$k=t$	100	200	240	320	<u>380</u>
$k=t+1$	<u>140</u>	240	<u>280</u>	410	
$k=t+2$	220	320			

Deliveries 65 in period 1, 80 in period 3, and 90 in period 5. Cost $380.

b) Two periods $(100 + 40)/2 = 70 \le 100$,
Three periods $(100 + 40 + 80)/3 = 73.3 > 70$, delivery in period 3.

Two periods $(100 + 40)/2 = 70 \le 100$,
Three periods $(100 + 40 + 180)/3 = 106.7 > 70$, delivery in period 5, i.e., the same solution.

c) Two periods $40 \le 100$,
Three periods $40 + 80 > 100$, delivery in period 3.

Two periods $40 \le 100$,
Three periods $40 + 180 > 100$, delivery in period 5, i.e., same solution.

4.17 a)

Period t	1	2	3	4	5	6
d_t	10	40	95	70	120	50
$k=t$	100	200	220	315	355	455
$k=t+1$	<u>120</u>	247.5	<u>255</u>	375	<u>380</u>	
$k=t+2$	215	317.5		425		

Deliveries 50 in period 1, 165 in period 3, and 170 in period 5. Cost $380.

b) Two periods $(100 + 20)/2 = 60 \le 100$,
Three periods $(100 + 20 + 95)/3 = 71.7 > 60$, delivery in period 3.

Two periods $(100 + 35)/2 = 67.5 \le 100$,
Three periods $(100 + 35 + 120)/3 = 85 > 67.5$, delivery in period 5,

Two periods $(100 + 25)/2 = 62.5 \le 100$, i.e., same solution.

4.18 a)

Period t	1	2	3	4	5	6	7
d_t	10	24	12	7	5	4	3
$k=t$	<u>100</u>	200	296	348	404	464	500
$k=t+1$	196	<u>248</u>	324	368	420	476	
$k=t+2$	292	304	364	400	444		
$k=t+3$	376	364	412	<u>436</u>			
$k=t+4$	456	428	460				
$k=t+5$	536	488					
$k=t+6$	608						

Deliveries 10 in period 1, 36 in period 2, and 19 in period 4. Cost $436.

b) Two periods $(100 + 96)/2 = 98 \le 100$,
Three periods $(100 + 96 + 96)/3 = 97.3 \le 98$,
Four periods $(100 + 96 + 96 + 84)/4 = 94 \le 97.3$,
Five periods $(100 + 96 + 96 + 84 + 80)/5 = 91.2 \le 94$,
Six periods $(100 + 96 + 96 + 84 + 80 + 80)/6 = 89.3 \le 91.2$,
Seven periods $(100 + 96 + 96 + 84 + 80 + 80 + 72)/7 = 86.9 \le 89.3$, i.e., just one delivery in period 1. Cost $608. Note the large difference due to the special demand structure.

5.1 Use the generating function.

5.2 Derive the generating function.

5.6 a) 68, 104, and 165.
b) 92.5, 98.3, and 99.2 percent.
c) 77.3, and 99.9 percent.

5.7 a) $SS = 135 - 100 = 35$. $S_1 = \Phi(0.7) = 75.8$ percent. $S_2 = 96.4$ percent.
b) Let the lead-time demand be uniform on $(100 - a, 100 + a)$. We obtain

$$(\sigma')^2 = 2500 = \int_{100-a}^{100+a} \frac{1}{2a}(u-100)^2 du = a^2/3,$$

i.e., $a = \sqrt{7500} = 86.60$. $S_1 = (135 - 100 + 86.60)/173.20 = 70.2$ percent. The average backordered quantity per cycle is

$$E(B) = \int_{135}^{186.6} \frac{(u-135)}{173.20} du = 7.686,$$

and $S_2 = 1 - 7.686/200 = 96.2$ percent.

5.8 a) Initially $\mu' = 800$ and $\sigma' = 100$. We get $S_2 = 96.7$ percent. After reducing the lead-time $\mu' = 400$ and $\sigma' = 50\sqrt{2} = 70.71$. We get $S_2 > 99.99$ percent. We obtain the average stock on hand as

$$E(IL^+) = R + Q/2 - \mu' + \frac{\sigma'^2}{Q}\left[H\left(\frac{R-\mu'}{\sigma'}\right) - H\left(\frac{R+Q-\mu'}{\sigma'}\right)\right]$$

The stock on hand is increasing from 351.7 to 750.

b) R can be reduced to 419. The stock on hand is 320.3.

5.11 The inventory position is uniform on 2, 3, 4.

a) We get $P(2) = (e^{-2}/3)(1 + 2 + 2) = 0.226$.
b) Furthermore, $P(1) = 0.241$, $P(3) = 0.135$ and $P(4) = 0.045$. $E(I)^+ = 1.278$. $E(I) = 3 - 2 = 1$. We get $E(I)^- = 1.278 - 1 = 0.278$. The average waiting time is obtained by Little's formula $0.278/2 = 0.139$.

5.12 Determine the average shortage corresponding to Q.

$$B = \int_R^{R+Q}(x-R)\frac{1}{m}e^{-\frac{x}{m}}dx + \int_{R+Q}^{\infty}Q\frac{1}{m}e^{-\frac{x}{m}}dx$$

$$= \left[-(x-R)e^{-\frac{x}{m}}\right]_R^{R+Q} + \int_R^{R+Q}e^{-\frac{x}{m}}dx + \left[-Qe^{-\frac{x}{m}}\right]_{R+Q}^{\infty}$$

$$= m(e^{-\frac{R}{m}} - e^{-\frac{R+Q}{m}}).$$

$$S_2 = 1 - B/Q = 1 - \frac{m}{Q}(e^{-\frac{R}{m}} - e^{-\frac{R+Q}{m}}).$$

5.13 a) $S_2 = 0.935$.
b) Set $\mu' = \lambda$ and $\sigma' = \lambda^{1/2}$. $S_2 = 0.932$.

5.21 The same probabilities 0.128, 0.16 and 0.2.

6.1 It is constant $S_2 = b_1/(h + b_1) = 10/11$. See (5.67).

ANSWERS AND HINTS   315

6.3   We obtain $\mu' = 200$ and $\sigma' = 2^{0.7}\sqrt{\pi/2} \cdot 40 = 81.44$.

   a) $SS = 1.28 \cdot 81.44 \approx 104$, and $R = 304$.
   b) $S_2 = 92$ percent, $S_2 = 96$ percent, and $S_2 = 99.7$ percent.

6.4   a) $\hat{x}_{t,t+\tau} = \hat{a}_t = 0.8 \cdot \hat{a}_{t-1} + 0.2 \cdot x_t$
       $MAD_t = 0.7 \cdot MAD_{t-1} + 0.3 \cdot |x_t - \hat{x}_{t-1,t}|$

x_t	112	96	84	106	110
$\hat{a}_t$	102.40	101.12	97.70	99.36	101.49
MAD	10.6	9.34	11.67	10.66	10.66

   $\sigma^2 \cong (1.25 \cdot MAD)^2$
   Expected demand 101.5 units per week with variance 177.6.

   b) $Q = \sqrt{2AD/h} = \sqrt{2 \cdot 100 \cdot 101.49/1} \cong 142$
      $SS = k \cdot \sigma$, $k = 1.64$
      $\sigma' = \sigma \cdot \sqrt{L} = 13.325 \cdot \sqrt{2} = 18.84$
      $SS = 1.64 \cdot 18.84 = 30.9$
      $R = 30.9 + 101.49 \cdot 2 = 234$

6.5   a) Simple exponential smoothing

Week		16	17	18	19	20
Demand		97	99	100	126	112
$\hat{a}$	100.00	99.40	99.32	99.46	104.77	106.21
MAD	7.00	6.20	5.04	4.17	8.64	8.36

   $\hat{a}_{t+1} = (1-\alpha)\hat{a}_t + \alpha \cdot x_{t+1}$
   $MAD_{t+1} = (1-\alpha) \cdot MAD_t + \alpha \cdot |\hat{x}_{t,t+1} - x_{t+1}|$
   $\sigma \approx 1.25 \cdot 8.36 = 10.45$

   Exponential smoothing with trend

Week		16	17	18	19	20
Demand		97	99	100	126	112
$\hat{a}$	100.00	99.40	99.13	99.10	104.35	107.48
$\hat{b}$	0.00	-0.24	-0.25	-0.16	2.00	2.45
MAD	7.00	6.20	4.99	4.22	8.79	8.16

   $\hat{a}_{t+1} = (1-\alpha) \cdot (\hat{a}_t + \hat{b}_t) + \alpha \cdot x_{t+1}$

$\hat{b}_{t+1} = (1-\beta) \cdot \hat{b}_t + \beta \cdot (\hat{a}_{t+1} - \hat{a}_t)$
$MAD_{t+1} = (1-\alpha) \cdot MAD_t + \alpha \cdot |\hat{x}_{t,t+1} - x_{t+1}|$
$\sigma \approx 1.25 \cdot 8.16 = 10.20$
$Q = (2 \cdot 2500 \cdot 106.21/10)^{1/2} \approx 230$
$\sigma' = 10.45 \cdot 3^{1/2} \approx 18.10$
$G(SS/\sigma') \approx Q \cdot (1-S_2) / \sigma' = 230 \cdot (1-0.95) / 18,10 = 0.635$
$SS/\sigma' = -0.4$ (from table)
$SS = -0.4 \cdot 18.10 \approx -7.24$
$R = -7.24 + 3 \cdot 106.21 = 311.39 \approx 312$

b) $S_1 = \Phi(SS/\sigma') \approx 35\%$

7.3   a) Applying (7.12) and (7.13) we obtain $T_1 = 27.037$, $T_2 = 39.528$, $T_3 = 43.759$, and the costs $C_1 = 59.178$, $C_2 = 25.298$, $C_3 = 45.705$. The sum of these costs $C = 130.182$ is a lower bound for the total costs.

    b) Assume that the independent solution is feasible and that we start production of product 3 at times t', $t' + 43.759$, $t' + 2 \cdot 43.759$ ... etc. Assume that the first production of product 2 after time t' starts at time $t' + \Delta_1$. Clearly $\Delta_1 \geq \sigma_3 = 15.003$. But the next time the time difference is $\Delta_2 = \Delta_1 + T_2 - T_3 = \Delta_1 - 4.231$. The time after that it is $\Delta_3 = \Delta_2 - 4.231$, etc. After some time we will obtain for some n, $0 < \Delta_n < \sigma_3$ which means that the solution is infeasible.

    c) Applying (7.17)-(7.19) we obtain $T_{min} = 7.292$, and $\hat{T} = 34.462$, i.e., $T_{opt} = \hat{T} = 34.462$. The resulting total costs per unit of time are $C = 133.481$.

    d) Starting with $W = 27.037$ we obtain $n_1 = 1$, $n_2 = n_3 = 2$. Next we get the final solution $W = 23.617$ with the same multipliers and the total costs $C = 131.259$. The solution is feasible if we produce items 2 and 3 in different basic periods. In basic periods where product 1 and 2 are produced, the time required is $\sigma_1 + \sigma_2 = 6.168 + 9.697 = 15.865 < 23.617$. In basic periods where product 1 and 3 are produced, the time required is $\sigma_1 + \sigma_3 = 6.168 + 16.115 = 22.283 < 23.617$.

7.5   a) $T^* = 5.020$, $T_{min} = 1.806$, $T_{opt} = 5.020$. $C = 9960.0$
$Q_I \cong 500$, $Q_{II} \cong 250$, $Q_{III} \cong 100$

    b) $T_I = 460/100 = 4.6$, $T_{II} = 230/50 = 4.6$, T_{III} $184/20 = 9.2$, $C_{DW} = 9877.7$. Product I and II are produced in each basic period, III in every second basic period. Feasible schedule. In a basic period with production of all three products the production time is $2.956 < 4.6$.

8.3   a) Change first to the equivalent $R_2^i = 2$ such that $IP_2^{i0} - R_2^i$ is a multiple of Q_1. Next we use (8.4) to obtain $R_1^e = 8$ and $R_2^e = 20$.

    b) For both policies we have: Installation 1 order at times 7, 17, 27,..., and installation 2 at times 7, 27, 47, ...

ANSWERS AND HINTS

c) At times 6, 26, 46, ... No equivalent installation stock policy exists, since the echelon stock policy is not nested.

8.7

Item A	Period	1	2	3	4	5	6	7	8
Lead-time = 1	Gross requirements	7	9	28	30	18	16	24	13
Order quantity = 30	Scheduled receipts								
	Projected inventory	22	15	6	8	8	20	4	10	27
	Planned orders			30	30	30		30	30	

Item B	Period	1	2	3	4	5	6	7	8
Lead-time = 1	Gross requirements		30	30	30		30	30	
Order quantity = 30	Scheduled receipts								
	Projected inventory	34	34	4	4	4	4	4	4	4
	Planned orders			30	30		30	30		

Item C	Period	1	2	3	4	5	6	7	8
Lead-time = 1	Gross requirements		90	90	60	30	90	60	
Order quantity = 90	Scheduled receipts								
	Projected inventory	12	12	12	12	42	12	12	42	42
	Planned orders		90	90	90		90	90		

8.8 a)

Item A	Period	1	2	3	4	5	6
Lead-time = 1	Gross requirements	5	26		13	12	
Order quantity = 10	Scheduled receipts						
Safety stock = 10	Projected inventory	27	22	16	16	13	11	11
	Planned orders	20		10	10			

Item B	Period	1	2	3	4	5	6
Lead-time = 1	Gross requirements	20		10	10		
Order quantity = 10	Scheduled receipts						
Safety stock = 10	Projected inventory	34	14	14	14	14	14	14
	Planned orders			10	10			

Item C	Period	1	2	3	4	5	6
Lead-time = 1	Gross requirements	40	10	30	20		
Order quantity = 20	Scheduled receipts						
Safety stock = 20	Projected inventory	75	35	25	35	35	35	35
	Planned orders		40	20				

No delayed orders.

b)

Item A	Period	1	2	3	4	5	6
Lead-time = 1	Gross requirements	5	26			13	12
Order quantity = 10	Scheduled receipts						
Safety time = 1	Projected inventory	27	22	6	16	13	1	1
	Planned orders		10	10	10		

Item B	Period	1	2	3	4	5	6
Lead-time = 1	Gross requirements	10	10	10			
Order quantity = 10	Scheduled receipts						
Safety time = 1	Projected inventory	34	24	14	4	4	4	4
	Planned orders						

Item C	Period	1	2	3	4	5	6
Lead-time = 1	Gross requirements	20	20	20			
Order quantity = 20	Scheduled receipts						
Safety time = 1	Projected inventory	75	55	35	15	15	15	15
	Planned orders						

The order for item A in period 1 is delayed one period.

9.1 Denote the final demand $d = 100$.

$$C = \frac{d}{p_1}\frac{Q}{2}1 + \frac{d}{p_2}\frac{Q(1-p_2/p_1)}{2}2 + \frac{Q(1-d/p_2)}{2}3 + \frac{d}{Q}(A_1 + A_2 + A_3).$$

We get $Q^* \approx 526$.

9.6 a) We have $e_1 = e_2 = 1$. Since $A_2/e_2 > A_1/e_1$, the constraint $Q_2 \geq Q_1$ will be satisfied automatically if we optimize the items separately in the relaxed problem. We get $Q_1^* = \sqrt{2A_1 d/e_1} \approx 44.72$, and $Q_2^* = \sqrt{2A_2 d/e_2} = 100$. A lower bound for the costs is $\sqrt{2A_1 de_1} + \sqrt{2A_2 de_2} \approx 144.72$. Without any lack of generality we can assume that $20\sqrt{5}/\sqrt{2} < q \leq 20\sqrt{5}\cdot\sqrt{2}$. This means that $Q_1 = q$. Furthermore, we should choose $Q_2 = 2q$ for $q \geq 50/\sqrt{2}$ and $Q_2 = 4q$ otherwise. We obtain

$$C(q) = 5\frac{q}{2} + \frac{2250}{q}, \quad 20\sqrt{5}/\sqrt{2} < q < 50/\sqrt{2},$$

ANSWERS AND HINTS 319

$$C(q) = 3\frac{q}{2} + \frac{3500}{q}, \quad 50/\sqrt{2} \le q \le 20\sqrt{5} \cdot \sqrt{2}.$$

We obtain the optimal $q \approx 48.30$, $Q_1 = q$, and $Q_2 = 2q$. The optimal costs are $C \approx 144.91$, i.e., only 0.13 percent above the lower bound.

b) From (9.13) we obtain $k^* = \sqrt{5}$. Since $\sqrt{5}/2 \le 3/\sqrt{5}$ it is optimal to have $k = 2$. Inserting in (9.10) and (9.11) we get the same solution as in a).

9.8 a) Let $Q_2 = k_2 Q_1$ and $Q_3 = k_3 Q_2$. We obtain

$$C = C_1^e + C_2^e + C_3^e = (e_1 + k_2 e_2 + k_3 k_2 e_3)\frac{Q_1}{2} + (A_1 + \frac{A_2}{k_2} + \frac{A_3}{k_3 k_2})\frac{d}{Q_1},$$

i.e., $A_1' = A_1 + \dfrac{A_2}{k_2} + \dfrac{A_3}{k_3 k_2}$, $h_1' = e_1 + k_2 e_2 + k_3 k_2 e_3$.

b) $p(2) = 3, p(1) = 2$. We obtain $A_3' = A_3$, $h_3' = e_3$, and

$$A_2' = A_2 + \frac{A_3}{k_3}, \quad h_2' = e_2 + k_3 e_3,$$

and

$$A_1' = A_1 + \frac{A_2'}{k_2} = A_1 + \frac{A_2}{k_2} + \frac{A_3}{k_3 k_2}, \quad h_1' = e_1 + k_2 h_2' = e_1 + k_2 e_2 + k_3 k_2 e_3.$$

10.2 Note first that $\mu_2' = \sigma_2' = 0$ and that S_1^e does not change. Consequently (10.9) degenerates to

$$\hat{C}_2(y_2) = h_2 y_2 + \hat{C}_1(S_1^e), \quad y_2 > S_1^e,$$

$$\hat{C}_2(y_2) = h_2 y_2 + \hat{C}_1(y_2), \quad y_2 \le S_1^e.$$

Since the costs for $y_2 > S_1^e$ are increasing with y_2, the optimal y_2 must occur for $y_2 \le S_1^e$. From (10.6) we obtain

$$\hat{C}_2(y_2) = h_1 y_2 - h_1 \mu_1'' + (h_1 + b_1)\sigma_1'' G\left(\frac{y_2 - \mu_1''}{\sigma_1''}\right).$$

This is the cost for a single-echelon system, since there is no stock at installation 2, and since the possible order-up-to level at installation 1 is bounded by y_2. We obtain the optimal y_2 from the condition

$$\Phi\left(\frac{y_2 - \mu_1''}{\sigma_1''}\right) = \frac{b_1}{h_1 + b_1} = \frac{10}{11.5} \approx 0.870,$$

i.e., $S_2^e = y_2^* \approx \mu_1'' + 1.13\sigma_1'' = 60 + 1.13 \cdot 5\sqrt{6} \approx 73.8$ and $\hat{C}_2(y_2^*) \approx 29.9$.

10.4 a) Using (10.18) we obtain $P(IL_0 = 1) = 0.2240$, $P(IL_0 = 2) = 0.1494$, $P(IL_0 = 3) = 0.0498$. The holding costs are determined from (10.19). For Poisson demand the fill rate is equivalent to the ready rate, i.e., $P(IL_0 > 0) = 0.2240 + 0.1494 + 0.0498 = 42.3$ percent.

b) From (10.21) and (10.22) we obtain $E(IL_0^-) = 0.672$ and $E(W_0) = 0.672/3 = 0.224$. Consequently, $\overline{L}_1 = \overline{L}_2 = 1.224$. For retailer 1 we obtain $P(IL_1 = 1) = 0.2203$, $P(IL_1 = 2) = 0.3599$, $P(IL_1 = 3) = 0.2941$. The holding costs are 1.822 and the backorder costs 0.463. The fill rate is 87.4 percent. For retailer 2 we get $P(IL_2 = 1) = 0.1294$, $P(IL_2 = 2) = 0.2114$, $P(IL_2 = 3) = 0.2591$, $P(IL_2 = 4) = 0.2117$, $P(IL_2 = 5) = 0.0865$. The holding costs are 2.608 and the backorder costs 0.565. The fill rate is 89.8 percent.

APPENDIX 2

NORMAL DISTRIBUTION TABLES

$$\varphi(x) = \frac{1}{\sqrt{2\pi}} e^{-\frac{x^2}{2}}, \quad \varphi(-x) = \varphi(x). \qquad \Phi(x) = \int_{-\infty}^{x} \varphi(v)dv, \quad \Phi(-x) = 1 - \Phi(x).$$

$$G(x) = \int_{x}^{\infty} (v-x)\varphi(v)dv = \varphi(x) - x(1 - \Phi(x)), \qquad G(-x) = G(x) + x.$$

$$H(x) = \int_{x}^{\infty} G(v)dv = \frac{1}{2}\left[(x^2 + 1)(1 - \Phi(x)) - x\varphi(x)\right], \quad H(-x) = -H(x) + \frac{1}{2}(x^2 + 1).$$

x	$\varphi(x)$	$\Phi(x)$	$G(x)$	$H(x)$	x	$\varphi(x)$	$\Phi(x)$	$G(x)$	$H(x)$
0.00	0.3989	0.5000	0.3989	0.2500	0.26	0.3857	0.6026	0.2824	0.1620
0.01	0.3989	0.5040	0.3940	0.2460	0.27	0.3847	0.6064	0.2784	0.1592
0.02	0.3989	0.5080	0.3890	0.2421	0.28	0.3836	0.6103	0.2745	0.1564
0.03	0.3988	0.5120	0.3841	0.2383	0.29	0.3825	0.6141	0.2706	0.1537
0.04	0.3986	0.5160	0.3793	0.2344	0.30	0.3814	0.6179	0.2668	0.1510
0.05	0.3984	0.5199	0.3744	0.2307	0.31	0.3802	0.6217	0.2630	0.1484
0.06	0.3982	0.5239	0.3697	0.2269	0.32	0.3790	0.6255	0.2592	0.1458
0.07	0.3980	0.5279	0.3649	0.2233	0.33	0.3778	0.6293	0.2555	0.1432
0.08	0.3977	0.5319	0.3602	0.2197	0.34	0.3765	0.6331	0.2518	0.1407
0.09	0.3973	0.5359	0.3556	0.2161	0.35	0.3752	0.6368	0.2481	0.1382
0.10	0.3970	0.5398	0.3509	0.2125	0.36	0.3739	0.6406	0.2445	0.1357
0.11	0.3965	0.5438	0.3464	0.2091	0.37	0.3725	0.6443	0.2409	0.1333
0.12	0.3961	0.5478	0.3418	0.2056	0.38	0.3712	0.6480	0.2374	0.1309
0.13	0.3956	0.5517	0.3373	0.2022	0.39	0.3697	0.6517	0.2339	0.1285
0.14	0.3951	0.5557	0.3328	0.1989	0.40	0.3683	0.6554	0.2304	0.1262
0.15	0.3945	0.5596	0.3284	0.1956	0.41	0.3668	0.6591	0.2270	0.1239
0.16	0.3939	0.5636	0.3240	0.1923	0.42	0.3653	0.6628	0.2236	0.1217
0.17	0.3932	0.5675	0.3197	0.1891	0.43	0.3637	0.6664	0.2203	0.1194
0.18	0.3925	0.5714	0.3154	0.1859	0.44	0.3621	0.6700	0.2169	0.1173
0.19	0.3918	0.5753	0.3111	0.1828	0.45	0.3605	0.6736	0.2137	0.1151
0.20	0.3910	0.5793	0.3069	0.1797	0.46	0.3589	0.6772	0.2104	0.1130
0.21	0.3902	0.5832	0.3027	0.1766	0.47	0.3572	0.6808	0.2072	0.1109
0.22	0.3894	0.5871	0.2986	0.1736	0.48	0.3555	0.6844	0.2040	0.1088
0.23	0.3885	0.5910	0.2944	0.1707	0.49	0.3538	0.6879	0.2009	0.1068
0.24	0.3876	0.5948	0.2904	0.1677	0.50	0.3521	0.6915	0.1978	0.1048
0.25	0.3867	0.5987	0.2863	0.1649					

x	$\varphi(x)$	$\Phi(x)$	$G(x)$	$H(x)$	x	$\varphi(x)$	$\Phi(x)$	$G(x)$	$H(x)$
0.51	0.3503	0.6950	0.1947	0.1029	1.01	0.2396	0.8438	0.0817	0.0368
0.52	0.3485	0.6985	0.1917	0.1009	1.02	0.2371	0.8461	0.0802	0.0360
0.53	0.3467	0.7019	0.1887	0.0990	1.03	0.2347	0.8485	0.0787	0.0352
0.54	0.3448	0.7054	0.1857	0.0972	1.04	0.2323	0.8508	0.0772	0.0345
0.55	0.3429	0.7088	0.1828	0.0953	1.05	0.2299	0.8531	0.0757	0.0337
0.56	0.3410	0.7123	0.1799	0.0935	1.06	0.2275	0.8554	0.0742	0.0329
0.57	0.3391	0.7157	0.1771	0.0917	1.07	0.2251	0.8577	0.0728	0.0322
0.58	0.3372	0.7190	0.1742	0.0900	1.08	0.2227	0.8599	0.0714	0.0315
0.59	0.3352	0.7224	0.1714	0.0882	1.09	0.2203	0.8621	0.0700	0.0308
0.60	0.3332	0.7257	0.1687	0.0865	1.10	0.2179	0.8643	0.0686	0.0301
0.61	0.3312	0.7291	0.1659	0.0849	1.11	0.2155	0.8665	0.0673	0.0294
0.62	0.3292	0.7324	0.1633	0.0832	1.12	0.2131	0.8686	0.0659	0.0287
0.63	0.3271	0.7357	0.1606	0.0816	1.13	0.2107	0.8708	0.0646	0.0281
0.64	0.3251	0.7389	0.1580	0.0800	1.14	0.2083	0.8729	0.0634	0.0275
0.65	0.3230	0.7422	0.1554	0.0784	1.15	0.2059	0.8749	0.0621	0.0268
0.66	0.3209	0.7454	0.1528	0.0769	1.16	0.2036	0.8770	0.0609	0.0262
0.67	0.3187	0.7486	0.1503	0.0754	1.17	0.2012	0.8790	0.0596	0.0256
0.68	0.3166	0.7517	0.1478	0.0739	1.18	0.1989	0.8810	0.0584	0.0250
0.69	0.3144	0.7549	0.1453	0.0724	1.19	0.1965	0.8830	0.0573	0.0244
0.70	0.3123	0.7580	0.1429	0.0710	1.20	0.1942	0.8849	0.0561	0.0239
0.71	0.3101	0.7611	0.1405	0.0696	1.21	0.1919	0.8869	0.0550	0.0233
0.72	0.3079	0.7642	0.1381	0.0682	1.22	0.1895	0.8888	0.0538	0.0228
0.73	0.3056	0.7673	0.1358	0.0668	1.23	0.1872	0.8907	0.0527	0.0222
0.74	0.3034	0.7704	0.1334	0.0654	1.24	0.1849	0.8925	0.0517	0.0217
0.75	0.3011	0.7734	0.1312	0.0641	1.25	0.1826	0.8944	0.0506	0.0212
0.76	0.2989	0.7764	0.1289	0.0628	1.26	0.1804	0.8962	0.0495	0.0207
0.77	0.2966	0.7794	0.1267	0.0615	1.27	0.1781	0.8980	0.0485	0.0202
0.78	0.2943	0.7823	0.1245	0.0603	1.28	0.1758	0.8997	0.0475	0.0197
0.79	0.2920	0.7852	0.1223	0.0591	1.29	0.1736	0.9015	0.0465	0.0193
0.80	0.2897	0.7881	0.1202	0.0578	1.30	0.1714	0.9032	0.0455	0.0188
0.81	0.2874	0.7910	0.1181	0.0567	1.31	0.1691	0.9049	0.0446	0.0184
0.82	0.2850	0.7939	0.1160	0.0555	1.32	0.1669	0.9066	0.0436	0.0179
0.83	0.2827	0.7967	0.1140	0.0543	1.33	0.1647	0.9082	0.0427	0.0175
0.84	0.2803	0.7995	0.1120	0.0532	1.34	0.1626	0.9099	0.0418	0.0171
0.85	0.2780	0.8023	0.1100	0.0521	1.35	0.1604	0.9115	0.0409	0.0166
0.86	0.2756	0.8051	0.1080	0.0510	1.36	0.1582	0.9131	0.0400	0.0162
0.87	0.2732	0.8078	0.1061	0.0499	1.37	0.1561	0.9147	0.0392	0.0158
0.88	0.2709	0.8106	0.1042	0.0489	1.38	0.1539	0.9162	0.0383	0.0155
0.89	0.2685	0.8133	0.1023	0.0478	1.39	0.1518	0.9177	0.0375	0.0151
0.90	0.2661	0.8159	0.1004	0.0468	1.40	0.1497	0.9192	0.0367	0.0147
0.91	0.2637	0.8186	0.0986	0.0458	1.41	0.1476	0.9207	0.0359	0.0143
0.92	0.2613	0.8212	0.0968	0.0449	1.42	0.1456	0.9222	0.0351	0.0140
0.93	0.2589	0.8238	0.0950	0.0439	1.43	0.1435	0.9236	0.0343	0.0136
0.94	0.2565	0.8264	0.0933	0.0430	1.44	0.1415	0.9251	0.0336	0.0133
0.95	0.2541	0.8289	0.0916	0.0420	1.45	0.1394	0.9265	0.0328	0.0130
0.96	0.2516	0.8315	0.0899	0.0411	1.46	0.1374	0.9279	0.0321	0.0127
0.97	0.2492	0.8340	0.0882	0.0402	1.47	0.1354	0.9292	0.0314	0.0123
0.98	0.2468	0.8365	0.0865	0.0394	1.48	0.1334	0.9306	0.0307	0.0120
0.99	0.2444	0.8389	0.0849	0.0385	1.49	0.1315	0.9319	0.0300	0.0117
1.00	0.2420	0.8413	0.0833	0.0377	1.50	0.1295	0.9332	0.0293	0.0114

NORMAL DISTRIBUTION

x	$\varphi(x)$	$\Phi(x)$	$G(x)$	$H(x)$	x	$\varphi(x)$	$\Phi(x)$	$G(x)$	$H(x)$
1.51	0.1276	0.9345	0.0286	0.0111	2.01	0.0529	0.9778	0.0083	0.0028
1.52	0.1257	0.9357	0.0280	0.0109	2.02	0.0519	0.9783	0.0080	0.0027
1.53	0.1238	0.9370	0.0274	0.0106	2.03	0.0508	0.9788	0.0078	0.0026
1.54	0.1219	0.9382	0.0267	0.0103	2.04	0.0498	0.9793	0.0076	0.0026
1.55	0.1200	0.9394	0.0261	0.0100	2.05	0.0488	0.9798	0.0074	0.0025
1.56	0.1182	0.9406	0.0255	0.0098	2.06	0.0478	0.9803	0.0072	0.0024
1.57	0.1163	0.9418	0.0249	0.0095	2.07	0.0468	0.9808	0.0070	0.0023
1.58	0.1145	0.9429	0.0244	0.0093	2.08	0.0459	0.9812	0.0068	0.0023
1.59	0.1127	0.9441	0.0238	0.0090	2.09	0.0449	0.9817	0.0066	0.0022
1.60	0.1109	0.9452	0.0232	0.0088	2.10	0.0440	0.9821	0.0065	0.0021
1.61	0.1092	0.9463	0.0227	0.0086	2.11	0.0431	0.9826	0.0063	0.0021
1.62	0.1074	0.9474	0.0222	0.0084	2.12	0.0422	0.9830	0.0061	0.0020
1.63	0.1057	0.9484	0.0216	0.0081	2.13	0.0413	0.9834	0.0060	0.0020
1.64	0.1040	0.9495	0.0211	0.0079	2.14	0.0404	0.9838	0.0058	0.0019
1.65	0.1023	0.9505	0.0206	0.0077	2.15	0.0396	0.9842	0.0056	0.0018
1.66	0.1006	0.9515	0.0201	0.0075	2.16	0.0387	0.9846	0.0055	0.0018
1.67	0.0989	0.9525	0.0197	0.0073	2.17	0.0379	0.9850	0.0053	0.0017
1.68	0.0973	0.9535	0.0192	0.0071	2.18	0.0371	0.9854	0.0052	0.0017
1.69	0.0957	0.9545	0.0187	0.0069	2.19	0.0363	0.9857	0.0050	0.0016
1.70	0.0940	0.9554	0.0183	0.0067	2.20	0.0355	0.9861	0.0049	0.0016
1.71	0.0925	0.9564	0.0178	0.0066	2.21	0.0347	0.9864	0.0047	0.0015
1.72	0.0909	0.9573	0.0174	0.0064	2.22	0.0339	0.9868	0.0046	0.0015
1.73	0.0893	0.9582	0.0170	0.0062	2.23	0.0332	0.9871	0.0045	0.0014
1.74	0.0878	0.9591	0.0166	0.0060	2.24	0.0325	0.9875	0.0044	0.0014
1.75	0.0863	0.9599	0.0162	0.0059	2.25	0.0317	0.9878	0.0042	0.0013
1.76	0.0848	0.9608	0.0158	0.0057	2.26	0.0310	0.9881	0.0041	0.0013
1.77	0.0833	0.9616	0.0154	0.0056	2.27	0.0303	0.9884	0.0040	0.0013
1.78	0.0818	0.9625	0.0150	0.0054	2.28	0.0297	0.9887	0.0039	0.0012
1.79	0.0804	0.9633	0.0146	0.0053	2.29	0.0290	0.9890	0.0038	0.0012
1.80	0.0790	0.9641	0.0143	0.0051	2.30	0.0283	0.9893	0.0037	0.0012
1.81	0.0775	0.9649	0.0139	0.0050	2.31	0.0277	0.9896	0.0036	0.0011
1.82	0.0761	0.9656	0.0136	0.0048	2.32	0.0270	0.9898	0.0035	0.0011
1.83	0.0748	0.9664	0.0132	0.0047	2.33	0.0264	0.9901	0.0034	0.0010
1.84	0.0734	0.9671	0.0129	0.0046	2.34	0.0258	0.9904	0.0033	0.0010
1.85	0.0721	0.9678	0.0126	0.0044	2.35	0.0252	0.9906	0.0032	0.0010
1.86	0.0707	0.9686	0.0123	0.0043	2.36	0.0246	0.9909	0.0031	0.0009
1.87	0.0694	0.9693	0.0119	0.0042	2.37	0.0241	0.9911	0.0030	0.0009
1.88	0.0681	0.9699	0.0116	0.0041	2.38	0.0235	0.9913	0.0029	0.0009
1.89	0.0669	0.9706	0.0113	0.0040	2.39	0.0229	0.9916	0.0028	0.0009
1.90	0.0656	0.9713	0.0111	0.0039	2.40	0.0224	0.9918	0.0027	0.0008
1.91	0.0644	0.9719	0.0108	0.0037	2.41	0.0219	0.9920	0.0026	0.0008
1.92	0.0632	0.9726	0.0105	0.0036	2.42	0.0213	0.9922	0.0026	0.0008
1.93	0.0620	0.9732	0.0102	0.0035	2.43	0.0208	0.9925	0.0025	0.0008
1.94	0.0608	0.9738	0.0100	0.0034	2.44	0.0203	0.9927	0.0024	0.0007
1.95	0.0596	0.9744	0.0097	0.0033	2.45	0.0198	0.9929	0.0023	0.0007
1.96	0.0584	0.9750	0.0094	0.0032	2.46	0.0194	0.9931	0.0023	0.0007
1.97	0.0573	0.9756	0.0092	0.0031	2.47	0.0189	0.9932	0.0022	0.0007
1.98	0.0562	0.9761	0.0090	0.0031	2.48	0.0184	0.9934	0.0021	0.0006
1.99	0.0551	0.9767	0.0087	0.0030	2.49	0.0180	0.9936	0.0021	0.0006
2.00	0.0540	0.9772	0.0085	0.0029	2.50	0.0175	0.9938	0.0020	0.0006

x	$\varphi(x)$	$\Phi(x)$	$G(x)$	$H(x)$	x	$\varphi(x)$	$\Phi(x)$	$G(x)$	$H(x)$
2.51	0.0171	0.9940	0.0019	0.0006	3.01	0.0043	0.9987	0.0004	0.0001
2.52	0.0167	0.9941	0.0019	0.0006	3.02	0.0042	0.9987	0.0004	0.0001
2.53	0.0163	0.9943	0.0018	0.0005	3.03	0.0040	0.9988	0.0003	0.0001
2.54	0.0158	0.9945	0.0018	0.0005	3.04	0.0039	0.9988	0.0003	0.0001
2.55	0.0154	0.9946	0.0017	0.0005	3.05	0.0038	0.9989	0.0003	0.0001
2.56	0.0151	0.9948	0.0017	0.0005	3.06	0.0037	0.9989	0.0003	0.0001
2.57	0.0147	0.9949	0.0016	0.0005	3.07	0.0036	0.9989	0.0003	0.0001
2.58	0.0143	0.9951	0.0016	0.0005	3.08	0.0035	0.9990	0.0003	0.0001
2.59	0.0139	0.9952	0.0015	0.0004	3.09	0.0034	0.9990	0.0003	0.0001
2.60	0.0136	0.9953	0.0015	0.0004	3.10	0.0033	0.9990	0.0003	0.0001
2.61	0.0132	0.9955	0.0014	0.0004	3.11	0.0032	0.9991	0.0003	0.0001
2.62	0.0129	0.9956	0.0014	0.0004	3.12	0.0031	0.9991	0.0002	0.0001
2.63	0.0126	0.9957	0.0013	0.0004	3.13	0.0030	0.9991	0.0002	0.0001
2.64	0.0122	0.9959	0.0013	0.0004	3.14	0.0029	0.9992	0.0002	0.0001
2.65	0.0119	0.9960	0.0012	0.0004	3.15	0.0028	0.9992	0.0002	0.0001
2.66	0.0116	0.9961	0.0012	0.0003	3.16	0.0027	0.9992	0.0002	0.0001
2.67	0.0113	0.9962	0.0012	0.0003	3.17	0.0026	0.9992	0.0002	0.0001
2.68	0.0110	0.9963	0.0011	0.0003	3.18	0.0025	0.9993	0.0002	0.0001
2.69	0.0107	0.9964	0.0011	0.0003	3.19	0.0025	0.9993	0.0002	0.0000
2.70	0.0104	0.9965	0.0011	0.0003	3.20	0.0024	0.9993	0.0002	0.0000
2.71	0.0101	0.9966	0.0010	0.0003	3.21	0.0023	0.9993	0.0002	0.0000
2.72	0.0099	0.9967	0.0010	0.0003	3.22	0.0022	0.9994	0.0002	0.0000
2.73	0.0096	0.9968	0.0010	0.0003	3.23	0.0022	0.9994	0.0002	0.0000
2.74	0.0093	0.9969	0.0009	0.0003	3.24	0.0021	0.9994	0.0002	0.0000
2.75	0.0091	0.9970	0.0009	0.0003	3.25	0.0020	0.9994	0.0002	0.0000
2.76	0.0088	0.9971	0.0009	0.0002	3.26	0.0020	0.9994	0.0001	0.0000
2.77	0.0086	0.9972	0.0008	0.0002	3.27	0.0019	0.9995	0.0001	0.0000
2.78	0.0084	0.9973	0.0008	0.0002	3.28	0.0018	0.9995	0.0001	0.0000
2.79	0.0081	0.9974	0.0008	0.0002	3.29	0.0018	0.9995	0.0001	0.0000
2.80	0.0079	0.9974	0.0008	0.0002	3.30	0.0017	0.9995	0.0001	0.0000
2.81	0.0077	0.9975	0.0007	0.0002	3.31	0.0017	0.9995	0.0001	0.0000
2.82	0.0075	0.9976	0.0007	0.0002	3.32	0.0016	0.9995	0.0001	0.0000
2.83	0.0073	0.9977	0.0007	0.0002	3.33	0.0016	0.9996	0.0001	0.0000
2.84	0.0071	0.9977	0.0007	0.0002	3.34	0.0015	0.9996	0.0001	0.0000
2.85	0.0069	0.9978	0.0006	0.0002	3.35	0.0015	0.9996	0.0001	0.0000
2.86	0.0067	0.9979	0.0006	0.0002	3.36	0.0014	0.9996	0.0001	0.0000
2.87	0.0065	0.9979	0.0006	0.0002	3.37	0.0014	0.9996	0.0001	0.0000
2.88	0.0063	0.9980	0.0006	0.0002	3.38	0.0013	0.9996	0.0001	0.0000
2.89	0.0061	0.9981	0.0006	0.0002	3.39	0.0013	0.9997	0.0001	0.0000
2.90	0.0060	0.9981	0.0005	0.0001	3.40	0.0012	0.9997	0.0001	0.0000
2.91	0.0058	0.9982	0.0005	0.0001	3.41	0.0012	0.9997	0.0001	0.0000
2.92	0.0056	0.9982	0.0005	0.0001	3.42	0.0012	0.9997	0.0001	0.0000
2.93	0.0055	0.9983	0.0005	0.0001	3.43	0.0011	0.9997	0.0001	0.0000
2.94	0.0053	0.9984	0.0005	0.0001	3.44	0.0011	0.9997	0.0001	0.0000
2.95	0.0051	0.9984	0.0005	0.0001	3.45	0.0010	0.9997	0.0001	0.0000
2.96	0.0050	0.9985	0.0004	0.0001	3.46	0.0010	0.9997	0.0001	0.0000
2.97	0.0048	0.9985	0.0004	0.0001	3.47	0.0010	0.9997	0.0001	0.0000
2.98	0.0047	0.9986	0.0004	0.0001	3.48	0.0009	0.9997	0.0001	0.0000
2.99	0.0046	0.9986	0.0004	0.0001	3.49	0.0009	0.9998	0.0001	0.0000
3.00	0.0044	0.9987	0.0004	0.0001	3.50	0.0009	0.9998	0.0001	0.0000

INDEX

ABC inventory analysis, 301
Abramowitz, M., 86, 124
Adan, I., 84, 124
Adaptive forecasting, 36
Advanced Planning System (APS), 170
Afentakis, P., 240, 242
Aggregate, 5, 21, 170, 178, 299
Aggregation, 179
Akella, R., 215
Alfredsson, P., 193, 215
Andersson, J., 280, 281, 287
Anupindi, R., 215
Arborescent system, *see* Distribution system
Archibald, A. W., 194, 215
Archibald, B. C., 119, 124
ARIMA, 27
ARMA, 28
Assembly system, 190
Atkins, D. R., 181
Axsäter, S., 52, 68, 69, 70, 135, 136, 145, 160, 171, 181, 193, 194, 200, 203, 204, 210, 215, 229, 241, 242, 243, 260, 267, 270, 272, 273, 278, 279, 280, 281, 287, 288, 307

Backlog, *see* Backorders
Backorder costs, 45, 59, 95 - 96

Backorders, 46, 59
Baganha, M. P., 214, 215
Baker, K. R., 68, 71, 213, 215
Balintfy, J. L., 181
Base stock policy, *see* S policy
Basic period, 150
Bassok, Y., 194, 215
Batch quantities, *see* Lot sizing
Benton, W. C., 56, 71
Berling, P., 44, 50
Berry, W. L., 6, 217, 307
Bertrand, J. W. M., 171, 181
Beyer, D., 119, 125, 140, 145
Bhaskaran, S., 65, 71
Billington, P. J., 170, 181
Bill of material (BOM), 191, 205
Bitran, G. R., 170, 181
Blackburn, J. D., 70, 71, 238, 239, 242
BOM, *see* Bill of material
Bomberger, E. A., 156, 159, 181
Bomberger's problem, 156
Bowman, A., 183
Bowman, R. A., 163, 181, 182
Box, G. E. P., 27, 37
Box-Jenkins techniques, 27-29
Browne, S., 85, 125
Brown, R. G., 5
Buchanen, D. J., 119, 125

Buffer stocks, *see* Safety stocks
Bullwhip effect, 214

Cachon, G. P., 280, 284, 288
Candea, D., 5
Can-order policy, 181
Capacity constraints, 155 - 170
Capacity Requirements Planning (CRP), 211, 212
Capital costs, *see* Holding costs
Carrying costs, *see* Holding costs
Chakravarty, A., 193, 216
Chen, F., 106, 125, 138, 145, 195, 214, 215, 256, 273, 274, 278, 280, 284, 288
Chikán, A., 5
Clark, A. J., 215, 288
Clark-Scarf model, 248 - 260
Classical economic order quantity (EOQ), 52 - 55
Cohen, M. A., 194, 214, 215, 217
Compound Poisson demand, 77 - 83
Constant model, 9
Continuous review, 47
Coordination and contracts, 283 - 287
Costs, 44 - 46, 95 - 96
Croston, J. D., 26, 37
CRP, *see* Capacity Requirements Planning
Customer service, 44 - 46, 94 - 95
Cycle counting, 298
Cycle time, 150
Cyclic schedule, 150

Dada, M., 193, 215
Dannenbring, D. G., 36, 37
Das, C., 194, 215
Dekker, R., 183, 195, 215, 217
De Kok, A. G., 5, 170, 182, 291
Del Vecchio, A., 214, 217
Demand forecasts, *see* Forecasting
Demand models, 8 - 11, 24 - 29, 77-88
Demand processes, *see* Demand models

De Matteis, J. J., 68, 71
Dependent demand, *see* Material Requirements Planning
Deuermeyer, B., 279, 288
Diks, E. B., 256, 260, 288, 291
Disaggregate, 267, 279 - 280
Distribution Requirements Planning (DRP), 211
Distribution system, 188-189
Dobson, G., 163, 182
Dogru, M. K., 260, 289
Doll, C. L., 160, 182, 184
DRP, *see* Distribution Requirements Planning
Dynamic programming, 63 - 66, 159 - 160

Echelon holding costs, 225-230
Echelon stock, 197 - 198, 213, 225 - 236, 278
Echelon stock reorder point policies, 197 - 204, 278
Economic Lot Scheduling Problem (ELSP) 155 - 163
Economic Order Quantity (EOQ), 52 - 61
  backorders allowed, 59 - 61
  classical economic order quantity, 52 - 55
  finite production rate, 55 - 56, 223 - 225
  quantity discounts, 56 - 58
Education, 306 - 307
Electronic Data Interchange (EDI), 187
Elmaghraby, S. E., 160, 163, 182
Enterprise Resource Planning (ERP), 170, 212, 299
EOQ, *see* Economic Order Quantity
Eppen, G. D., 170, 182, 256, 260, 289
Erkip, N. K., 247, 260, 281, 289
Erlang distribution, 87
ERP, *see* Enterprise Resource Planning

INDEX 327

Exponential smoothing, 12 - 16
Exponential smoothing with trend, 16 - 18

Federgruen, A., 65, 71, 112, 125, 130, 132, 145, 146, 163, 181, 182, 249, 256, 260, 289
Feller, W., 77, 125
Fill rate, 94 - 95
Finite production rate, 55 - 56, 223 - 225
Fisher, M., 280, 288
Fleischmann, B., 170, 182
Fleischmann, M., 215, 217
Forecast errors, 29 - 34
Forecasting, 7 - 37
    adaptive forecasting, 36
    ARIMA, 27
    ARMA, 28
    Box-Jenkins techniques, 27 - 29
    correlated stochastic variations, *see* Box-Jenkins techniques
    demand models, 8 - 11, 24 - 29
    exponential smoothing, 12 - 16
    exponential smoothing with trend, 16 - 18
    forecast errors, 29-34
    initial forecast, 15, 17, 31, 35
    manual forecasts, 36 - 37
    Mean Absolute Deviation (MAD), 29 - 34
    monitoring forecasts, 34 - 36
    moving average, 11 - 12
    regression analysis, 21 - 26
    sporadic demand, 26 - 27
    Winters' trend-seasonal method, 18 - 21
Forrester, J. W., 214, 215
Forsberg, R., 270, 280, 289
Fransoo, J. C., 170, 182
Frenk, J. B. G., 183

Gallego, G., 163, 182, 281, 289
Gamma distribution, 86 - 87
Gardiner, J. S., 38
Gardner, E. S., 36, 37

Gavish, B., 242
General system, 190 - 192
Geometric distribution, 82
Goyal, S. K., 176, 182
Grahovac, J., 193, 216
Graves, S. C., 5, 71, 125, 163, 181, 182, 183, 215, 240, 242, 260, 266, 282, 287, 288, 289
Groenevelt, H., 182
Groff, G., 70, 71
Gross requirements, 206, 208
Grouping of items, 141, 301
Grubbström, R. W., 44, 50
Guaranteed service model, 281 - 283
Guide, Jr., V. D. R., 216
$G(x)$, 91- 92, 321 - 324

Hadley, G., 5, 119, 125, 145
Hariharan, R., 283, 289
Harris, F. W., 52, 71
Hausman, W. H., 247, 281, 289
Hax, A., 5
Heyman, D. P., 125, 145
Hill, R. M., 6, 38, 50, 72, 119, 125, 146, 216
Holding costs, 44, 225 - 230
Holt, C. C., 5, 16, 37
Hopp, D. L., 217
Hsu, W., 163, 182
$H(x)$, 104, 321 - 324
Hyndman, R. J., 38

Iglehart, D., 140, 145
Implementation, 295 - 307
    ABC inventory analysis, 301
    bar coding, 296
    cycle counting, 298
    education, 306 - 307
    fixed locations, 297
    grouping of items, 301
    inventory records, 296 - 298
    performance evaluation, 298 - 299
    periodic counting, 297
    random locations, 297
    short-run consequences, 305 - 306
    simulation, 304 - 305

step-by-step implementation, 303 - 304
Independent lead-times, 120
Inderfurth, K., 215, 282, 290, 291
Initial forecast, 15, 17, 31, 35
Installation stock reorder point, 196 - 197
Intermittent demand, *see* Sporadic demand
Inventory level, 46 - 47
Inventory level distribution, 88 - 94
Inventory position, 46 - 47
Inventory position distribution, 88 - 89
Inventory records, 296 - 298
Inventory turnover, 299
Iyogun, P., 181

Jackson, P. L., 179, 182, 183, 260, 290
Jenkins, G. M., 27, 37
Johansen, S. G., 119, 125, 194, 216
Johnson, L. A., 5, 38
Joint optimization of order quantity and reorder point, 129 - 137
Joint replenishments, 172 - 181
Juntti, L., 203, 204, 215

KANBAN policy, 49, 196 - 197
Kapur, S., 216
Karmarkar, U. S., 171, 182, 242
Katalan, Z., 163, 182
Kimball, G. E., 282, 290
Kolen, A., 71
Kukreja, A., 193, 216
Kunreuther, H., 65, 71

Lagrangean optimization, 163 - 170, 176 - 180, 230 - 236
Lateral transshipments, 192 - 194
Lead-time, 47, 119 - 124
Least Unit Cost (LUC), 68
Lee, H. L., 71, 125, 193, 214, 216, 280, 290
Little's formula, 59, 101, 263, 279
Logarithmic distribution, 80

Loss function, *see* $G(x)$
Lost sales, 34, 46, 117 - 119, 135
Lot sizing, 51 - 70, 129 - 137, 149 - 181, 221 - 242
  backorders allowed, 59 - 61
  coordinated ordering, 149 - 181
  finite production rate, 55 - 56
  joint optimization of order quantity and reorder point, 129 - 137
  multi-echelon, constant demand, 221 - 236
  multi-echelon, time-varying demand, 236 - 242
  Roundy's 98 percent approximation, 230 - 236
  quantity discounts, 56 - 58
  Silver-Meal heuristic, 66 - 68
  single-echelon, constant demand, 52 - 61
  single-echelon, time-varying demand, 61 - 70
  Wagner-Whitin algorithm, 63 - 66
Love, R. F., 119, 125
Love, S. F., 6
Lundin, R., 65, 71

MAD, *see* Mean Absolute Deviation
Makridakis, S., 29, 38
Manne, A. S., 168, 183
Manual forecasts, 36 - 37
Manufacturing Resource Planning, 211 - 212
Marklund, J., 280, 288, 290
Martin, R. K., 170, 182
Master Production Schedule (MPS), 205
Material Requirements Planning (MRP), 8, 204 - 213
  bill of material, 205
  Capacity Requirements Planning (CRP), 211 - 212
  Distribution Requirements Planning (DRP), 211
  Enterprise Resource Planning (ERP), 212
  gross requirement, 206, 208

INDEX

lot-for-lot, 206
Manufacturing Resource Planning, 211 - 212
Master Production Schedule (MPS), 205
MRP II, 211
nervousness, 211
net requirement, 206
peg requirements, 210
planned orders, 207
safety time, 209
Maxwell, W., 182
McClain, J. O., 6, 181
McGavin, E. J., 260, 290
Meal, H. C., 66, 68, 71, 74
Mean Absolute Deviation (MAD), 7, 29 - 32
Mendoza, A. G., 68, 71
METRIC, 261 - 266, 278 - 279, 281
Meyr, H., 170, 182
Millen, R. A., 70, 71, 238, 239, 242
Miller, D. M., 216
Minner, S., 194, 216, 282, 290
MOD-METRIC, 265 - 266
Moinzadeh, K., 194, 216, 280, 290
Monitoring forecasts, 34 - 36
Montgomery, D. C., 5, 38
Moon, I., 163, 182
Morton, T., 65, 71
Moving average, 11 - 12, 13 - 14, 16, 23, 25
MPS, see Master Production Schedule
MRP, see Material Requirements Planning
MRP II, 211
Muckstadt, J. A., 163, 179, 181, 182, 183, 193, 216, 230, 242, 247, 260, 265, 281, 290
Muharremoglu, A., 195, 216, 280, 290
Muller, M., 6, 307
Multi-echelon systems, 187 - 293
Multi-stage systems, see Multi-echelon systems

Naddor, E., 6
Nahmias, S., 6, 34, 38, 71, 125, 192, 194, 216
Negative binomial distribution, 81 - 82
Nervousness, 211
Nested policy, 200
Net requirements, 206
Net stock, see Inventory level
Newsboy model, 114 - 116, 283 - 287
Normal distribution, 35, 85 - 86, 92 - 94
Nuttle, H. L. W., 229, 241, 242

Olsson, F., 194, 216
Optimality of ordering policies, 137 - 140
Optimization, 129 - 137, 249 - 256, 271 - 273, 283 - 287
Ordering costs, 44 - 45, 173
Ordering policies, see Ordering systems
Ordering system dynamics, 213 - 214
Ordering systems, 46 - 50, 195 - 214
  comparison of echelon stock and installation stock polices, 198 - 204
  echelon stock reorder point policies, 197 - 198
  installation stock reorder point policies and KANBAN policies, 196 - 197
  Material Requirements Planning, 204 - 213
  ordering system dynamics, 213 - 214
  (R, Q) policy, 48 - 49
  (s, S) policy, 49 - 50
Order points, see Reorder points
Order quantities, see Lot sizing
Order-up-to-S policy, see S policy
Orlicky, J., 6, 213, 216
Overage cost, 115

Padmanabhan, P., 216

Palm, C., 122, 125
Park, S., 56, 71
Performance evaluation, 298 - 299
Periodic counting, 297 - 298
Periodic review, 47, 109 - 114, 248 - 261, 273 - 276
Peterson, R., 6, 38, 50, 71, 125, 146, 183, 217, 307
Planche, R., 181, 183
Planned orders, 207 - 208, 213
Plossl, G. W., 6
Poisson demand, 77 - 78, 85
Porteus, E. L., 6, 125, 140, 145, 146, 300, 307
Powers-of-two policies, 150 - 153
Probabilistic demand, *see* Demand models
Production smoothing, 154 - 172
Product structure, 191
Projected inventory, 206
Pull system, 210
Push system, 210
Pyke, D. F., 6, 38, 50, 71, 125, 146, 183, 216, 217, 307

(Q, R) policy, *see* (R, Q) policy
Quantity discounts, 56 - 58, 172

Ready rate, 94 - 95
Reinsell, G. C., 37
Regression analysis, 21 - 26
Remanufacturing, 194 - 195
Renberg, B., 181, 183
Reorder points, multi-echelon, 247 - 287
   batch-ordering policies, 273 - 281
   Clark-Scarf model, 248 - 260
   METRIC approach, 261 - 266
   optimization, 271 - 273
   other assumptions, 281 - 287
   two exact techniques, 266 - 270
Reorder points, single-echelon, 48 - 50, 54 - 55, 77 - 124, 129 - 137
   continuous review (s, S) policy, 107 - 109
   given service level, 96 - 101

   inventory level distribution, 88 - 94
   inventory position distribution, 88 - 89
   joint optimization of order quantity and reorder point, 129 - 137
   lost sales, 117 - 119
   newsboy model, 114 - 116
   periodic review - fill rate, 109 - 114
   shortage cost per unit, 106 - 107
   shortage cost per unit and time unit, 101 - 105
   stochastic lead-times, 119 - 124
   updating in practice, 140 - 145
Repairable items, 192, 265
Replenishment quantities, *see* Lot sizing
Reservations, 46
Resing, J., 124
RFID, 187
Rinnooy Kan, A., 5
Robb, D. J., 216
Robinson, L. W., 194, 216
Rolling horizon, 65, 70
Rosling, K., 106, 117, 118, 125, 133, 135, 140, 146, 200, 203, 204, 210, 215, 240, 243, 256, 290
Rough Cut Capacity Planning (RCCP), 211 - 212
Roundy, R., 163, 176, 179, 182, 183, 230, 242, 243
Roundy's 98 percent approximation, 176 - 180, 230 - 236
(R, Q) policy, 48 - 49
Rudi, N., 194, 216
Rustenburg, W. D., 266, 290

Safety capacity, 248
Safety factor, 96 - 97, 282 - 283
Safety stocks, 93 - 94, *see also* Reorder points
Safety time, 205, 209, 248
Salomon, M., 217
Samroengraja, R., 214, 215
Sassen, S. A., 215

INDEX

Satir, A. T., 176, 182
Scarf, H., 195, 215, 248, 249, 256, 261, 271, 288, 291
Scheduled receipts, 206
Schmidt, C. P., 216
Schrage, L., 256, 260, 289
Schwarz, L. B., 216, 242, 279, 288, 289, 290
Seasonal index, 10
Selective inventory control, 301
Sensitivity analysis, 54
Sequential deliveries, 119
Serial system, 188 - 189
Service constraints, see Service levels
Service levels, 94 - 95
Sethi, S. P., 65, 71, 140, 145
Setup costs, see Ordering costs
Shang, K., 256, 290
Shapiro, J. F., 170, 183
Sherbrooke, C. C., 6, 192, 193, 216, 261, 290, 291
Shortage costs, 45, 95 - 96
Silver, E. A., 6, 38, 44, 50, 66, 68, 69, 71, 74, 125, 146, 176, 181, 183, 213, 216, 217, 288, 300, 307
Silver-Meal heuristic, 66 - 68
Simon, R. M., 266, 291
Simpson, K. F., 282, 291
Simulation, 247, 304 - 305
Single period problem, see Newsboy model
Smoothing constant, 12, 16, 19, 27, 30, 31
Sobel, M. J., 111, 125, 145
Song, J. S., 120, 125, 256, 288, 290
Sox, C. R., 163, 183
Spearman, M. L., 197, 217
Spence, A. M., 300, 307
S policy, 49
Sporadic demand, 26 - 27
$(S-1, S)$ policy, see S policy
(s, S) policy, 49 - 50
Standard deviation, 29
Stegun, I., 86, 124
Step-by-step implementation, 303 - 304

Sterman, J., 214, 217
Stochastic lead-times, 119 - 124
Stockout costs, see Shortage costs
Supply Chain Management, 1, 187
Svoronos, A., 279, 291

Tagaras, G., 194, 217
Tersine, R. J., 6, 307
Teunter, R. J., 216
Thomas, L. C., 215
Thomas, L. J., 6, 181, 247, 281, 290
Thorstenson, A., 119, 125, 194, 216
Tijms, H. C., 88, 111, 125, 182
Time horizon, see Rolling horizon
Time-varying demand, 61 - 70, 163 - 170, 236 - 242
Tirupati, D., 170, 181
Towill, D., 214, 217
Trend model, 9 - 10
Trend-seasonal model, 10
Tsitsiklis, J. N., 195, 216, 280, 290
Tzur, M., 65, 71

Underage cost, 115

Van der Heijden, M. C., 260, 291
Van der Laan, E. A., 195, 217
Van Hoesel, S., 71
Van Houtum, G. J., 249, 256, 260, 290, 291
Variance, 29
Veatch, M. H., 197, 217
Veinott, A., 6, 140, 145, 146
Vendor Managed Inventory (VMI), 188, 284
Verrijdt, J. H. C. M., 193, 215, 260, 291
Vidgren, S., 193, 217
Viswanathan, S., 180, 183
VMI, see Vendor Managed Inventory
Vollman, T. E., 6, 213, 217, 307

Wagelmans, A., 65, 71
Wagner, H. M., 6, 63, 65, 66, 68, 70, 71, 74, 75, 145, 237, 238

Wagner-Whitin algorithm, 63 - 66, 70, 237 - 238
Ward, J. E., 290
Wein, L. M., 197, 217
Whang, S., 216
Wheelwright, S. C., 38
Whitin, T. M., 5, 63, 65, 70, 71, 119, 125, 145
Whybark, D. C., 6, 160, 182, 184, 217, 307
Wight, O. W., 6
Wildeman, R.E., 176, 183
Willems, S. P., 282, 289
Williams, J. F., 230, 243
Wilson, R. H., 52, 71

Winters, P. R., 18, 38
Winters' trend-seasonal method, 18 - 21
Woodruff, D. L., 217

Zangwill, W. I., 65, 72
Zhang, W. F., 270, 288
Zheng, Y. S., 52, 72, 106, 112, 125, 130, 132, 140, 145, 146, 273, 274, 278, 280, 288
Zijm, W. H. M., 256, 290, 291
Zipkin, P. H., 5, 6, 44, 50, 72, 85, 120, 125, 146, 163, 171, 183, 249, 256, 260, 279, 281, 283, 289, 291

Early Titles in the
INTERNATIONAL SERIES IN
OPERATIONS RESEARCH & MANAGEMENT SCIENCE
Frederick S. Hillier, Series Editor, *Stanford University*

Saigal/ *A MODERN APPROACH TO LINEAR PROGRAMMING*
Nagurney/ *PROJECTED DYNAMICAL SYSTEMS & VARIATIONAL INEQUALITIES WITH APPLICATIONS*
Padberg & Rijal/ *LOCATION, SCHEDULING, DESIGN AND INTEGER PROGRAMMING*
Vanderbei/ *LINEAR PROGRAMMING*
Jaiswal/ *MILITARY OPERATIONS RESEARCH*
Gal & Greenberg/ *ADVANCES IN SENSITIVITY ANALYSIS & PARAMETRIC PROGRAMMING*
Prabhu/ *FOUNDATIONS OF QUEUEING THEORY*
Fang, Rajasekera & Tsao/ *ENTROPY OPTIMIZATION & MATHEMATICAL PROGRAMMING*
Yu/ *OR IN THE AIRLINE INDUSTRY*
Ho & Tang/ *PRODUCT VARIETY MANAGEMENT*
El-Taha & Stidham/ *SAMPLE-PATH ANALYSIS OF QUEUEING SYSTEMS*
Miettinen/ *NONLINEAR MULTIOBJECTIVE OPTIMIZATION*
Chao & Huntington/ *DESIGNING COMPETITIVE ELECTRICITY MARKETS*
Weglarz/ *PROJECT SCHEDULING: RECENT TRENDS & RESULTS*
Sahin & Polatoglu/ *QUALITY, WARRANTY AND PREVENTIVE MAINTENANCE*
Tavares/ *ADVANCES MODELS FOR PROJECT MANAGEMENT*
Tayur, Ganeshan & Magazine/ *QUANTITATIVE MODELS FOR SUPPLY CHAIN MANAGEMENT*
Weyant, J./ *ENERGY AND ENVIRONMENTAL POLICY MODELING*
Shanthikumar, J.G. & Sumita, U./*APPLIED PROBABILITY AND STOCHASTIC PROCESSES*
Liu, B. & Esogbue, A.O./ *DECISION CRITERIA AND OPTIMAL INVENTORY PROCESSES*
Gal, T., Stewart, T.J., Hanne, T. / *MULTICRITERIA DECISION MAKING: Advances in MCDM Models, Algorithms, Theory, and Applications*
Fox, B.L. / *STRATEGIES FOR QUASI-MONTE CARLO*
Hall, R.W. / *HANDBOOK OF TRANSPORTATION SCIENCE*
Grassman, W.K. / *COMPUTATIONAL PROBABILITY*
Pomerol, J-C. & Barba-Romero, S. / *MULTICRITERION DECISION IN MANAGEMENT*
Axsäter, S. / *INVENTORY CONTROL*
Wolkowicz, H., Saigal, R., & Vandenberghe, L. / *HANDBOOK OF SEMI-DEFINITE PROGRAMMING: Theory, Algorithms, and Applications*
Hobbs, B.F. & Meier, P. / *ENERGY DECISIONS AND THE ENVIRONMENT: A Guide to the Use of Multicriteria Methods*
Dar-El, E. / *HUMAN LEARNING: From Learning Curves to Learning Organizations*
Armstrong, J.S. / *PRINCIPLES OF FORECASTING: A Handbook for Researchers and Practitioners*
Balsamo, S., Personé, V., & Onvural, R./ *ANALYSIS OF QUEUEING NETWORKS WITH BLOCKING*
Bouyssou, D. et al. / *EVALUATION AND DECISION MODELS: A Critical Perspective*
Hanne, T. / *INTELLIGENT STRATEGIES FOR META MULTIPLE CRITERIA DECISION MAKING*
Saaty, T. & Vargas, L. / *MODELS, METHODS, CONCEPTS and APPLICATIONS OF THE ANALYTIC HIERARCHY PROCESS*
Chatterjee, K. & Samuelson, W. / *GAME THEORY AND BUSINESS APPLICATIONS*
Hobbs, B. et al. / *THE NEXT GENERATION OF ELECTRIC POWER UNIT COMMITMENT MODELS*
Vanderbei, R.J. / *LINEAR PROGRAMMING: Foundations and Extensions, 2nd Ed.*
Kimms, A. / *MATHEMATICAL PROGRAMMING AND FINANCIAL OBJECTIVES FOR SCHEDULING PROJECTS*
Baptiste, P., Le Pape, C. & Nuijten, W. / *CONSTRAINT-BASED SCHEDULING*
Feinberg, E. & Shwartz, A. / *HANDBOOK OF MARKOV DECISION PROCESSES: Methods and Applications*
Ramík, J. & Vlach, M. / *GENERALIZED CONCAVITY IN FUZZY OPTIMIZATION AND DECISION ANALYSIS*
Song, J. & Yao, D. / *SUPPLY CHAIN STRUCTURES: Coordination, Information and Optimization*
Kozan, E. & Ohuchi, A. / *OPERATIONS RESEARCH/ MANAGEMENT SCIENCE AT WORK*
Bouyssou et al. / *AIDING DECISIONS WITH MULTIPLE CRITERIA: Essays in Honor of Bernard Roy*

Early Titles in the
INTERNATIONAL SERIES IN
OPERATIONS RESEARCH & MANAGEMENT SCIENCE
(Continued)

Cox, Louis Anthony, Jr. / *RISK ANALYSIS: Foundations, Models and Methods*
Dror, M., L'Ecuyer, P. & Szidarovszky, F. / *MODELING UNCERTAINTY: An Examination of Stochastic Theory, Methods, and Applications*
Dokuchaev, N. / *DYNAMIC PORTFOLIO STRATEGIES: Quantitative Methods and Empirical Rules for Incomplete Information*
Sarker, R., Mohammadian, M. & Yao, X. / *EVOLUTIONARY OPTIMIZATION*
Demeulemeester, R. & Herroelen, W. / *PROJECT SCHEDULING: A Research Handbook*
Gazis, D.C. / *TRAFFIC THEORY*
Zhu/ *QUANTITATIVE MODELS FOR PERFORMANCE EVALUATION AND BENCHMARKING*
Ehrgott & Gandibleux/ *MULTIPLE CRITERIA OPTIMIZATION: State of the Art Annotated Bibliographical Surveys*
Bienstock/ *Potential Function Methods for Approx. Solving Linear Programming Problems*
Matsatsinis & Siskos/ *INTELLIGENT SUPPORT SYSTEMS FOR MARKETING DECISIONS*
Alpern & Gal/ *THE THEORY OF SEARCH GAMES AND RENDEZVOUS*
Hall/*HANDBOOK OF TRANSPORTATION SCIENCE - 2nd Ed.*
Glover & Kochenberger/ *HANDBOOK OF METAHEURISTICS*
Graves & Ringuest/ *MODELS AND METHODS FOR PROJECT SELECTION: Concepts from Management Science, Finance and Information Technology*
Hassin & Haviv/ *TO QUEUE OR NOT TO QUEUE: Equilibrium Behavior in Queueing Systems*
Gershwin et al/ *ANALYSIS & MODELING OF MANUFACTURING SYSTEMS*

* A list of the more recent publications in the series is at the front of the book *